ELECTROPOLYMERIZATION

Edited by **Ewa Schab-Balcerzak**

Electropolymerization
Edited by Ewa Schab-Balcerzak

Published by InTech
Janeza Trdine 9, 51000 Rijeka, Croatia

Publishing Process Manager Martina Durovic

Technical Editor Teodora Smiljanic

Cover Designer InTech Design Team

Additional hard copies can be obtained from orders@intechopen.com

Electropolymerization, Edited by Ewa Schab-Balcerzak
p. cm.
ISBN 978-953-307-693-5

For EU product safety concerns:
IN TECH d.o.o., Prolaz Marije Krucifikse Kozulić 3, 51000 Rijeka, Croatia,
info@intechopen.com or visit our website at intechopen.com.

Meet the editor

Ewa Schab-Balcerzak is a professor at the University of Silesia in Katowice, Poland, and head of the Department of Polymer Chemistry. She is also a professor at the Polish Academy of Sciences in the Centre of Polymer and Carbon Materials in Zabrze, where she began her research career in 1991. She received her MS in 1991, and her PhD in 1999 in polymer chemistry from Silesian University of Technology in Gliwice. In 2010, she received her DSc in chemistry from Warsaw University of Technology. In 1999, Ewa Schab-Balcerzak was a visiting researcher at LEMP/MAO at the University of Montpellier, France. After, she went on to work in the Department of Organic and Polymeric Materials in Tokyo Institute of Technology in Japan from 2001 to 2002. In 2003, she worked at Fraunhofer Institute of Applied Polymer Research in Golm, Germany. Her expertise and main research interests are in the design, synthesis, and characterization of new processable polymers and low molecular weight compounds for optoelectronic and photonic applications. Her scientific achievements contain over 75 papers and a few book chapters.

Contents

Preface XI

Chapter 1 **Electropolymerization of Polysilanes with Functional Groups 1**
Lai Chen, Xianfu Li and Jinliang Sun

Chapter 2 **Electropolymerization of Some Ortho-Substituted Phenol Derivatives on Pt-Electrode from Aqueous Acidic Solution; Kinetics, Mechanism, Electrochemical Studies and Characterization of the Polymer Obtained 21**
S.M. Sayyah, A.B. Khaliel, R.E. Azooz and F. Mohamed

Chapter 3 **Oxidation of Porphyrins in the Presence of Nucleophiles: From the Synthesis of Multisubstituted Porphyrins to the Electropolymerization of the Macrocycles 53**
Delphine Schaming, Alain Giraudeau, Laurent Ruhlmann, Clémence Allain, Jian Hao, Yun Xia, Rana Farha, Michel Goldmann, Yann Leroux and Philippe Hapiot

Chapter 4 **Electrochemical Polymerization of Aniline 77**
Milica M. Gvozdenović, Branimir Z. Jugović, Jasmina S. Stevanović, Tomislav Lj. Trišović and Branimir N. Grgur

Chapter 5 **Electrochemical Preparation and Properties of Novel Conducting Polymers Derived from 5-Amino-2naphtalensulfonic Acid, Luminol and from Mixtures of Them 97**
Luz María Torres-Rodríguez, María Irene López-Cázares, Antonio Montes-Rojas, Olivia Berenice Ramos-Guzmán and Israel Luis Luna-Zavala

Chapter 6 **Electrosynthesis and Characterization of Polypyrrole in the Presence of 2,5-di-(2-thienyl)-Pyrrole (SNS) 119**
Nasser Arsalani, Amir Mohammad Goganian, Gholam Reza Kiani, Mir Ghasem Hosseini and Ali Akbar Entezami

Chapter 7 **Polypyrrole Composites: Electrochemical Synthesis, Characterizations and Applications 131**
R. N. Singh, Madhu and R. Awasthi

Chapter 8 **Polypyrrole Soft Actuators 159**
Yasushiro Nishioka

Chapter 9 **Electrochemical Sensors Based on Electropolymerized Films 187**
Xu Qin, Hu Xiao-Ya and Hao Shi-Rong

Chapter 10 **Stability of Peptide in Microarrays: A Challenge for High-Throughput Screening 199**
Marie-Bernadette Villiers, Carine Brakha, Arnaud Buhot, Christophe Marquette and Patrice N. Marche

Preface

In recent years, great focus has been placed upon polymer thin films. These polymer thin films are important in many technological applications, ranging from coatings and adhesives to organic electronic devices, including sensors and detectors. Polymers can be prepared using chemical and/or electrochemical methods of polymerization. There are a few advantages of electrosynthesis over chemical methods. These include, but are not limited to: the wide range of oxidation and reduction reactions possible, the possibility of reaching very high levels of product purity and selectivity, and significantly less energy requirement. The process of electropolymerization leads to simple and reproducible formations of polymer films, which led to a broad material diversity of applications.

Nowadays, electrochemistry plays an important role in a wide number of fundamental research and applied areas. The great development in electropolymerization is due to the growing interest in synthesis of conducting polymers, which can be synthetically tailored to produce the desired electronic answer. The electropreparation of electroactive polymers is of major importance for their use in different electrochemical systems, such as batteries, electrocatalysis materials, sensor devices. Additionally, they can be used in fields of organic electronic devices, that is, light emitting diodes (OLEDs), field transistors (OFETs), and solar cells (OSCs). Electrochemical polymerization is preferable, especially if the polymeric product is intended for use as polymer thin films, because electrogeneration of films has the advantage of convenient one-step polymer synthesis. This allows control over the film thickness, an important parameter for fabrication of optoelectronic devices. Moreover, it was demonstrated that it is possible to modify the material properties by parameter control of the electrodeposition process, including the kind of solution in which the polymerization process takes place. It should be stressed that electrochemistry is an excellent tool, not only for synthesis, but also for characterization and application of various types of materials.

This book provides a timely overview of a current state of knowledge regarding the use of electropolymerization for new materials preparation, including conducting polymers and various possibilities of applications. The chapters of this book have been contributed by experienced and internationally recognized scientists. The material is presented in ten chapters.

The first chapter is devoted to electropreparation of two kinds of polysilanes, that is, one with a double bond and the second with attached zirconium. Such modified polymers let to obtain C/C composite with significantly enhanced anti-oxidative ablation. Chapter 2 concerns the study of electropolymerization, kinetics, and mechanisms of ortho-chloro-substituted phenol and catechol. Moreover, the chosen properties and possibility of the application of synthesized polymers as pH sensors, and as absorbent of methylene blue dye for purification of waste water were examined. Chapter 3 presents an overview of original methods of porphyrins electropolymerization, as a promising approach to elaborate new functional materials. Chapter 4 deals with all important aspects of electrochemical polymerization of aniline. Chapter 5 is devoted to electrosynthesis of novel conductive polymers based on 5-amino-2-naphthalenesulfonic acid, and luminol and evaluation of obtained polymers properties. Chapters 6, 7 and 8 concern polypyrroles (PPy). The effect of 2,5-di-(2-thenyl)-pyrrole on electropolymerization and electrochemical behavior of pyrrole was investigated in Chapter 6, while in Chapter 7, electrochemical methods for PPy composites preparation together with their properties and applications, and electrochemical devices are described. On the other hand, Chapter 8 presents polypyrrole films prepared by electropolymerization as a new material for organic soft actuators. Chapter 9 is a review devoted to electrochemical sensors based on electrogenerated films mainly for analytical purpose, that is, for clinical and environmental monitoring. Chapter 10 concerns the investigations of peptides stability in microarrays in the case that is grafted on the gold surface of the biochip. This is done by electrochemical copolymerization of pyrrole-peptide conjugates, or by electrodeposition of diazonium-peptide adducts. Thus, the topics addressed in this book present wide range of utilization of electrochemistry for both materials synthesis and their applications.

Finally, I would like to thank all the contributors who helped make this work possible.

Dr. Ewa Schab-Balcerzak
Centre of Polymer and Carbon Materials
Polish Academy of Sciences
Poland

1

Electropolymerization of Polysilanes with Functional Groups

Lai Chen, Xianfu Li and Jinliang Sun
School of Materials Science and Engineering, Shanghai University, Shanghai
China

1. Introduction

Polysilane (PS) is a kind of polymers with backbone consisting of only silicon atoms. It arouses great attention for its special electronic and optical properties[1], as well as SiC ceramic precursor[2]. The main method to synthesize polysilane is Wurth coupling reaction which rules out the production of many polysilanes with functional groups for the harsh reaction conditions[3]. Among the various methods for the synthesis of polysilanes[4], the electrosynthesis technique is gaining importance for its mild and safe condition and aptness of making functional molecules[5]. Shono[6] studied the effects of electrode materials, monomer concentration, amount of supplied electricity, and ultrasonic on the synthesis of polysilanes and related polymers. The concerned monomers bear inert side groups, such as methyl, ethyl, butyl, phenyl. The first attempt of attaching reactive vinyl groups to polysilane by electrosynthesis failed[7]. Our purpose to get high ceramic yield SiC precursor has prompted us searching for ways to introduce cross-linkable groups into polysilane molecules. Using allyl chloride and methyltrichlorosilane($MeSiCl_3$) as monomers, we managed to synthesize polysilane with double bonds[8], an excellent SiC precursor[9]. To increase the anti-oxidative ablation property of SiC, one way is to adulterate refractory metals M, such as Ti, Zr, Hf, et al., into SiC ceramic[10]. Coordination of M into polysilane molecule needs ligands. Cyclopentadiene(Cp), which is served as ligand for transitional metals, was attached on PS chains, thus zirconium atom was chemically combined with polysilane chain. And a new route to make composite ceramic MC-SiC was invented.

2. Electropolymerization of polysilanes

2.1 The setup of electrosynthesis

To take advantage of electrosynthesis, many efforts are needed to design the electrolytic apparatus. The paper[6] briefly introduces the setup of electropolymerization. Here we present the details of the apparatus.

The setup consists of four parts (Scheme1). The DC power can operate under two modes: the constant current mode and the constant voltage mode. We usually choose the constant current mode. The commutator can change the polarity of the electrodes in the range of 5 ~ 60 seconds. The supplied electricity is counted by a coulometer. When a predetermined electric volume is reached, it will cut off the electric supply to the electrolytic cell. A detailed description of the cell is plotted in Scheme 2.

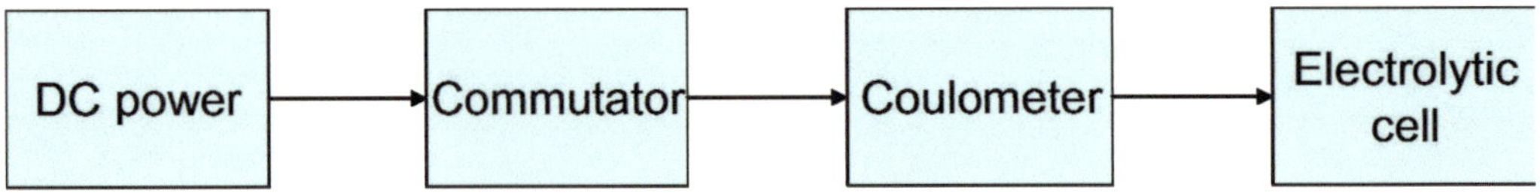

Scheme 1. The structure of the setup

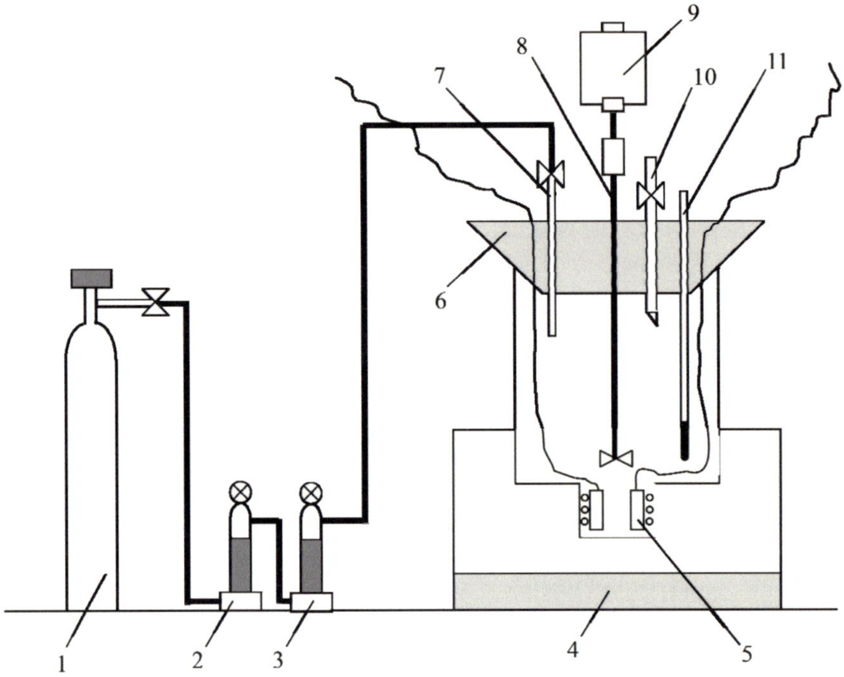

Scheme 2. The scheme of the electrolytic cell 1. N_2 bottle (99.99%) 2.drying tower (molecular sieve) 3.drying tower ($CaCl_2$) 4.ultrasonic 5.electrode 6. rubber plug 7. N_2 inlet 8.stirrer 9.motor 10.feeding hole 11.thermometer

The rectangle electrodes are 40mm long, 30mm wide and 10mm thick. It is rubbed with emery cloth before use. The space between the two poles is fixed by inserting two hollow polyethylene tubes (Φ4mm), and the electrodes and the tubes are binded together with nylon wire. The power of the ultrasonic (20 kHz) is 100W.

2.2 Electropolymerization of polysilanes with double bonds

The introduction of double bonds will increase the ceramic yield, shorten the production period of the C/C-SiC composites, and diminish the mechanical damage to the composites. We will discuss in detail the influential factors in the reaction and characterize the reaction product.

2.2.1 Electropolymerization and isolation

The $MeSiCl_3$ and allyl chloride were put in a compartment cell equipped with Mg cathodes, mechanic stirrer and thermograph. Then THF and $LiClO_4$ (0.05M) were added. All reactions were carried out under dry nitrogen atmosphere in order to eliminate the oxygen. Constant current electrolysis was applied at room temperature. During the electroreduction the electrolysis cell was sonicated and the polarity of electrodes was periodically alternated. The supplied amount of electricity was counted by a coulometer. Electrolysis was terminated at the desired amount of electricity. Then a quantity of toluene was poured in the cell and adequate gaseous ammonia was introduced to react with the residual Si-Cl. The solution was filtrated to remove $MgCl_2$ precipitate, and then was vacuum distilled to remove THF in the solvent. Subsequently, the residue was filtrated again to remove insoluble material. Finally, toluene was evaporated at reduced pressure, in each case to yield the light yellow liquid, polysilane with double bonds (EPS1).
The reaction scheme is shown as follows:

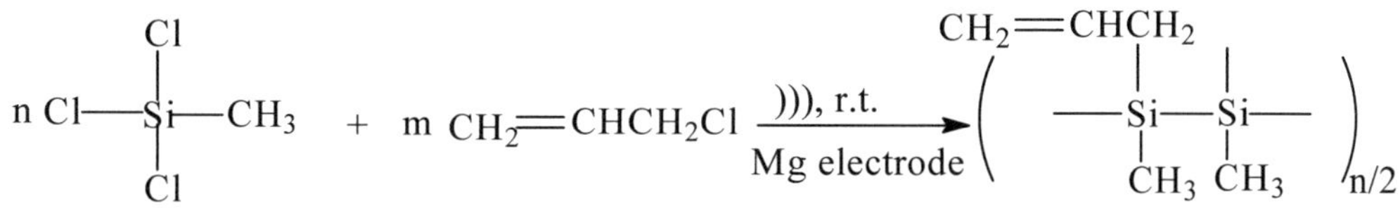

Scheme 3. Electrosynthesis of polysilane with double bonds

Polysilane without double bonds (EPS2) was synthesized in the same way using only $MeSiCl_3$ as monomer for comparison.
The products were identified using GPC (Agilent1100 American) with THF as solvent, ^1H NMR (AV-500), XRD (D/Max-rc Japan) and FTIR (Thermo Nicolet-AVRTAR370FT-IR). FTIR spectra were measured using KBr method.
The double bond content was determined by bromine addition method[11].
The measurement of monomer reaction rate is as follows:
Half an hour later after the monomers are plunged into the reaction bottle, the electrodes are charged. A small amount of sample is drawn out at intervals. It is hydrolyzed, and then titrated with sodium hydroxide solution. Therefore, the content of Si-Cl bonds in the bottle is determined. By following the variation of the Si-Cl bonds, we can analyze the reaction rate of the electropolymerization.

2.2.2 Influential factors of electrosynthesis

In the following episodes, the various factors influencing the reaction will be discussed in detail.

2.2.2.1 Effect of molar ratios of monomers

The influence of the $MeSiCl_3$/allyl chloride ratio on the yield, remaining double bonds, and Mw of polysilanes is shown in Table 1.
The optimum molar ratio of $MeSiCl_3/H_2C{=}CHCH_2Cl$ was 3/1 where the highest yield, double bonds concentration and molecular weight were reached. The relatively low yield and molecular weight at ratio of 10/1 and 5/1, where allyl chloride was deficient, might be due to the intramolecular reaction of $MeSiCl_3$. This reaction led to highly cross-linked structure, which was insoluble in solvent. Only the low molecular weight fraction was

soluble and obtainable as product. As the molar ratio surpassed 3/1, the excessive allyl chloride, acting as end capping agent, blocked the propagation of polysilane chains, resulting in lower molecular weight.

Ratio of ($MeSiCl_3/H_2C{=}CHCH_2Cl$)	10/1	5/1	3/1	2/1	3/2
Yield, %	49	51	84	71	79
Retaining ratio of double bonds, %	7.4	9.1	12.6	11.3	8.2
M_w, Dalton	1625	1974	3749	2293	834
M_w/M_n	3.19	3.84	8.26	4.52	1.64

[a] $MeSiCl_3$ monomer concentration was 1M , 10% theoretical charge was applied, and cathodes were Mg ingots.

Table 1. Effect of molar ratios of monomers[a]

The IR spectra of polysilanes are showed in Fig.1. The polysilane with and without double bonds both have C-H stretching (2960 and 2920cm^{-1}) groups, C-H bending (1450 and 1410cm^{-1}) groups, Si-CH_3 (1260 and 790cm^{-1}), Si-Si (460cm^{-1}). These absorptions imply the presence of Si-Si and Si-CH_3 bonds in both polysilanes. The presence of Si-O-Si (1080cm^{-1}) linkages was the consequence of the polymers' reactivity toward oxygen during work-up. The absorptions at 1640cm^{-1} of C=C stretching groups in the curve (1) indicate the presence of double bonds[12].

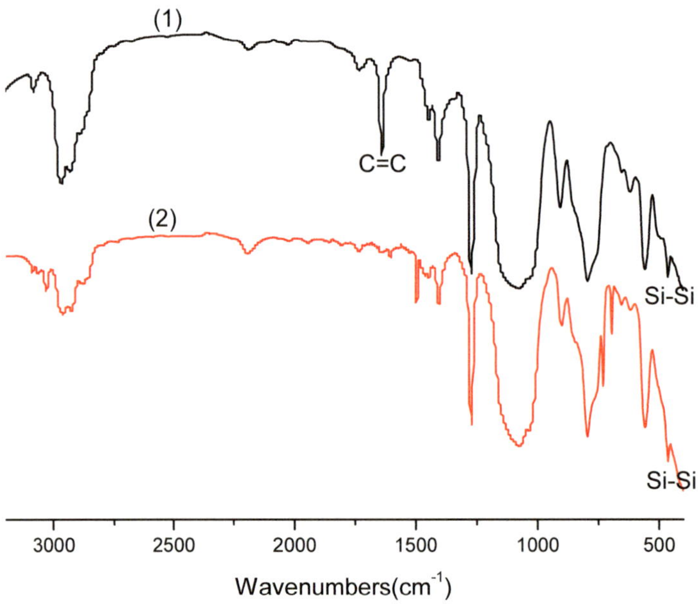

Fig. 1. FT-IR spectra of polysilanes with and without double bonds

1. EPS1:polysilane with double bonds synthesized with the $MeSiCl_3/H_2C=CHCH_2Cl$ ratio of 3:1
2. EPS2:polysilane without double bonds synthesized with the $MeSiCl_3$

^1H NMR spectrum of polysilane with double bonds (Fig.2) has four major peaks near δ 0.09, 0.25, 0.55 and 0.85 ppm, these are chemical shifts of $Si\text{-}CH_3$. The broad and complication of these peaks are due to the complex structures around the $Si\text{-}CH_3$ groups. The observed peaks near δ 5.8, 5.1-5.2, and 3.8-3.9 are the characteristic chemical shifts for Si-allyl moiety[13]. This proves the presence of double bonds in polysilane.

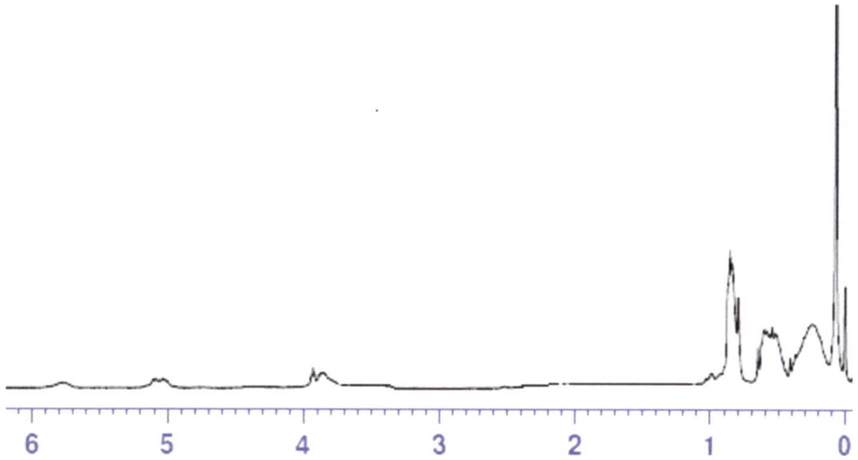

Fig. 2. ^1H NMR spectrum of polysilane with double bonds

The following experiments were performed on $MeSiCl_3$/Allyl Chloride ratio of 3:1.

2.2.2.2 Effect of monomer concentration

Under different monomer concentrations, the polysilanes were synthesized with 5% theoretical electricity. As shown in Table 2, the yield and molecular weight of the polysilanes became lower with increase of concentration. When the concentration was higher, $MeSiCl_3$ was apt to polymerize with itself, forming giant three dimensional networks, which would precipitate from solution.

Monomer concentration	1M	2M	3M
Yield, %	93	72	49
M_w, Dalton	1723	861	474
M_w/M_n	4.89	2.29	1.33

Table 2. Effect of monomer concentration

The following experiments were performed at 1M monomer concentration.

2.2.2.3 Effect of electrode materials

The electrode materials have a profound influence on the formation of polysilanes. As Table 3 shows, the entry using Mg anode and cathode gave the highest molecular weight and yield. Aluminum was less effective than Mg, and Cu the worst. It was said that Mg played some important roles in the formation of the Si-Si bond[6]. Although details of the role of Mg in the mechanism of formation of the Si-Si bond are not clear now, the unique reactivity of Mg electrode is undoubtedly shown in this reaction.

Electrode materials	Mg-Mg	Mg-Al	Al-Al	Mg-Cu
Yield, %	85	78	34	18
M_w, Dalton	3952	1099	898	512
M_w / M_n	8.67	3.36	2.12	1.63

[a] The polysilane were synthesized with 10% theoretic electric charge and interval time 17s.

Table 3. Effect of electrode materials [a]

The following experiments were performed with Mg electrodes.

2.2.2.4 Effect of the amount of electricity

The molecular weight was monitored at the different amount of electricity during the reaction process (Fig. 3). The molecular weight went up rapidly before 3% theoretical electricity (1440C*mol^{-1}) and entered a linear increasing period afterwards. This suggests that electrochemical synthesis of polysilane is a step-growth polymerization.

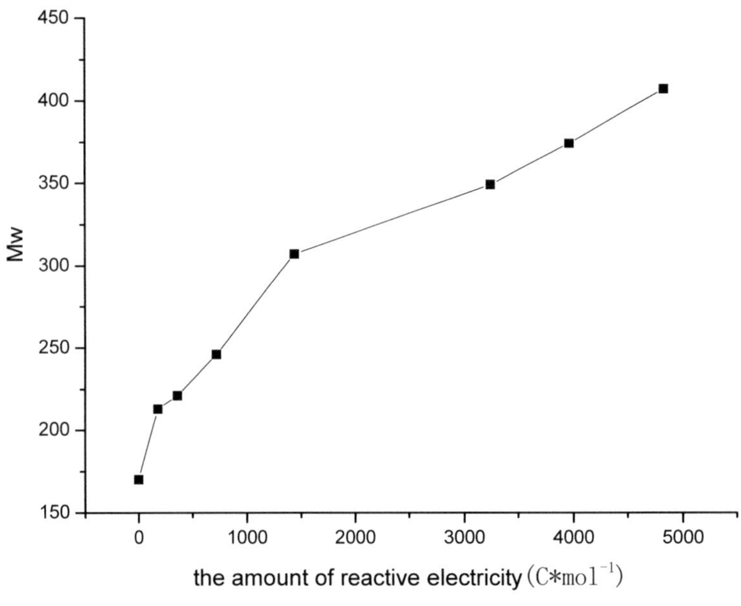

Fig. 3. The change of molecular weight of polysilane with the amount of electricity increasing

The amount of electricity is one of the most important factors to control the formation of Si-Si bond. The polymers, synthesized with different proportion of theoretical electricity, are listed in Table 4. With increase of the amount of electricity, the molecular weight rose stepwise and the molecular weight distributions became broader. The broader Mw /Mn might be due to the heterogeneous reaction condition and multiple reaction paths[14]. The results indicate that the polysilanes with different molecular weight can be obtained by controlling the amount of electricity.

The amount of electricity	5% theoretic electricity	7.5% theoretic electricity	10% theoretic electricity
Yield, %	87	92	85
M_w, Dalton	1723	2687	3952
M_w /M_n	4.89	6.58	8.67

Table 4. Effect of the amount of electricity

2.2.2.5 Effect of interval time

The electroreduction was carried out under ultrasound with different interval times of 8s, 17s and 26s, respectively. As shown in Fig. 4, the molecular weights of the products became higher with decreasing of the interval time. It could be explained that the alternation of anode and cathode might overcome the difficulty of keeping the electric current at a suitable level due to the increase of the voltage between anode and cathode with progress of the reaction[6]. The maximum voltage was relatively lower at shorter interval time. The lower voltage reduced the possibility of side reactions, thus benefiting the propagation of Si-Si bonds.

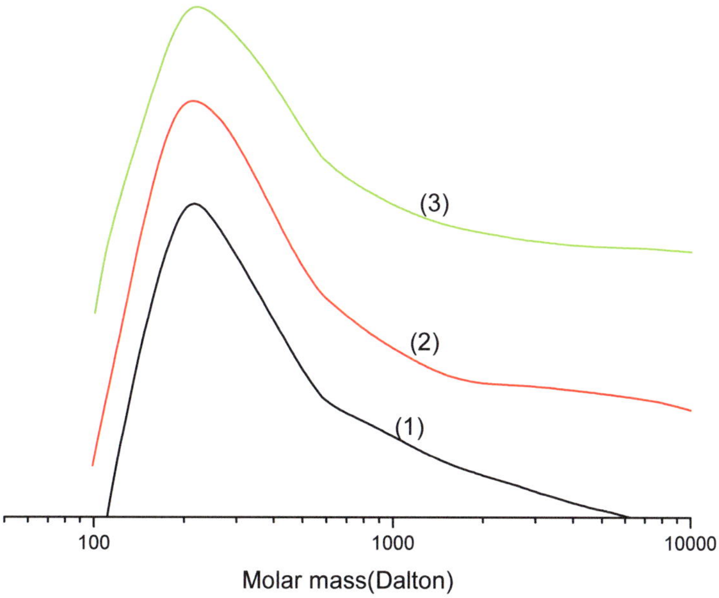

Fig. 4. Effect of interval time on the products' GPC spectra (1) $\overline{M_w}$ =2432, interval time 26s; (2) $\overline{M_w}$ =3031, interval time 17s; (3) $\overline{M_w}$ =6980, interval time 8s;

2.2.2.6 Effect of the concentration of supporting electrolyte

The increase in the concentration of supporting electrolyte resulted in the growth of molecular weight and yield (Table 5). Higher concentration of supporting electrolyte brought on lower electrode voltage to maintain constant current, thus reducing side reaction. The 0.05M concentration seemed reasonable because the acquired molecular weight and yield were satisfying, and even higher concentration leads to higher cost as the recovery of costly $LiClO_4$ is difficult.

Concentration of supporting electrolyte ($LiClO_4$)	0.05M	0.03M	0.005M
Yield, %	91	81	72
M_w, Dalton	3031	1309	665
M_w/M_n	7.89	2.56	1.42

Table 5. Effect of concentration of supporting electrolyte

2.2.2.7 The rate of electrosynthesis of polysilane

As shown in Fig.5, the EPS1 with double bonds has similar reaction tendency as the EPS2 without double bonds. About 30% ~ 40% Si-Cl had consumed at 0 mA*h of electricity because of the Grignard reaction between monomers and Mg in THF. The rate of electrosynthesis was relatively quicker in 0~100 mA*h range. The reaction finished as the amount of electricity reached 400 mA*h (30% of the theoretical amount of electricity). The EPS1 with double bonds reacted faster than the EPS2 without double bonds. At the same electricity volume, the Si-Cl content of EPS1 was 10% ~ 15% less than that of EPS2. The reason for this is that the allyl chloride might change into allyl anion by gaining electrons, then the obtained allyl anion initiates the anion polymerization of polysilane, thus increasing the reaction rate.

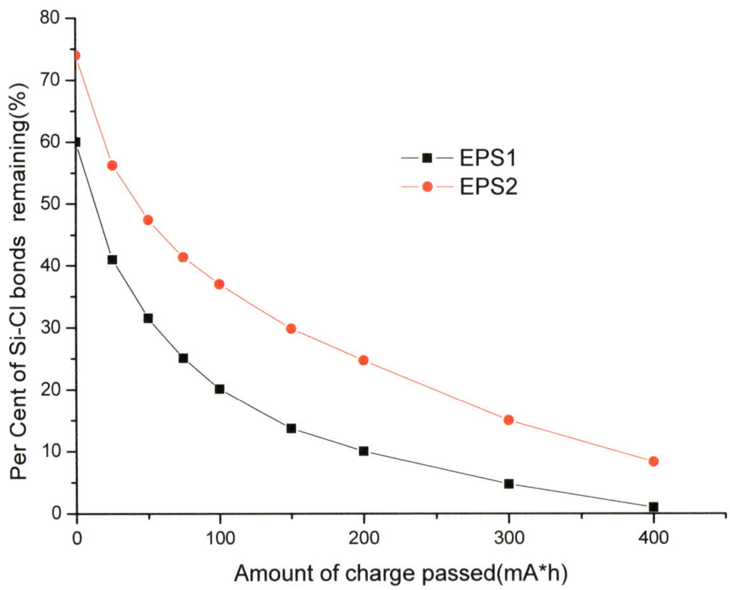

Fig. 5. Reaction Rate of EPS1 and EPS2 by Electroreduction

2.2.2.8 Comparison of the rates of electrosynthesis with and without electricity

The result above indicates polymerization can happen without electricity. To evaluate the role of electricity in the synthesis, the synthetic reactions of EPS1 and EPS2 with and without electricity were pursued (Fig. 6). The Si-Cl content in the case of electricity (curve b and d) is 5% ~ 13% less than that without electricity (curve a and c). In the whole electrosynthesis, the Grignard reaction between monomers and Mg ingot is in the majority, its ratio being as high as 80% ~ 90%. This result implies that the amount of charge passed could be much lower than that of the theoretical one to fulfill the electrosynthesis.

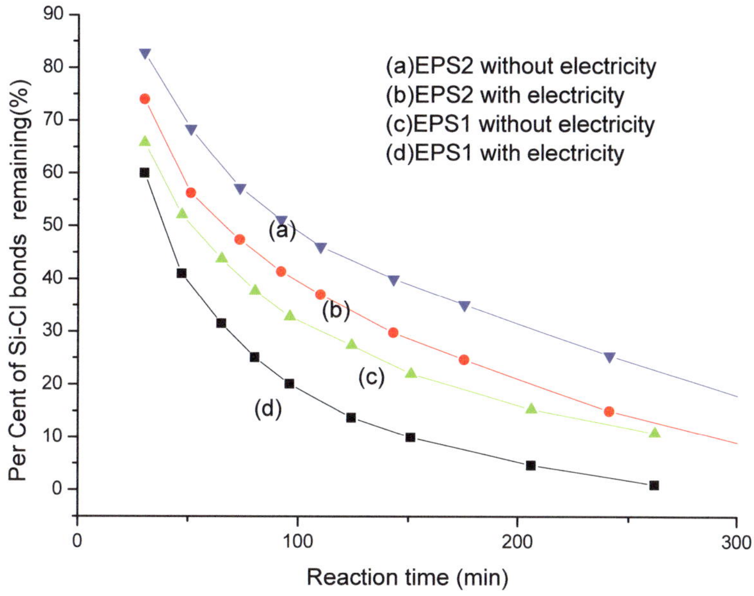

Fig. 6. Reaction rate of EPS1 and EPS2 with and without electricity

2.2.3 On the reaction mechanism

The reaction mechanism for synthesis of polysilane through Wurtz route has been studied in the past few decade[15]. However, the mechanism by electrochemical route is scarcely reported[16]. According to above-mentioned, we speculate the mechanism might be divided into two parts. The first part is the Grignard reaction mechanism (Scheme 4). The monomers can react with Mg to form Grignard reagent, which again react with CH_3SiCl_3 to produce a large molecule. The process repeats to form polysilane.

Another part is the electropolymerization mechanism (Scheme 5). The reactions include the electrode reaction and thermal reaction. First, the monomer gains an electron, being reduced to anion. The anion changes to silicon radical by eliminating chlorine anion, which can react with magnesium ion from sacrificing anode. The formed silicon radical might gain another electron to change into silicon anion intermediate. This intermediate reacts with other monomers by nucleophilic substitution to form oligomers, which can also be reduced to silicon anions. In the process of chain elongation, the delocalization of the electron along the chain makes it easier for the reduction of the oligomer. The thermal process includes the nucleophilic substitution of silicon anion with monomer or oligomer, and the coupling reaction between radicals. The nucleophilic substitution may happen intramolecularly or intermolecularly. The intramolecular one leads to cross-linking, while the intermolucular makes the chain to grow.

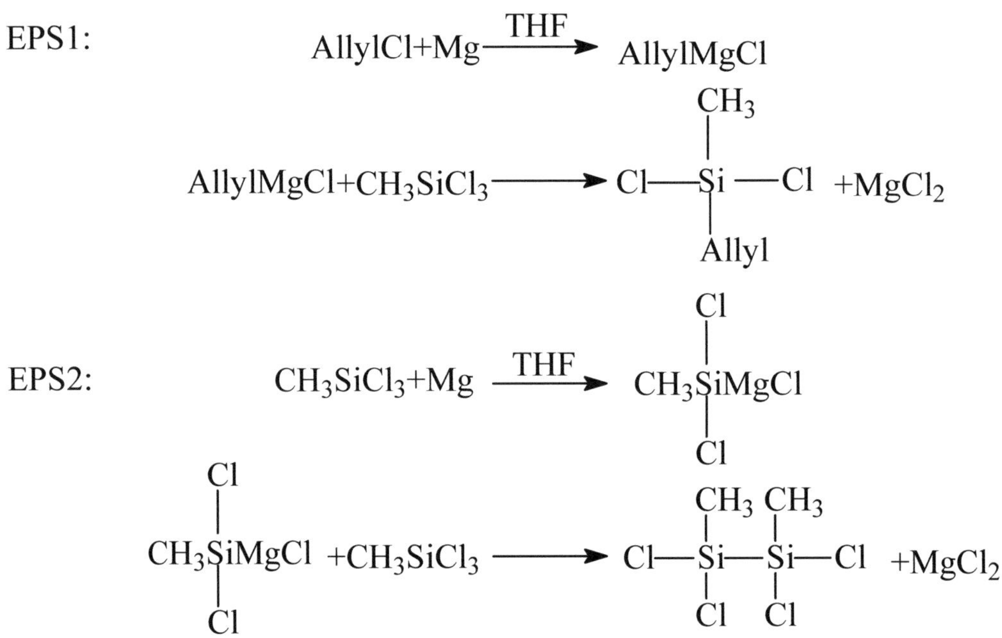

Scheme 4. The proposed reaction mechanism for Grignard reaction of CH_3SiCl_3

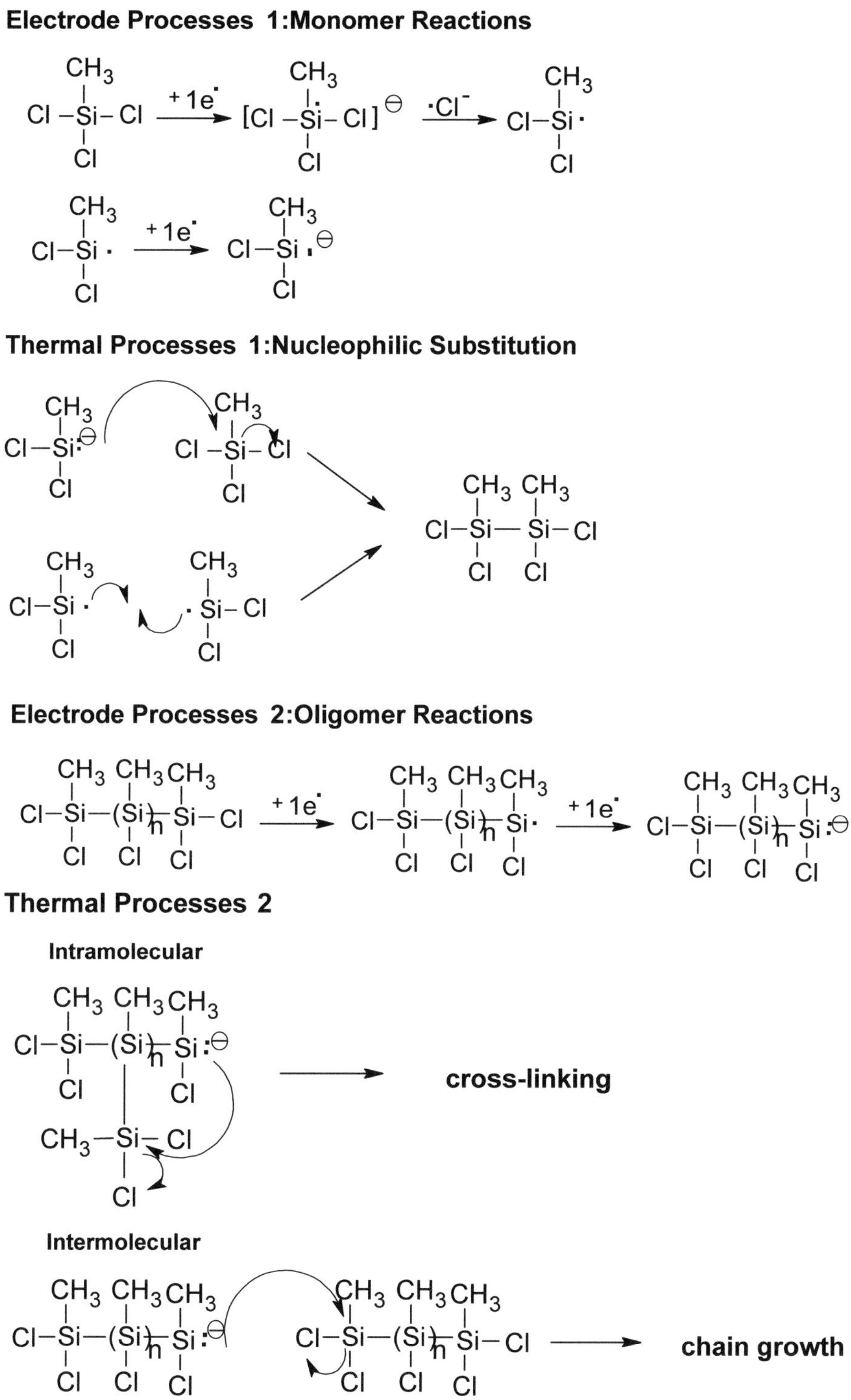

Scheme 5. Proposed reaction mechanism for electropolymerization of CH_3SiCl_3

2.3 Electropolymerization of polysilanes with refractory metals

The study of refractory-metals-containing SiC through polymer-derived ceramic (PDC) method is of interest[17], because the introduction of refractory metals M (such as Ta, Ti, Zr, Hf, Nb, Mo et al.) can enhance the heat-resistance property and the anti-oxidative ablation property of SiC ceramic. The methods of introducing M metals can be divided into two classes. One way to obtain M containing ceramic precursors is by pyrolysis of polycarbosilane (PCS) or polysilacarbosilane (PSCS) with organometallic compounds, such as $M(acac)_n$, $M(OR)_n$, Cp_2MCl_2[18-20]. Another way is by condensation of Si-H containing polymethylsilane with MCl_n[21], or by copyrolysis of polydimethylsilane with $M(OR)_n$[22]. The polymers used in above methods are synthesized by Wurtz reaction, which is featured by the relatively high cost and harsh reaction conditions. Our success in synthesis of polysilane with double bonds has prompted us to fabricate polysilane with refractory metals. By attaching cyclopentadienyl (Cp) ligand to the side chains of PS, the M atoms can be chemically combined with the Cp through η^5 π bonds.

2.3.1 Electrosynthetic procedure of polysilane containing zirconium

Into the electrolytic cell were plunged $MeSiCl_3$, allyl chloride, and cyclopentadiene in the molar ratio of 1:1/3:1/5. THF was used as solvent, $LiClO_4$ as supporting electrolyte, and magnesium ingot as electrodes. After deairing and inputing nitrogen three times, start stirring and ultrasonic. Set the interval time 10 seconds. The reaction was carried out at constant current mode. $ZrCl_4$/THF solution was charged into the cell as the passed electricity reached 90% of the predetermined electrical amount. It took about 8 hours to finish the electrosynthesis. Then 200ml toluene was poured into the cell, and ammonia gas was introduced to eliminate the remaining Si-Cl groups. After pressure filtration plus vacuum distillation two times, the Zr-containing polysilane EPS3, a black liquid, was obtained. As a comparison, the polysilane EPS4 was synthesized in the same procedure without adding $ZrCl_4$/THF solution.

The reaction is schemed as follows:

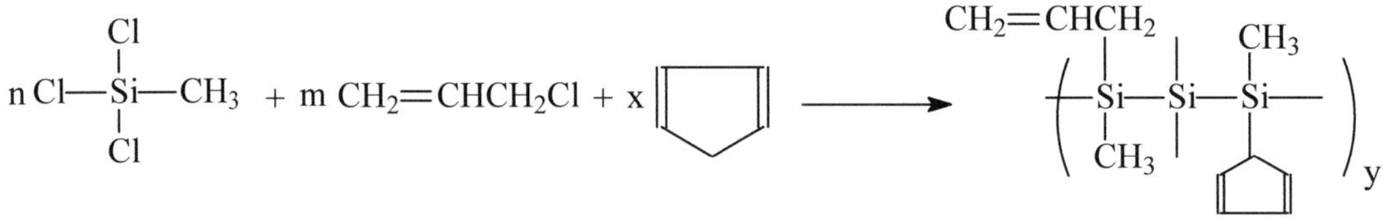

EPS4

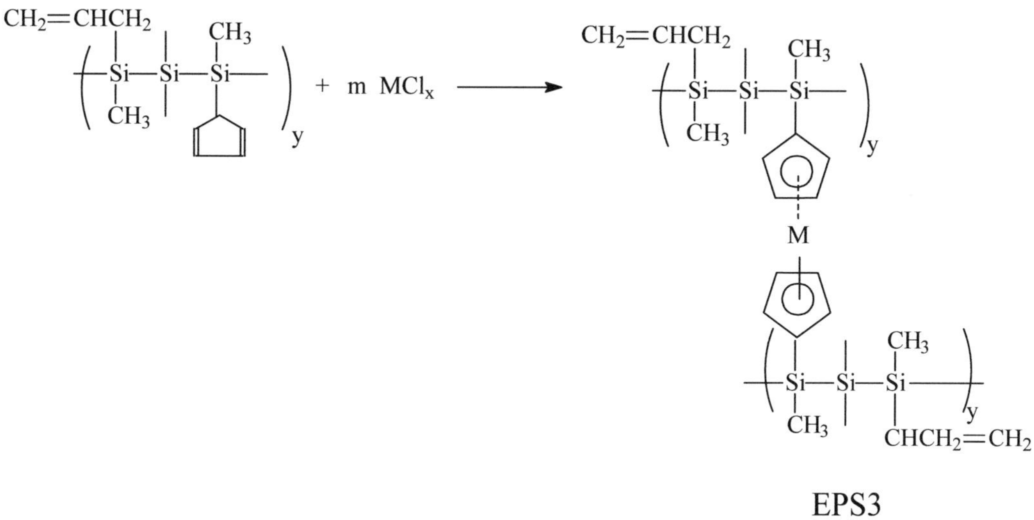

Scheme 6. Electrosynthesis of polysilane containing refractory metals

2.3.2 Characterization of the Zr-containing polysilane

The EPS3, being synthesized with the monomers of $MeSiCl_3$, allyl chloride, cyclopentadiene and $ZrCl_4$ in the ratio of 1:1/3:1/5:1/5, had the element contents (Wt%): 48.8%Si, 42.6%C, 4.1%Zr and a little oxygen. The retaining ratio of double bonds was 10.8% and the product yield was 71.2%.

The GPC analysis result and viscosity of EPS2 and EPS3 is listed in Table 6. As a result of the introduction of Zr increased the size of the polysilane. Both the EPS2 and EPS3 can dissolve in toluene, THF and chloroform.

Sample	$\overline{M_w}$ /(Dalton)	$\overline{M_w}$ / $\overline{M_n}$	Viscosity/(mPa.s/25℃)
EPS2	1723	1.89	75
EPS3	3749	3.24	125

Table 6. Molecular weight and Shear viscosity of EPS2 and EPS3

The IR absorption of EPS3 at 1405cm^{-1} corresponds to Si-Cp[23] and the 1640cm^{-1} peak corresponds to the C=C stretching(Fig.7).

The UV maximum absorptions of EPS3 and EPS4 shift to red, being broadening and higher, compared with EPS2 (Fig.8). The first reason for this phenomenon is the delocalization of the double bonds of the allyl groups, which increases the σ-π conjugation between the backbone and the side groups. Next the Cp group can conjugate with the main chain, making it bigger the whole conjugation system of the molecule. Thus the electron-transfer energy is lowered, causing significant spectral red shifts into the accessible UV region. Furthermore, the big π bond formed between Zr and Cp makes the conjugation system even wider, leading to still broader adsorption peak.

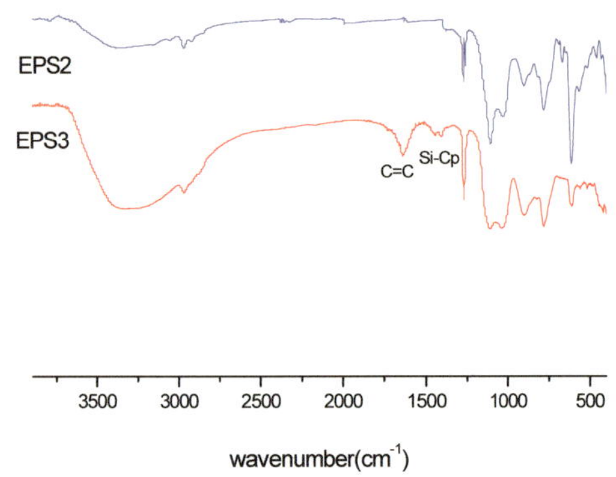

Fig. 7. FT- IR spectrum of EPS2 and EPS3

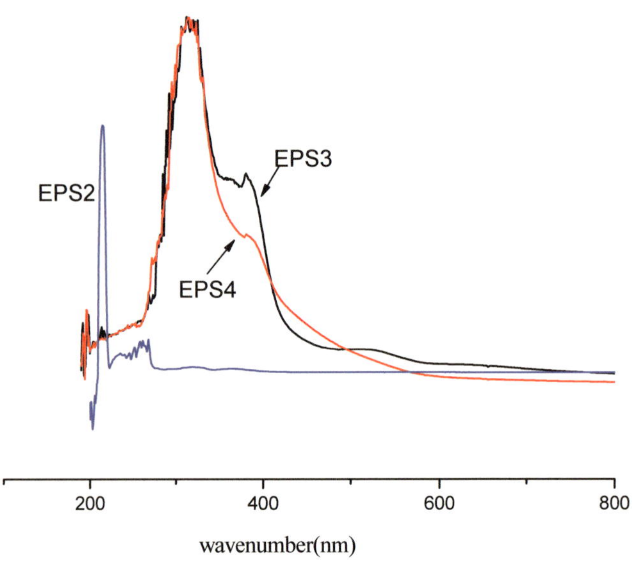

Fig. 8. UV spectra of EPS2, EPS3 and EPS4

The chemical shift δ=4.9-5.1, 5.8 and 1.5 on the ^{1}H-NMR spectrum of EPS3 (Fig.9) correspond to the primary, secondary and tertiary H of the allyl group[13]. The spectral peak at δ=5.5 and 6.0 belongs to the H of Cp group attaching to Zr elements[24, 25].

The above results prove that EPS3 has the predesigned structure. With Cp groups as a bridge, the refractory metals can be chemically attached to the backbone of the chain. As a result, the polysilane with both double bonds and hetero-elements has been invented.

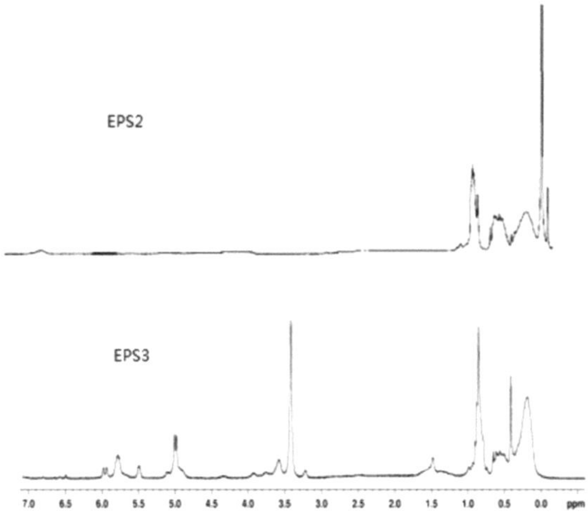

Fig. 9. ^{1}H-NMR spectra of EPS2 and EPS3

3. Ceramization of polysilanes with functional groups

3.1 The procedure of cross-linking and ceramization

The samples were loaded onto the sample holders. The cross-linking was carried out under dry nitrogen atmosphere at 130°C/12h plus 200°C/3h. The samples' weights before (m_0) and after (m) cross-linking were measured so that the mass retaining ratio (m/m_0) can be obtained. Ceramization was carried out in a sealed alumina tube furnace in flowing argon gas at 25-1300°C, with a heating rate 1°C/min and a dwell time of 3h. The crucible loaded with sample was weighed before and after pyrolysis so that ceramic yield may be calculated.

3.2 The cross-linking and pyrolysis of polysilanes with functional groups

The mass retaining ratio of EPS1 was much higher than that of EPS2 (Table 7), and was close to that of EPS1/DVB (1/0.5). It could be inferred that the double bonds on EPS1 had played important role in the cross-linking processes and were the cause of high mass retaining ratio. The ceramic yields of self-cross linking samples were dramatically higher, compared with specimens having DVB as curing agent. DVB might form inhomogeneous structure during curing process, leading to lower ceramic yield[26]. Therefore, it is not necessary to add curing agent to polysilane with double bonds, which is on its own a good ceramic precursor. The Zr-containing EPS3 also had a high mass retaining ratio and a moderate ceramic yield.

Sample	Mass retaining ratio	Ceramic yield
EPS1(self-cross linking)	97.5%	73.4%
EPS1+DVB (1:0.5)	99.1%	47.3%
EPS2(self-cross linking)	66.1%	70%
EPS2+DVB (1:0.5)	76.7%	40%
EPS3(self-crosslinking)	86.5%	65.5%
EPS3+DVB (1:0.5)	89.1%	60.1%

DVB: Divinylbenzene , as curing agent.

Table 7. Mass retaining ratio and Ceramic yield of EPS1, EPS2 and EPS3

XRD pattern of the pyrolyzed product of EPS1 at 1300℃(Fig. 10) shows broad peaks of β-SiC (2θ =35.6°,60°,71.8°). That of EPS3 (Figure 11) shows the existence of β-SiC, ZrC, Si and SiO_2.

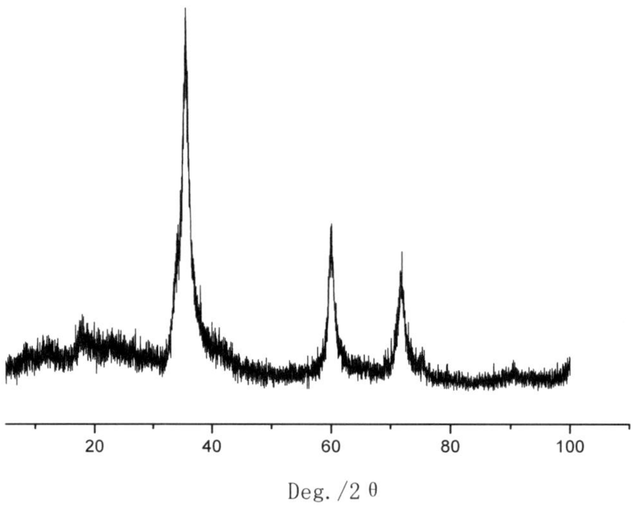

Fig. 10. XRD spectrum of EPS1's pyrolyzed product

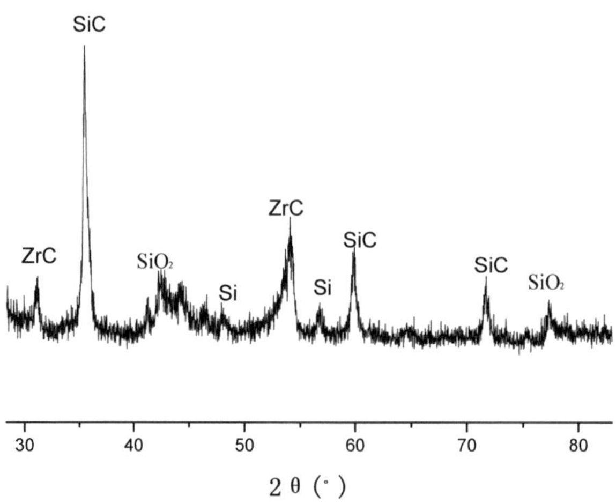

Fig. 11. XRD pattern of ceramic from EPS3

3.3 Improvement of the anti-oxidative ablation property of C/C composites by using polysilanes with functional groups

The poor oxidation resistance of C/C composites restricts its usage in oxidized environments. To improve its oxidative ablation resistance, the matrix of C/C composites is modified with SiC ceramic[27]. The introduction of SiC can be achieved by the chemical vapor infiltration (CVI) and the polymer infiltration and pyrolysis (PIP)[28]. PIP has aroused much interest due to its low manufacturing temperature, ability for the design of precursor molecules, simplicity of formation and feasibility of fabricating complex components. The main precursors for SiC at present are polycarbosilane and polysilane (PS)[4]. The C/C-SiC composites were manufactured using PS with double bonds by the PIP technique[9]. The C/C-ZrC-SiC composites were prepared using Zr-containing polysilane. The ablation behavior of the C/C-SiC composites and C/C-ZrC-SiC composites were measured.

The used C/C composites with a density of 1.29 g/cm^3 were made by CVI of bulk needled carbon felt and temperature treatment at 2 300 °C. The C/C-SiC composites and C/C-ZrC-SiC composites were densified by the following procedure: C/C composites → vacuum + pressure infiltration of precursor → cross-linking → pyrolysis → sintering. The process was repeated several times. The maximum density of the obtained C/C-SiC and C/C-ZrC-SiC composites were 1.75 g/cm^3 and 1.69 g/cm^3, respectively.

The specimens, with a size of ø30 mm×10 mm, were flushed vertically with a H_2-O_2 torch flame. The inner diameter of the nozzle tip was 4.68 mm. The distance between the nozzle tip and specimen surface was 15.8 mm. The pressure and flux of oxygen were 1.55 MPa and 2.1 L/min, and for hydrogen, they were 0.18 MPa and 1.68 L/min, respectively. The ablation time was 180 s.

The thickness of the specimen before ablation (d1) was measured with a micrometer (precision 0.01 mm), and the mass (m1) was measured with an analytical balance (precision 0.1 mg). After ablation, the thickness at the lowest point (d2) and mass (m2) were measured. For the ablation time of t, the linear ablation rate was calculated by (d1-d2)/t, and the mass ablation rate was calculated by (m1-m2)/t.

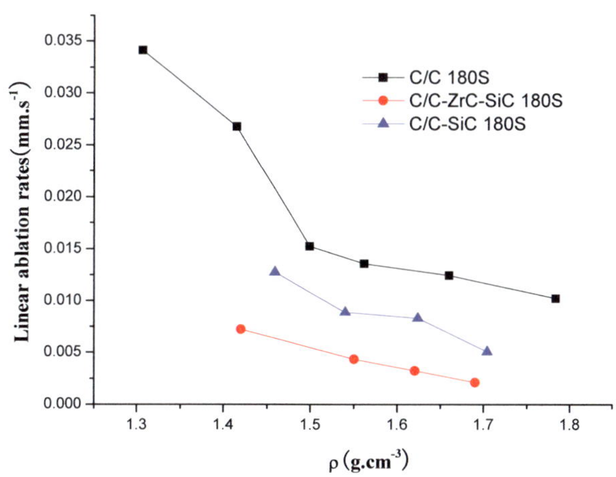

Fig. 12. Linear ablation rate of specimens

The ablation test results are depicted in Fig.12 and Fig.13, which show that the linear and mass ablation rate of the C/C-SiC and C/C-ZrC-SiC were lower than that of the C/C composites at the same density. With the increase of infiltration times, the density of C/C-SiC specimens increased, and the ablation rate decreased. The linear ablation rates of the 1.75 g/cm^3 C/C-SiC and 1.69 g/cm^3 C/C-ZrC-SiC specimen were 21.7% and 20.6% those of the 1.78 g/cm^3 C/C specimen, respectively. And the mass ablation rates of the 1.75 g/cm^3 C/C-SiC and 1.69 g/cm^3 C/C-ZrC-SiC specimen were 78.6% and 31.6% those of the 1.78 g/cm^3 C/C specimen, respectively. Therefore, introduction of SiC into C/C composite greatly improved its oxidative ablation-resistive property, and the adulteration of Zr element enhances the effect even more. It is worthwhile to investigate the impact of attaching refractory metals, such as Ta, Zr, Hf, Th, et al. solely or combinatorially to polysilanes on the anti-oxidative ablation property of the C/C composites.

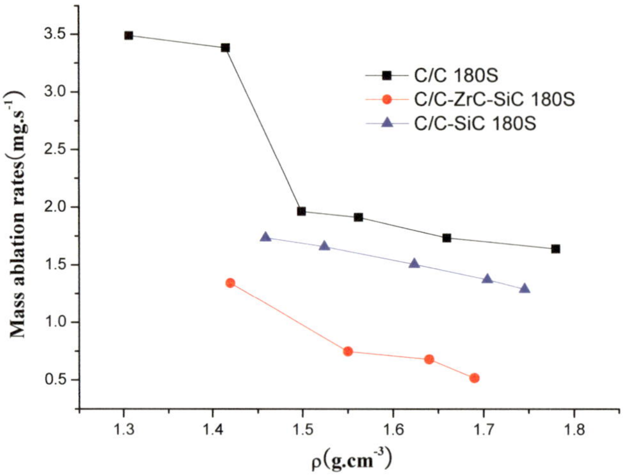

Fig. 13. Mass ablation rate of specimens

4. Conclusions

The electroreduction $MeSiCl_3$ and Allyl chloride monomers carried out with Mg electrodes in a single compartment cell gave polysilane with double bonds. The preferred $MeSiCl_3$/Allyl chloride ratio was 3:1. The molecular weight and yield of the products were affected by the concentration of monomer, the amount of electricity, the electrode's interval time and concentration of supporting electrolyte. By introducing cyclopendienyl ligands as side groups of polysilanes, the polysilanes with refractory metals were obtained by the combination of Cp with the metal elements. The synthesized polysilanes with self-curable ability and high ceramic yields are excellent ceramic precursors.

5. References

[1] Hayase, S., Polysilanes for semiconductor fabrication. *Progress in Polymer Science* 2003, 28, (3), 359-381.

[2] Roewer, G.; Herzog, U.; Trommer, K.; Muller, E.; Fruhauf, S., Silicon carbide - A survey of synthetic approaches, properties and applications. In *High Performance Non-Oxide Ceramics I*, 2002; Vol. 101, pp 59-135.

[3] Hayase, S., Polysilanes with functional groups. *Endeavour* 1995, 19, (3), 125-131.

[4] Chandrasekhar, V., Polysilanes and Other Silicon-Containing Polymers. In *Inorganic and Organometallic Polymers*, Springer Berlin Heidelberg: 2005; pp 249-292.

[5] Subramanian, K., A review of electrosynthesis of polysilane. *Journal of Macromolecular Science-Reviews in Macromolecular Chemistry and Physics* 1998, C38, (4), 637-650.

[6] Kashimura, S.; Ishifune, M.; Yamashita, N.; Bu, H. B.; Takebayashi, M.; Kitajima, S.; Yoshiwara, D.; Kataoka, Y.; Nishida, R.; Kawasaki, S.; Murase, H.; Shono, T., Electroreductive synthesis of polysilanes, polygermanes, and related polymers with magnesium electrodes. *Journal of Organic Chemistry* 1999, 64, (18), 6615-6621.

[7] Elangovan, M.; Muthukumaran, A.; Kulandainathan, M. A., Novel electrochemical synthesis and characterisation of poly(methyl vinylsilane) and its co-polymers. *European Polymer Journal* 2005, 41, (10), 2450-2460.

[8] Wu, S.; Chen, L.; Zhang, F.-j., Polysilane with double bonds by electrochemical synthesis. *Gaofenzi Cailiao Kexue Yu Gongcheng* 2007, 23, (5), 64-67.

[9] Wu, S.; Chen, L.; Qian, L.; Zhang, J.; Pan, J.; Zhou, C.; Ren, M.; Sun, J., Ablation properties of C/C-SiC composites by precursor infiltration and pyrolysis process. *Guisuanyan Xuebao* 2008, 36, (7), 973-977,984.

[10] Cao, S.-w.; Xie, Z.-f.; Wang, J.; Wang, H.; Tang, Y.; Li, W.-h., Research progress in hetero-elements containing SiC ceramic precursors. *Gaofenzi Cailiao Kexue Yu Gongcheng* 2007, 23, (3), 1-5.

[11] Chen, Y.; Du, L., *Organic Quantitative Microanalysis*. Science Press, China: Beijing, China, 1978.

[12] Gozzi, M.; Yoshida, I., Structural evolution of a poly(methylsilane)/tetra-allylsilane mixture into silicon carbide. *European Polymer Journal* 1997, 1301-1306.

[13] Yang, G.; Gao, S.; Gu, Q.; Xu, N., Purification of carbosilane dendrimers by column chromatography. *Nanjing Gongye Daxue Xuebao, Ziran Kexueban* 2010, 32, (1), 19-22.

[14] Vermeulen, L. A.; Smith, K.; Wang, J. B., Electrochemical polymerization of alkyltrichlorosilane monomers to form branched Si backbone polymers. *Electrochimica Acta* 1999, 45, (7), 1007-1014.

[15] Krzysztof, M.; Dorota, G.; S., H. J.; Kyu, K. H., Sonochemical Synthesis of Polysilylenes by Reductive Coupling of Disubstituted Dichlorosilanes with Alkali Metals. *Macromolecules* 1995, 28, (1), 59–72.

[16] Okano, M.; Takeda, K.; Toriumi, T.; Hamano, H., Electrochemical synthesis of polygermanes. *Electrochimica Acta* 1998, 44, (4), 659-666.

[17] Colombo, P.; Mera, G.; Riedel, R.; Soraru, G. D., Polymer-Derived Ceramics: 40 Years of Research and Innovation in Advanced Ceramics. *Journal of the American Ceramic Society* 2010, 93, (7), 1805-1837.

[18] Yamaoka, H.; Ishikawa, T.; Kumagawa, K., Excellent heat resistance of Si-Zr-C-O fibre. *Journal of Materials Science* 1999, 34, (6), 1333-1339.

[19] Li, X.; Edirisinghe, M. J., Evolution of the ceramic structure during thermal degradation of a Si-Al-C-O precursor. *Chemistry of Materials* 2004, 16, (6), 1111-19.

[20] Amoros, P.; Beltran, D.; Guillem, C.; Latorre, J., Synthesis and characterization of SiC/MC/C ceramics (M = Ti, Zr, Hf) starting from totally non-oxidic precursors. *Chemistry of Materials* 2002, 14, (4), 1585-1590.

[21] Xing, X.; Liu, L.; Gou, Y.; Li, X., Research progress of polymethylsilane in precursor of silicon carbide ceramics. *Guisuanyan Xuebao* 2009, 37, (5), 898-904.

[22] Shibuya, M.; Yamamura, T., Characteristics of a continuous Si-Ti-C-O fibre with low oxygen content using an organometallic polymer precursor. *Journal of Materials Science* 1996, 31, (12), 3231-3235.

[23] Deng, J.; Shi, B.; Liao, X., Synthesis and characterization of polymethylcyclopentadienylsilane and its copolysilane. *Youjigui Cailiao* 2001, 15, (6), 1-5.

[24] Grimmond, B. J.; Rath, N. P.; Corey, J. Y., Enantiopure siloxy-functionalized group 4 metallocene dichlorides. Synthesis, characterization, and catalytic dehydropolymerization of PhSiH3. *Organometallics* 2000, 19, (16), 2975-2984.

[25] Braunschweig, H.; von Koblinski, C.; Wang, R. M., Synthesis and structure of the first 1 boratitanocenophanes. *European Journal of Inorganic Chemistry* 1999, (1), 69-73.

[26] Li, J.-j.; Chen, L.; Sun, J.-l.; Zhou, C.-j.; Zhang, J.-b.; Ren, M.-s., Cross-linking and pyrolysis of polycarbosilane/divinylbenzene and polysilane/divinylbenzene. *Shanghai Daxue Xuebao, Ziran Kexueban* 2004, 10, (5), 479-483.

[27] Lamouroux, F.; Bourrat, X.; Naslain, R.; Thebault, J., Silicon-carbide infiltration of porous C-C composites for improving oxidation resistance. *Carbon* 1995, 33, (4), 525-535.

[28] Fan, X.-q.; Chen, L.; Wu, S.; Zhang, F.-j.; Zhang, J.-b.; Ren, M.-s.; Sun, J.-l., Crosslinking and pyrolysis of polymethylsilane/divinylbenzene. *Cailiao Kexue Yu Gongcheng Xuebao* 2006, 24, (6), 920-922.

2

Electropolymerization of Some Ortho-Substituted Phenol Derivatives on Pt-Electrode from Aqueous Acidic Solution; Kinetics, Mechanism, Electrochemical Studies and Characterization of the Polymer Obtained

S.M. Sayyah*, A.B. Khaliel, R.E. Azooz and F. Mohamed
Polymer Research Laboratory, Chemistry Department, Faculty of Science, Beni-Suef University 62514, Beni-Suef City, Egypt

1. Introduction

One of the promising methods for waste water remediation is The electrochemical oxidation of hazardous organic species [Fleszar & Jolanta, *1985;* Comninellis, 1994]. Phenols due to their slow degradation, bioaccumulation and toxicity constitute a large group of organic pollutants. The quantitation of phenolic compounds in environmental, industrial and food samples is currently of great interest, which can be found in soils and groundwater [Wang et al., 1998]. Also, these compounds are important synthesis intermediates in chemical industry such as resins, preservatives, pesticides, etc. Another, the main sources of phenolic waste are in glass fiber insulation manufacture on petroleum refineries. Phenol and substituted phenolic compounds such as catechol, chlorophenol are hydroquinones and discharged in the effluent from a number of chemical process industries. Today, these compounds are found in relatively high amount in domestic and industrial wastewater, discharged mainly from the mechanical industries. Many treatment technologies are in use or have been proposed for phenol recovery or destruction.
The electrooxidation of phenolic compounds can be occurs as follows: in the first step of electrooxidation of phenols, phenoxy radicals are generated, then these species can be either oxidized further or be coupled, forming ether and oligomeric or polymeric compounds [Wang et al ,1991; Iotov, & Kalcheva, 1998]. Electropolymerization of phenol beings with the formation of the phenoxy radical, or it can react with a molecule of phenol to give predominantly a para-linked dimeric radical. This radical may be further oxidized to form a neutral dimmer or it may attach another molecule. The dimer may be further oxidized create oligomers to polymers. Formation of the insoluble polyphenol results in deactivation of electrode surface. The relative rates of the two pathways (polymerization and forming quinonic structure) depend on the phenols concentration, the nature of electrode, pH, solvent, additives, electrode potential and current density [Gattrell & Kirk,1993]. Electropolymerization of phenols occur on different electrodes, such as Fe, Cu, Ni, Ti, Au, Pt

and other type of electrodes [Iotov, & Kalcheva, 1998; Ezerskis & Jusys, 2002]. Deactivation of electrode due to the phenol polymerization is more characteristic in alkaline medium. Insoluble high molecular weight species block the electrode surface and prevent effective electrooxidation of phenol.

In the present work, we seek to contrast the electro-oxidation of OCP and OHP from aqueous H_2SO_4 medium as electrolyte using cyclic voltammetry technique. The kinetic study of the oxidation processes will be useful for optimize the parameters control it. Mechanisms of the electrochemical polymerization will be discussed using electrochemical data. Also, the characterization of the obtained polymer films were carried out by elemental analysis, TGA, SEM, XRD, IR, UV-vis., ^{1}H-NMR spectroscopy. We hope to have films with good characters to be used in applications (i.e. dye removal and pH sensor).

2. Expermental

2.1 Materials

OCP was obtained from Hopkin & Williams (Dagenham, Essex, UK), Sulfuric acid, K_2HPO_4, KHphthalate, Borax, NaOH, Hydrochloric acid and $NaHCO_3$ were provided by Merck (Darmstadt, Germany). OHP and MB dye was provided by Aldrich. All chemicals are of analytical pure grade and used as received. All solutions were prepared by using freshly double-distilled water

2.2 Electropolymerization cell

Electropolymerization of the monomers and formation of the polymer films was carried out using potentiodynamic technique. The cell used is shown in Fig. (1). This cell is made from transparent Prespex trough, which has the inside dimensions of 8cm length, 2.5cm width and 3cm height.

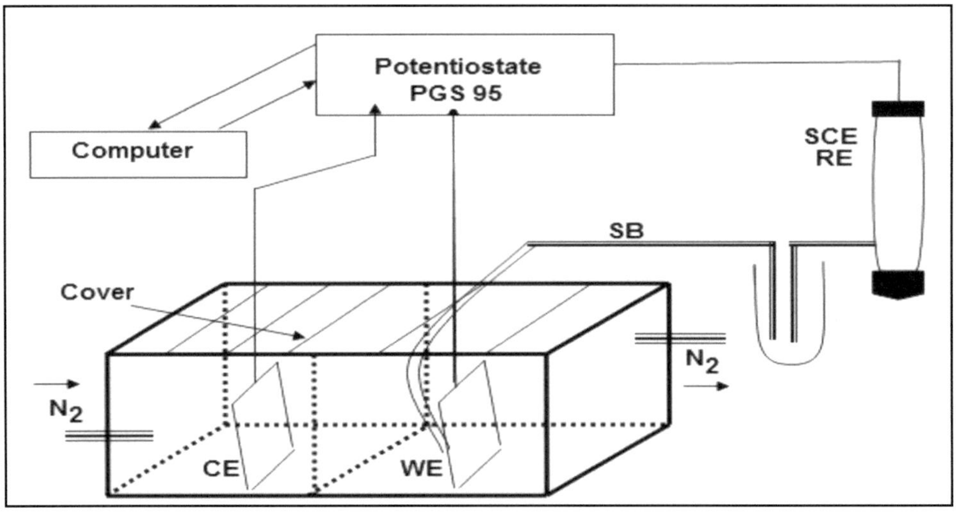

Fig. 1. Three electrode cell used for electropolymerization and cyclic voltammetry measurements. Were CE is counter electrode, WE is working electrode, SB is salt bridge, SCE is the standard calomel electrode and RE is the reference electrode.

2.2.1 Electrodes

2.2.1.1 Working electrode

The working electrode (WE) was a platinum sheet with dimensions of 1cm length and 0.5 cm width

2.2.1.2 Auxiliary electrode

The auxiliary (counter) electrode (CE) was a platinum foil with the same dimensions as the WE. Before each run, both the WE and the CE were cleaned and washed thoroughly with water, double distilled water, rinsed with ethanol and dried.

2.2.1.3 Reference electrode

A saturated calomel electrode (SCE) was used as a reference electrode. The values of the electrode potential in the present work are given relative to this electrode. The potential value for the SCE is 0.242 V vs. NHE at 25 °C. SCE was periodically calibrated and checked.
Electrochemical experiments were performed using the Potentiostat / Galvanostat Wenking PGS 95. i-E curves were recorded by computer software from the same company (Model ECT). Except otherwise stated, the potential was swept linearly from starting potential into the positive direction up to a certain anodic potential with a given scan rate and then reversed with the same scan rate up to the starting cathodic potential.
For each run, freshly prepared solutions as well as a cleaned set of electrodes were used. All experiments were conducted at a given temperature (± 0.5 °C) with the help of circular water thermostat. After polymer film formation, the working electrode was withdrawn from the cell, rinsed thoroughly with a doubly distilled water to remove any traces of the formed constituents in the reaction medium. The deposited polymer film was subjected to different experimental tests to characterize it.

2.2.2 Procedure

Potentiodynamic cyclic voltammetry measurements during the formation of the polymer films on the surface of the working electrode was carried out in the electrochemical cell shown in Fig.(1).The cell was filled with the test solution (aqueous solution containing H_2SO_4 as supporting electrolyte, and monomer). The working and counter electrodes were introduced in the cell. The reference electrode was attached to the cell by U-shaped salt bridge (SB) ended with a fine capillary tip (Luggin –Harber probe)wherein the reference electrode was positioned very closed to the working electrode to minimize the over potential due to electrolyte resistance .The bridge was filled with the test solution. Before and during measurements a current of pure nitrogen gas was bubbled in the test solution to remove dissolved oxygen.
Electrochemical experiments were performed using the potentiostat / Galvanostat Wenking PGS 95. i-E curves were recorded by computer software from the company (Model ECT).
Except otherwise stated ,the potential was swept linearly from the starting potential vs. (SCE) into the positive direction up to a certain anodic potential with a given scan rate and then reversed with the same scan rate up to the starting cathodic potential.
For each run, freshly prepared solutions as well as a cleaned set of electrodes were used. All experiments were conducted at a given temperature (± 0.5 °C) with the help of circular water thermostat. After polymer film formation, the working electrode was withdrawn from the cell, rinsed thoroughly with doubly distilled water to remove any traces of the medium

constituents. The deposited polymer film was subjected to different experimental tests to characterize it.

2.3 Characterization of the electro-prepared polymers

2.3.1 UV-vis, IR and ^{1}H-NMR spectroscopy, TGA and elemental analysis

UV-vis. absorption spectra of the prepared polymer sample was measured using Shimadzu UV spectrophotometer (M160 PC) at room temperature in the range 200-400 nm using dimethylformamide (DMF) as a solvent and reference. IR measurements were carried out using shimadzu FTIR-340 Jasco spectrophotometer (Japan) by KBr pellets disk technique. ^{1}H-NMR measurements were carried out using a Varian EM 360 L, 60-MHz NMR spectrometer. NMR signals of the electropolymerized sample were recorded in dimethylsulphoxide (DMSO) using tetramethylsilane as internal standard. TGA of the obtained polymer was performed using a Shimadzu DT-30 thermal analyzer (Shimadzu, Kyoto, Japan). The weight loss was measured from ambient temperature up to 600 °C, at the rate of 20 °C min^{-1} and nitrogen 50cc min^{-1} to determine the degradation rate of the polymer. Elemental analysis was carried out in the micro-analytical center at Cairo University (Cairo, Egypt) by oxygen flask combustion and dosimat E415 titrator (Metrohm).

2.3.2 Scanning electron microscopy and X-ray diffraction

Scanning electron microscopic (SEM) analysis was carried out on the as-prepared polymer film deposited on Pt-working electrode surface using a JSM-T20 Electron Probe Microanalyzer (JEOL, Tokyo, Japan). The X-ray diffraction analysis (XRD) (Philips 1976 Model 1390, Netherlands) was operated under the following conditions that were kept constant for all the analysis processes: X-ray tube, Cu; scan speed, 8 deg min^{-1}; current, 30 mA; voltage, 40 kV; preset time, 10 s.

2.4 Determination of the kinetic rate law of the electropolymerization reaction

The amount of polymer electrodeposited on the electrode surface can be determined directly from the peak current density (i_p) [Sayah et al, 2010] therefore, The peak current density (i_p) is proportional to the electropolymerization rate ($R_{P,E}$) at a given concentration of the monomer and H_2SO_4. The kinetic equation was calculated from the value of anodic peak current density (i_p) measured at each concentration during the electroformation of polymer. In this case, we used the value of (i_p) instead of ($R_{P,E}$). Therefore, the kinetic rate law can be expressed as follows

$$R_{P,E} = i_p = k_E\,[\text{Acid}]^a[\text{Monomer}]^b \tag{1}$$

where a and b are the reaction orders with respect to acid and monomer concentrations respectively, and k_E is the kinetic rate constant calculated from the electrochemical measurements.

2.5 Potentiometric data and pH measurements

The electropolymerization were performed with a three-electrode system mention above in section. Using a potential range -0.2 ~ +0.9V (vs.SCE) with a scan rate of 25 mVs^{-1}. Finally, a POCP modified Pt-electrode was ready for further experiments. the thickness of the polymer films were controlled by varying the no of repetitive cycles from 3 to 15 cycles as the thickness of polymer films were positively correlated with the no of repetitive cycles.

Potentiometric data were recorded by using (two electrodes system) shown in Fig.(2). Where the working electrode is the prepared POCP modified Pt-electrode with different thickness and the reference electrode is SCE. The potential of system (the electrode and the tested buffer solution) were recorded after immersion for 5 minute by using Avometer DT3900, then the polymer electrode removed from the solution and rinsed with bidistilled water before immersing in anther buffer solution for the next measurement All pH measurements were performed using the modified electrode as the pH sensor and SCE is the reference electrode. The actual pH of solution was determined by pH meter (HANNA Instruments pH 211).

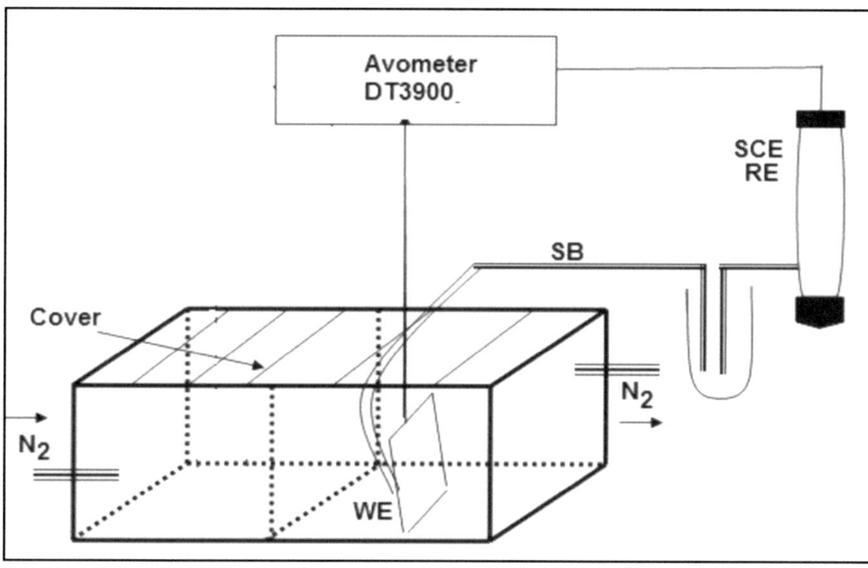

Fig. 2. Two electrode cell used foe potentiometric measurements.

2.6 Adsorption of methylene blue (MB) dye

Different concentrations of MB solution was added to 0.05gm of POHP previously deposited potentiodynamically on Pt electrode surface in 50 ml measuring flask and with continuous stirring for 2 h and then filtration. The concentration of dye in the filtrate was determined at different time intervals by using UV-Vis spectrophotometer at 664 nm for MB dye, the equilibrium uptake was calculated according to the following equation:

$$Q_e = (C_o - C_e)\ V / W \tag{2}$$

Where Qe is the amount adsorbed at equilibrium, C_o is the initial concentration of dye, C_e is the equilibrium conc. of the dye solution, V is volume of solution (L) and W is the mass of polymer taken for the experiment (mg)

The percentage removal of dye was calculated as

$$\text{Percentage removal} = 100\ (C_o - C_e) / C_o \tag{3}$$

3. Results and discussion

3.1 Electropolymerization kinetics and mechanisms of of OCP and OHP

3.1.1 Electropolymerization

Electropolymerization of OCP or OHP on platinum electrode from aqueous solution containing 0.6M H_2SO_4 at 303 K in the absence and presence of monomer, was studied by cyclic voltammetry at potential between -366 and +1780 mV(vs. SCE) with scan rate of 25 mVs^{-1}.

The obtained voltammograms in absence and presence of monomer is represented in Fig. 1(a-c). The voltammogram in the absence of monomer exhibit an oxidation peak (I) which developed at -300 mV vs. SCE, which is a result of hydrogen adsorption on Pt electrode [Arslan et al, 2005]. While, the voltammograms in the presence of OCP exhibit two oxidation peaks (I and II) that progressively developed at -300 and +863 mV (vs. SCE) repetitively and one reduction peak (II') at +248 mV (vs. SCE). While in the presence of OHP ; two oxidation peaks (I and II) that progressively developed at --200 and +622 mV (vs. SCE) repetitively and one reduction peak (II') at +0.20 mV (vs. SCE). One one hand, the first oxidation peaks (I) are a result of hydrogen reduction [Arslan et al, 2005] as mentioned above where, the second oxidation peaks (II) correspond to oxidation of monomer to give phenoxy radical which adsorbed on Pt-electrode [Gattrell & Kirk,1993]. The adsorbed radicals react with

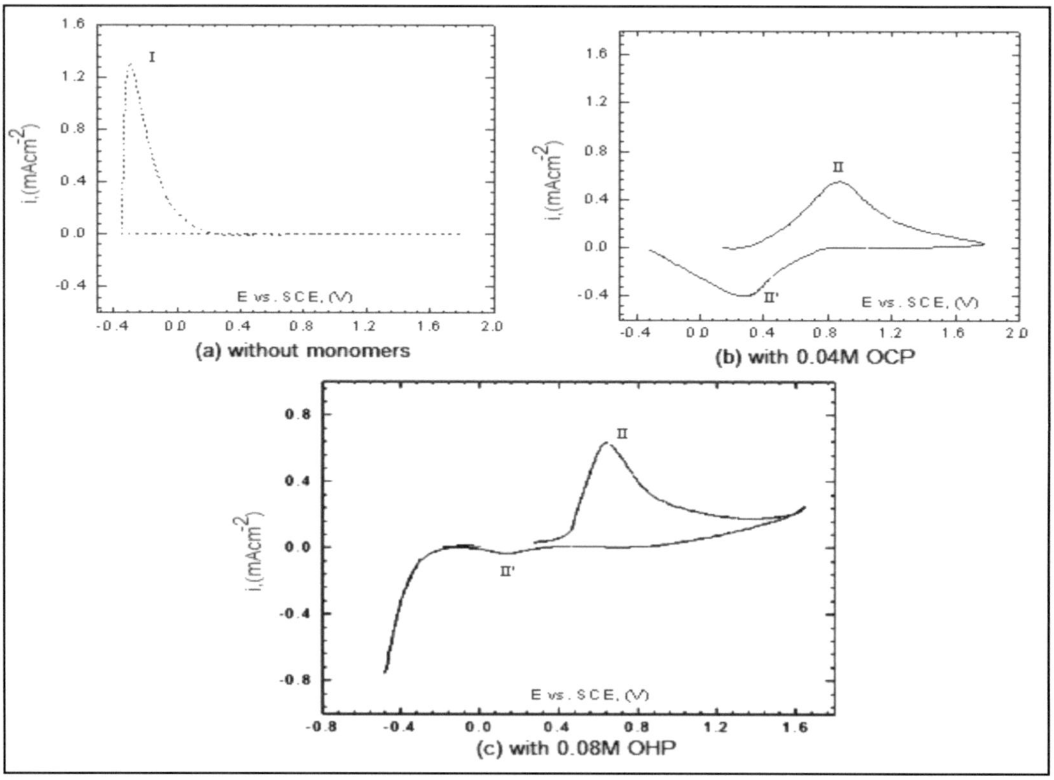

Note: peak (I) in both (b) and (c) is not shown in the figure.

Fig. 3. Cyclic voltammograms of solutions containing 0.06M H_2SO_4 at 303 K with scan rate 25mVs^{-1}.

other monomer molecule via head-to-tail coupling to form predominantly a para-linked dimeric radical and so on to form oligomer and polymer film; this film is a chain of isolated aromatic rings (polyethers) without п-electrons delocalization between each unit as shown in schemes (1 and 2). The oxidation occurs at more positive values ~ + 863 and +622 mV vs. SEC where the presence of (Cl and OH) make the oxidation process difficult.

On reversing the potential scan from, the reversing anodic current is very small indicating the presence of polymer layer adhered to the Pt-surface [Sayyah et al, 2010]. One cathodic peak (II') was found which could be ascribed to the reduction of the formed polymer films.

The effects of repetitive cycling on Pt- electrode in aqueous solution containing 0.6M H_2SO_4 with and without monomer at 303 K are shown in Fig. 4 (a-c). The data reveal that, in absence of monomer, the repetitive cycling show the oxidation peak (I) only. The current of this peak (i_{pI}) is almost the same and not affected with cycling up to 6 cycles (c.f. fig. 4 (a)).

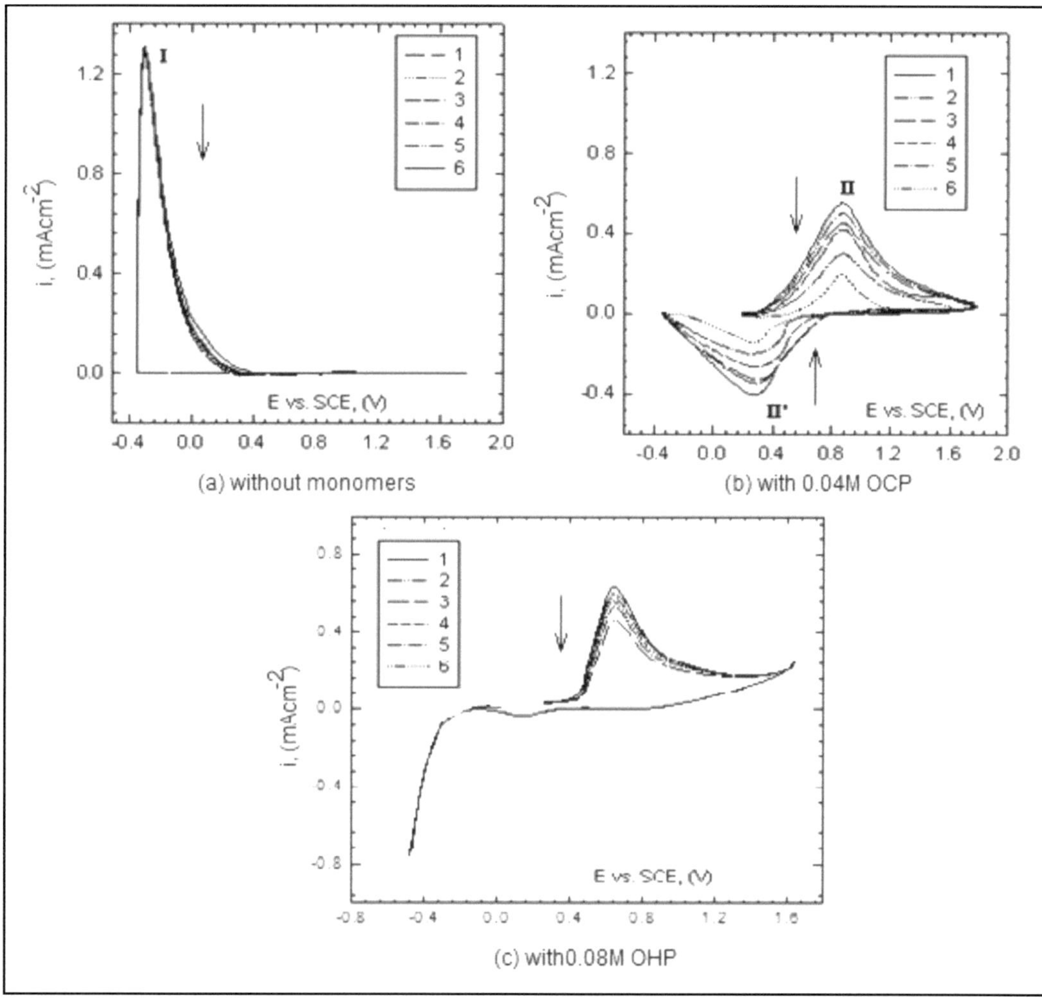

Fig. 4. Repetitive cycling of electropolymerization from solution containing 0.6M H_2SO_4 at 303 K with scan rate 25 mVs^{-1}

This means that, surface area of electrode is not affected by the H_2 adsorption where, in presence of monomers (OCP or OHP) the data reveal that, during the second cycle both the oxidation and the reduction peak currents decrease significantly with repetitive cycling. This behavior is observed elsewhere as a result of fouling of electrode [Arslan et al, 2005] where phenolic products block the electrode surface and the formed film hinders diffusion of further phenoxide ions to the electrode surface, thereby causing a significant decrease in the anodic peak current and also decrease the cathodic peak current. The potential position of the redox peaks does not shift with increasing number of cycles, indicating that the oxidation and the reduction reactions are independent on the polymer thickness [Sayyah et al, 2010].
Figure 5 (a and b) illustrates the influence of the scan rate (15 - 45 mV s^{-1}) on the potentiodynamic anodic polarization curves for OCP and OHP from aqueous solution containing 0.6 M H_2SO_4 at 303 K on platinum electrode. It is obvious that both the anodic and cathodic peak current densities (i_{pII} and $i_{pII'}$) in the two cases increases with the increasing of the scan rate. This behavior may be explained as follows, when an enough potential is applied at Pt- surface causing oxidation of species in solution, a current arises due to the depletion of the species in the vicinity of the Pt- surface. As a consequence, a concentration gradient appears in the solution. The current (i_p) is proportional to the gradient slope, dc/dx, imposed (i = dc/dx). As the scan rate increase the gradient increase and consequently the current (i_p).

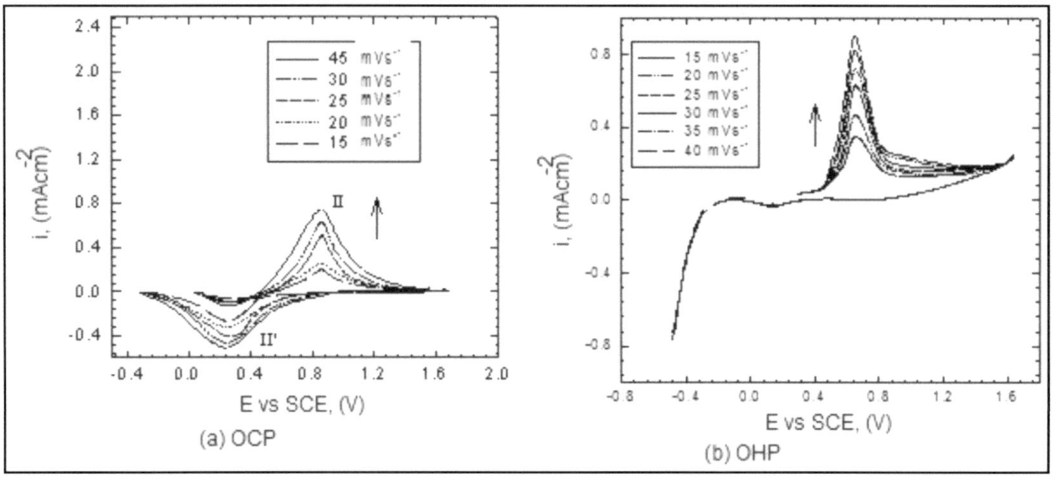

Fig. 5. Effect of Scan rate on electropolymerization on Pt electrode from solution containing 0.6M H_2SO_4 at 303 K.

Figure 6 (a and b) shows the linear dependency of the anodic peak current densities (i_{pII}) - which is corresponding to the formation of the polymer films POCP and POHP repetitively- versus the square root of scan rate ($v^{1/2}$). This linear relation suggests that the oxidation of OCP to POCP and/or OHP to POHP may be described by a partially diffusion-controlled process (diffusion of reacting species to the polymer film / solution interface) where the correlation coefficients (r^2) is higher than 0.9 but not equal to 1.0 suggesting the non-ideal simulation relation (i.e: the process is not completely diffusion control but it is exactly a partially diffusion control. Values of i_p are proportional directly to $v^{1/2}$ according to Randless [Randless, 1984] and Sevick [Sevick, 1948] equation:

$$i_{pII} = 0.4463\, n\, F\, A\, C\, (\, n\, F\, \upsilon\, D\, /\, R\, T\,)^{1/2} \qquad (4)$$

where n is the number of exchanged electron in the mechanism, F is Faraday's constant (96485 $Cmol^{-1}$), A is the electrode area (cm^2), C is the bulk concentration, D is the analyst diffusing coefficient (cm^2s^{-1}), and v is the scan rate (Vs^{-1}). R is the universal gas constant (8.134 $Jmol^{-1}K^{-1}$), T is the absolute temperature (K).The calculated values of D (at 0.6 M H_2SO_4 at 303 K with scan rate from 15 to 45 mV s^{-1}) are shown in Table 1;

Scan rate, (Vs^{-1})	Diffusing coefficient, (m^2s^{-1})	
	OCP→POCP	OHP→POHP
0.015	1.83676×10^{-11}	4.41282×10^{-11}
0.020	2.15245×10^{-11}	8.26886×10^{-11}
0.025	7.16612×10^{-11}	1.09351×10^{-10}
0.030	9.11262×10^{-11}	1.15739×10^{-10}
0.045	1.07765×10^{-10}	1.59405×10^{-10}

Table 1. Calculated values of Diffusion coefficients.

The values of D are seen to be constant for both cases over the range of sweep rates, which again shows that the oxidation process is diffusion-controlled [sayyah et al, 2010].
Figure 6 (a and b) shows the linear dependence of the anodic current peak, (i_{pII}) versus $v^{1/2}$. These linear regression equations are;
For OCP; i_{pII} *(mA)* $= 0.207\, \upsilon^{1/2}\, (mV\, s^{-1})^{1/2} - 0.60$: $r^2=0.90$ *and,*
For OHP; i_{pII} *(mA)* $= 0.228\, \upsilon^{1/2}\, (mV\, s^{-1})^{1/2} - 0.53$: $r^2=0.99$
From Fig. 6 and the above equations we notice that; 0.9 < correlation coefficient (r^2) <1. So we suggest that the electroformation of both POCP and POHP may be described partially by a diffusion-controlled process (diffusion of reacting species to the polymer film/solution interface). [Ardakani et al, 2009] It seems that, initially the electroformation of radical cations

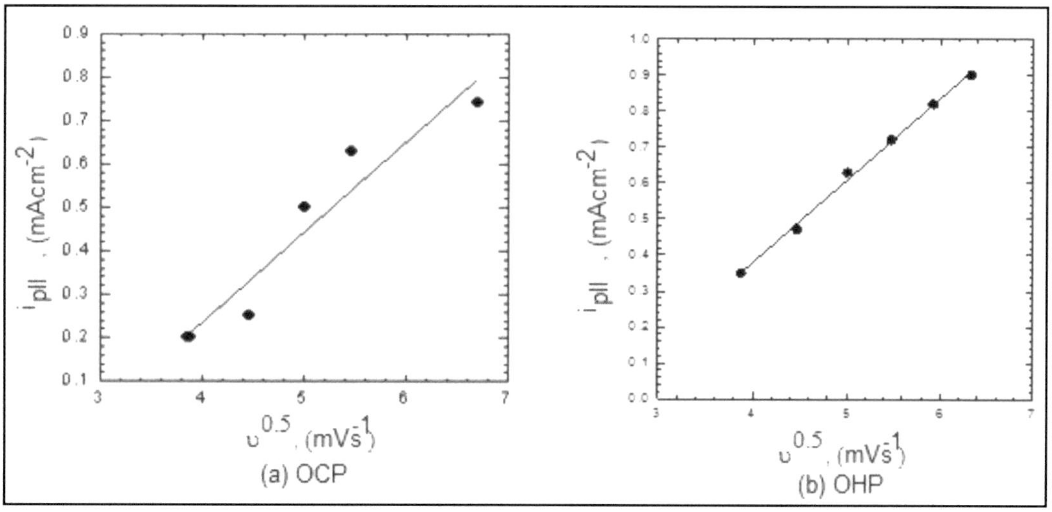

Fig. 6. Relation between ipII and square root of scan rate $\upsilon^{0.5}$

is controlled by charge transfer process. When the polymers become thick, the diffusion of reactant inside the film becomes the slowest step, the process changed to diffusion transfer, which confirms the data in Figure 4.

The intercepts in Figure 6 are small and negative, -0.60 and -0.53 for OCP and OHP respectively, which could be attributed to a decrease of the active area of the working electrode during the positive scan [Zanartu et al , 2002] or the increase of the covered area of working electrode by the adhered polymer sample, which confirm the data of Figure 4 .

3.1.2 Kinetic studies

The electropolymerization kinetics was investigated by using aqueous solution containing (different monomer concentrations in the range between 0.01 and 0.04M in case of OCP and in the range between 0.3 and 0.6M in case of OHP where H_2SO_4 concentration in the range between 0.3 and 0.6M at 303 K in both case. The cyclic voltammogram for each monomeric systems and the relation between the log i $_{pII}$ vs log [monomer] or log i $_{p11}$ vs log [H_2SO_4] are plotted from which linear relations were obtained.

3.1.2.1 Effect of monomer concentration on the electropolymerization processes

The influence of OCP and OHP concentrations on the CV behavior was studied using scan rate of 25 mVs^{-1} is shown in Figure7 (a and b). The voltammograms show that, the anodic peak current densities (i_{pI1}) increase with the increasing of the monomer concentration. This is obvious due to the increased availability of the electroactive species, OCP and OHP in solution, which is again in accordance with Eq.(4).

At higher monomer concentrations (i.e. concentration > 0.04 M for OCP and concentration > 0.08 M for OHP), no noticeable increase in peak currents was observed. This suggests that, at higher concentration, the oxidation reaction is not limited by diffusion alone

A double logarithmic plot of the current density related to oxidation peaks (II) against monomers concentrations are graphically represented in Figure 7 (c and d). Straight lines with slope of 1.1 for OCP and 0.96in case of OHP were obtained. Therefore, the reactions order with respect to both monomers concentration is a first-order reaction.

3.1.2.2 Effect of H2SO4 concentration on the polymerization process

The influence of acid concentration in the range between (0.3 and 0.6M) on the CV using of OCP or OHP using scan rate of 25 mVs^{-1} is represents in Figure 8 (a and b). Voltammograms show that, the anodic peak current densities (i_{pI1}) increase with the increasing of the acid concentration in both cases. At higher acid concentrations for both cases (i.e. concentration > 0.6 M), no noticeable increase in peak currents were observed but, it began to decrease as a result of degradation and the solubility of the polymer film from the platinum surface. A double logarithmic plot of the current density related to oxidation peaks (II) against acid concentrations in the range between 0.3 and 0.6M is graphically represented in Figure8 (c and d). Straight lines with slope of 0.98 in case of OHP and of 0.74in case of OHP were obtained. Therefore, the reactions may be considered as first-order with respect to H_2SO_4 concentration in both cases. Depending upon the above results, the kinetic rate laws obtained from this method can be written as:

For OCP; $R_{P,E} = k_E\ [monomer]^{1.1}\ [acid]^{0.98}$ *and*

For OHP; $R_{P,E} = k_E\ [monomer]^{0.96}\ [acid]^{0.74}$

Where, $R_{P,E}$ is the electropolymerization rate and k_E is the kinetic rate constant.

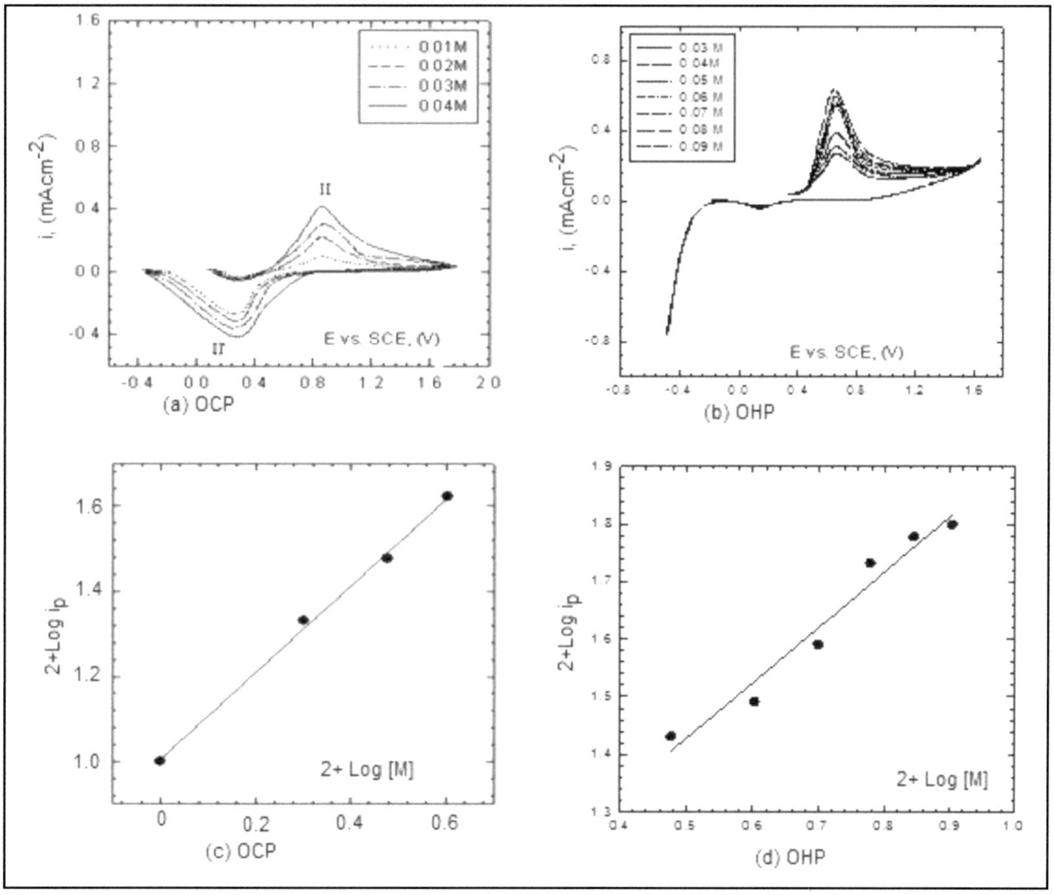

Fig. 7. Cyclic voltammetry curves showing the effect of monomers concentration on electropolymerization and its related double logarithmic plot.

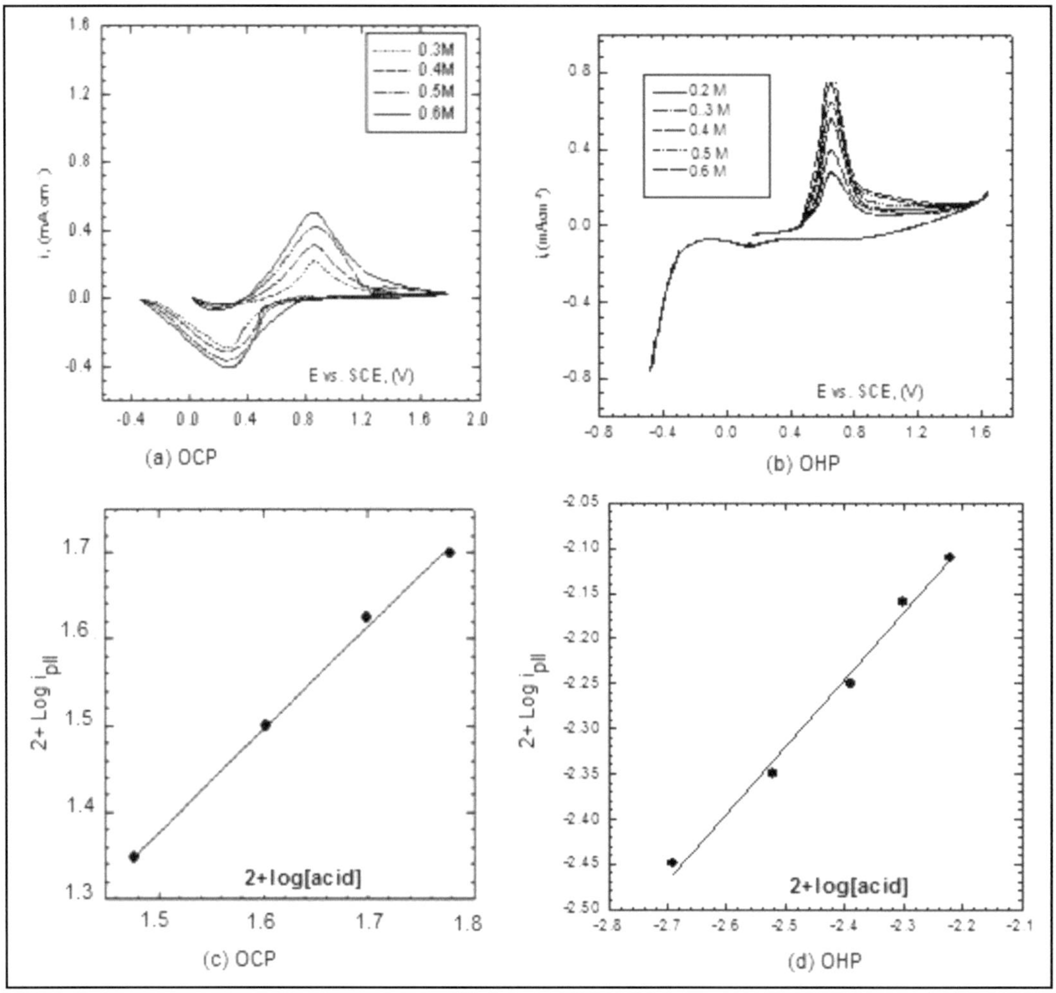

Fig. 8. Cyclic voltammetry curves showing the effect of acid concentration on electropolymerization and its related double logarithmic plot.

3.1.2.3 Effect of temperature and calculation of thermodynamic parameters

The potentiodynamic polarization curves as a function of the solution temperature in the range between 293 and 313 K under the optimum experimental conditions as mentioned above (i.e. 0.04M monomer and 0.6M acid in case of OCP or 0.08M monomer and 0.6M acid in case of OHP) were illustrated in Figure 9(a and b). From the figure, it is clear that, an increase of the reaction temperature resulted in a progressive increase of the charge included in the anodic peaks. The plot of the log (i_{p1}) versus 1/T is represented in Figure 9 (c and d), straight lines are obtained with slopes equal to -1.44 in case of OCP and -1.17 in case of OHP. The apparent activation energies were calculated using Arrhenius equation [sayyah et al, 2010] and it is found to be 27.57kJ mol^{-1} for OCP and 22.39 kJ mol^{-1} for OHP. The enthalpy ΔH^* and entropy ΔS^* of activation for the electropolymerization reaction can be calculated from Eyring equation plot at different temperatures (Fig. 9(e and f)) we obtained linear relationships with slopes of $-\Delta H^*/2.303R$ and intercepts of log {(R/Nh) +

ΔS*/2.303R}. From slopes and intercepts the values of ΔH* and ΔS* were found to be 31.54 k J mol^{-1} and -99.68 JK^{-1} mol $^{-1}$, respectively for OCP and 19.87k J mol^{-1} and -286.69 JK^{-1} mol $^{-1}$, respectively for OHP.

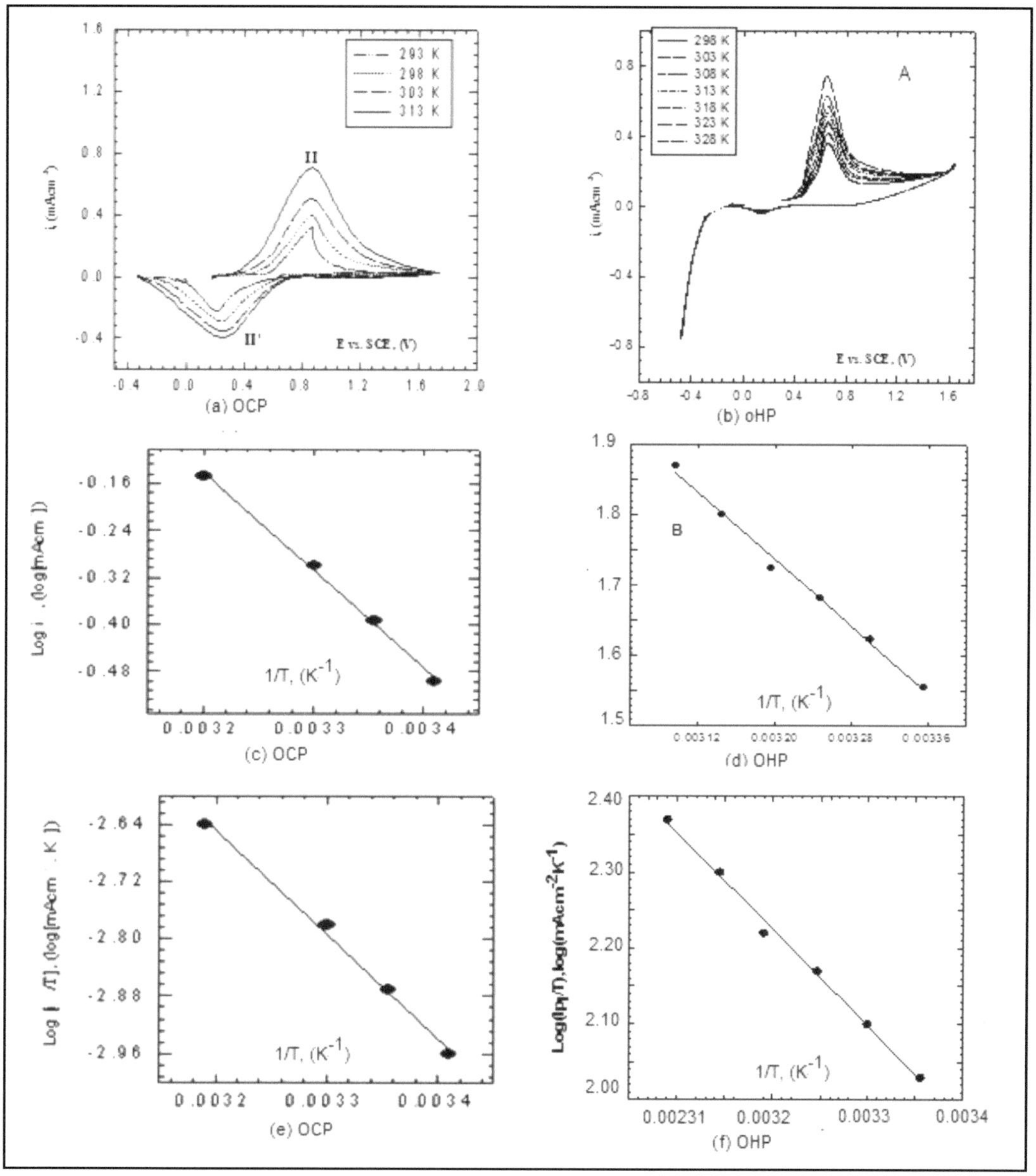

Fig. 9. Cyclic voltammetry curves for the effect of temperature on the electropolymerization, Arrhenius plot and Eyring equation plot

3.2 Electroactivity of the Pt-polymer film

The obtained polymer films adhere Pt-electrode -after preparation at the optimum conditions of preparation (0.04M monomer and 0.6M Acid for OCP and 0.08M monomer and 0.6M Acid for OHP at 303 K using 25 mV s^{-1})- were transferred to another cell with only 0.6M H_2SO_4 and cycled in the same range as the above work. The obtained date shows that, the oxidation peaks that was developed at +863 SCE and +622 mV vs. SCE for OCP and OHP respectively were disappear completely but the H_2 peak (developed at -300 and -200 mV vs. SCE for OCP and OHP respectively) are still found and a broad reduction peaks is observed (at ~ 300 and 200mV vs. SCE for OCP and OHP respectively) which is attributed to the reduction of the formed polymer films. The lack of oxidation peaks confirm the insulating properties of the films and lake of active species inside the polymer films. By repetitive cycling, the current of peak I still the same where the current of peak II decreases as a result of decreasing the polymer amount on Pt surface.

3.3 Mechanisms of electropolymerization

The anodic oxidative polymerization of OCP is preceded in different steps as follows:

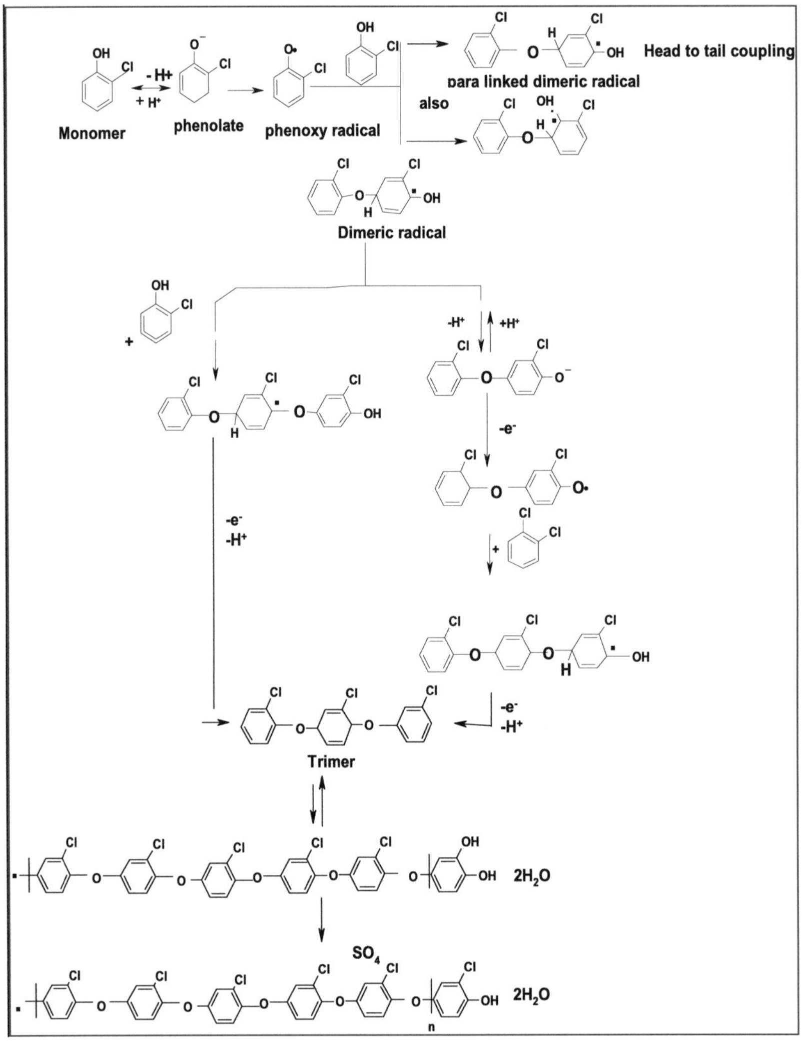

Scheme 1. The mechanism of electropolymerization of OCP.

The anodic oxidative polymerization of OHP is preceded in different steps as follows;

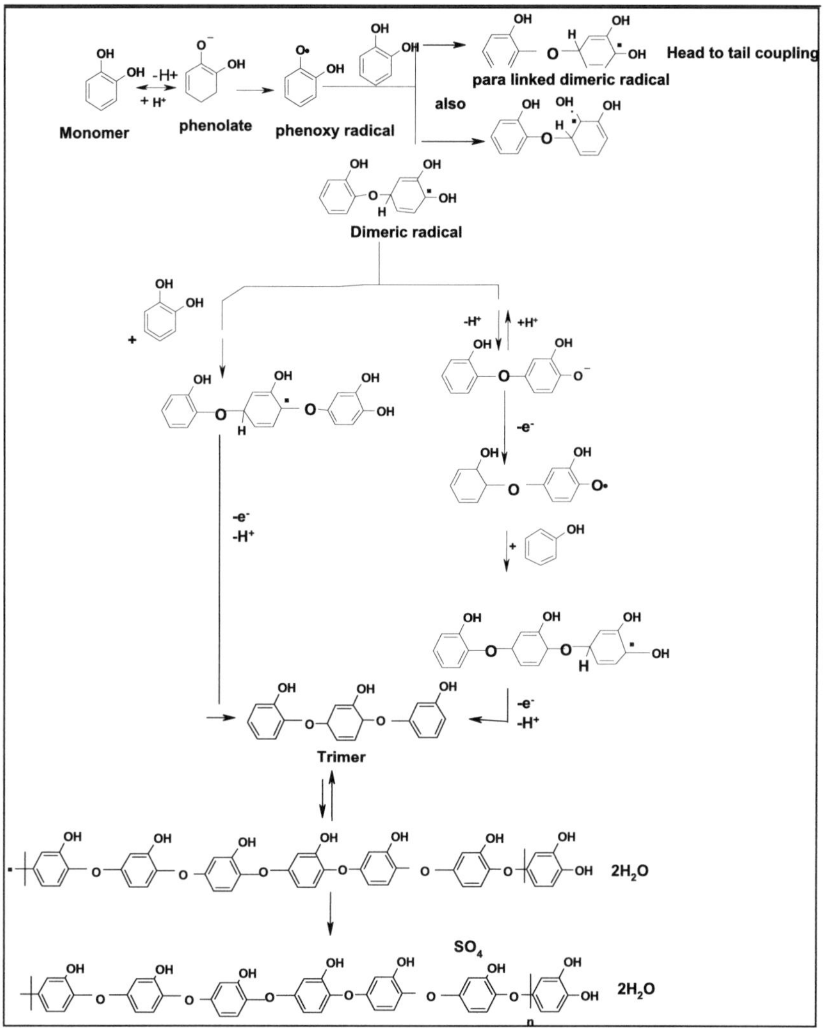

Scheme 2. The mechanism of electropolymerization of OHP.

3.4 Structure determination of the obtained polymers by elemental and spectroscopic an

3.4.1 Elemental analysis

Elemental analysis of the obtained polymers by anodic oxidative Electropolymerization of OCP and OHP were carried out in the micro- analytical laboratory at Cairo University. The percentage C, H, Cl and S for all investigated polymers samples are summarized in Table 2

The found elemental analyses are in a good agreement with the calculated data for the above suggested structures in schemes 1and 2.

No	Structure of polymer	Elemental analysis			
		C%	H%	Cl%	S%
		Cal/found	Cal/found	Cal/found	Cal/found
1	POCP	47.1 47.5	2.48 2.10	23.22 23.01	4.18 4.20
2	POHP	55.24 54.25	3.83 4.15	-- --	4.00 3.48

Table 2. Elemental analytical data of the prepared homopolymers

3.4.2 Spectroscopic analysis

3.4.2.1 Ultraviolet spectroscopic studies

The UV vis. spectra of OCP and its homopolymer POCP and OHP and its homopolymer POHP are shown in Fig 10 (a-d)

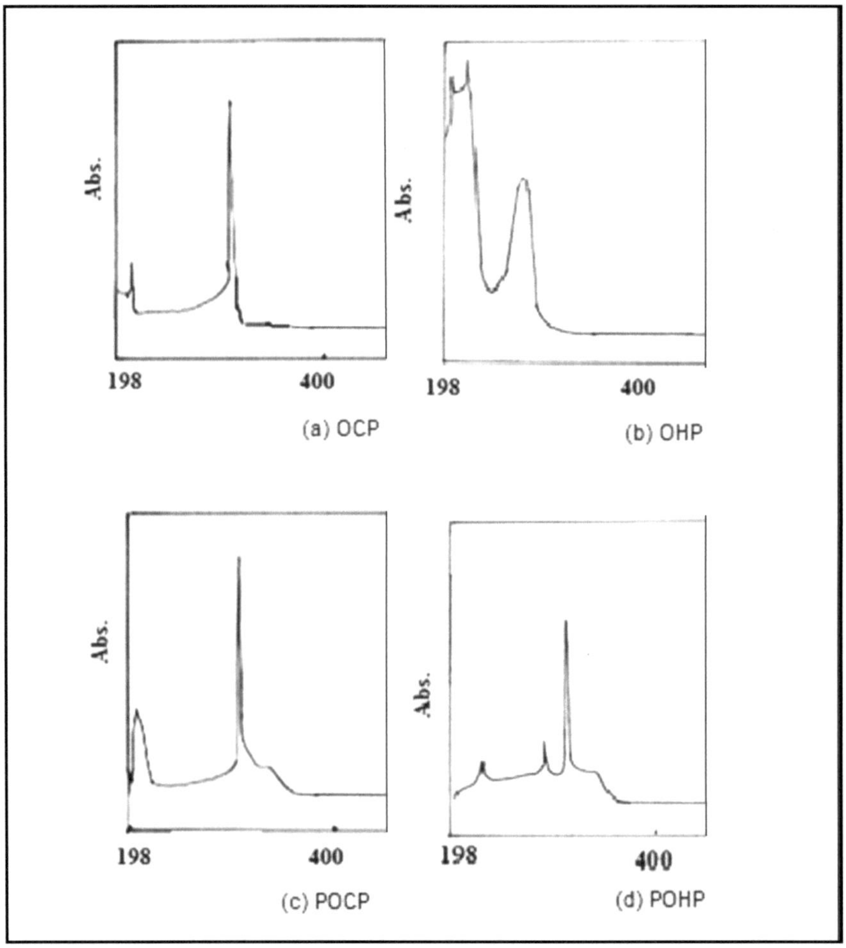

Fig. 10. Uv-vis. Spectra of OCP, OHP, POCP and POHP.

The absorption bands shown in the spectra are summarized in Table 3.

λ_{max}				Assignment
OCP	POCP	OHP	POHP	
---	---	200	---	π- π* transition
212	208	---	---	
---	---	225	223	
---	---	---	---	
---	---	---	273	
306	307	302	305	

Table 3. The Uv-vis. assignments.

3.4.2.2 Infrared spectroscopic studies

The infrared spectra of OCP monomer and its prepared homopolymer POCP are represented in Fig. 11(a) where the infrared spectra of OHP monomer and its prepared homopolymer POHP are represented in Fig. 11(b). The IR absorption bands and their assignments are given in Tables 4 and 5.

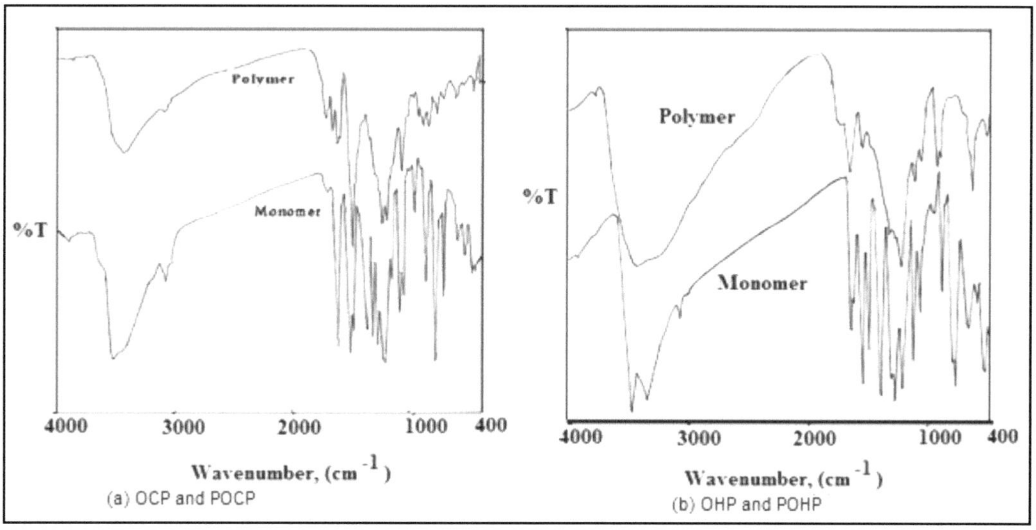

Fig. 11. IR spectra of OCP, POCP, OHP and POHP.

Wave number(cm^{-1}) Name		Assignments
OCP	POCP	
750^{s}	--	CH out of plane bending for 1,2 di-substituted benzene ring
833^{s}	--	
--	823^{s}	CH out of plane bending for 1,2,4 tri-substituted benzene ring
925^{s}	920^{m}	Stretching vibration for C-Cl group
1028^{s}	1058^{s}	Stretching vibration for C-O group
1055^{s}		
--	1187^{m}	Stretching for incorporation in the polymer sulphate group
1456^{s}	--	Stretching vibration of C=C in benzene ring
1482^{s}	1475^{s}	
1589^{s}	1490^{s}	
1645^{m}	1698^{s}	
3073^{m}	3070^{w}	Stretching vibration for CH aromatic
3520^{b}	3419^{b}	Stretching vibration intermolecular hydrogen solvated OH group or end group OH of polymeric chain

Table 4. IR adsorption bands and their assignments for OCP and POCP.(*where; s: strong, w: weak, b: broad, m:medium*}

Wave number(cm^{-1}) Name		Assignments
OHP	POHP	
746^{s}	--	CH out of plane bending for 1,2 di-substituted benzene ring
762^{m}	--	
851^{s}	--	
--	857^{s}	CH out of plane bending for 1,2,4 tri-substituted benzene ring
--	1176.4^{m}	SO_4 incorporation in the polymer
1188^{s}	1197^{s}	C-O stretching vibration
1268^{s}	1251^{s}	
1468^{s}	1475.3^{s}	C=C stretching vibration in benzene ring
--	1508^{s}	
1597^{s}	--	
1619^{s}	--	
--	1642^{s}	
3049^{w}	3100^{w}	CH stretching vibration in benzene ring
--	3382^{b}	intermolecular hydrogen solvated OH stretching vibration
3451^{m}	3743^{b}	Free OH stretching vibration
3328^{m}		

Table 5. IR adsorption bands and their assignments for OHP and POHP.(*where; s: strong, w: weak, b: broad, m:medium*)

3.4.2.3 ^{1}HNMR spectroscopic studies

The ^{1}HNMR spectrum of the prepared POCP and POHP are represented in Fig.12 (a and b). The figure shows two solvent signals at δ=2.55 ppm and δ=3.55 ppm in case of POCP where in case of POHP the solvent signals appear at δ=2.15 ppm and δ=2.55 ppm. The protons of benzene rings in the polymeric structures appear in the region from δ=6.04 to δ=8 ppm in both cases. The singlet signal appears at δ=4.3 ppm in case of POCP and at 5.5 ppm in case of POHP is attributed to OH protons for water of solvation. The singlet signal appears at δ=10.1ppm in case of POCP and at δ=9.5 is attributed to OH proton attached to for benzene ring. The signals of different (OH) are disappeared when deuterated water was added to the investigated sample.

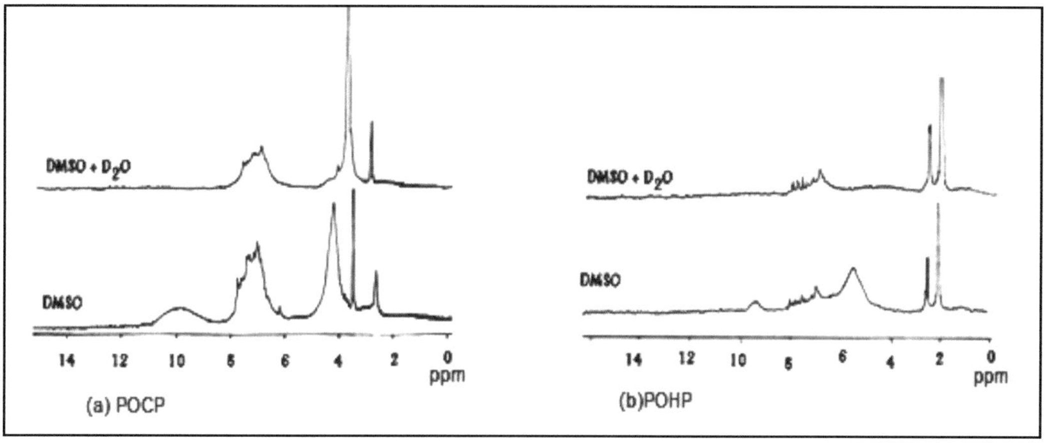

Fig. 12. ^{1}HNMR spectra of POCP and POHP in DMSO and DMSO+D_2O.

3.4.2.4 Thermogravimetric analysis(TGA)

Thermogravimetric analysis (TGA) for the electrochemically prepared POCP and POHP samples have been investigated and the TGA-curve is represented in Figure 13 (a and b).
The TGA steps of the prepared POCP are shown in Fig 13(a) from which it is shown that there are three stages during thermolysis of the POCP sample
The first stage : including the loss of $2H_2O$ molecules in the temperature range between 25°C and 120°C , The weight loss was found to be (5.30%) which is in good agreement with the calculated value (5.70%)this is in good agreement with what was found in the literature for water release [Sayyah et al 2010].
second stage: includes the loss of the dopant species , SO_3 and one benzenoide ring in the temperature range between 120 °C and 400°C , the weight loss for this step was found to be 24.04% which is in good agreement with the calculated value (24.8%)
The third stage: in the range of temperature >490 °C, the remaining part of the polymer decomposition,is equal to 70.70. The found data equal to 69.5.
The TGA steps of the prepared POHP are shown in Fig. 13 (b) from which it is shown that there are four stages during thermolysis of the polymer sample
The first stage: including the loss of $2H_2O$ molecules in the temperature range between 25 °C and 100°C , The weight loss was found to be(4.35%) which is in good agreement with the calculated value(4.6%) .

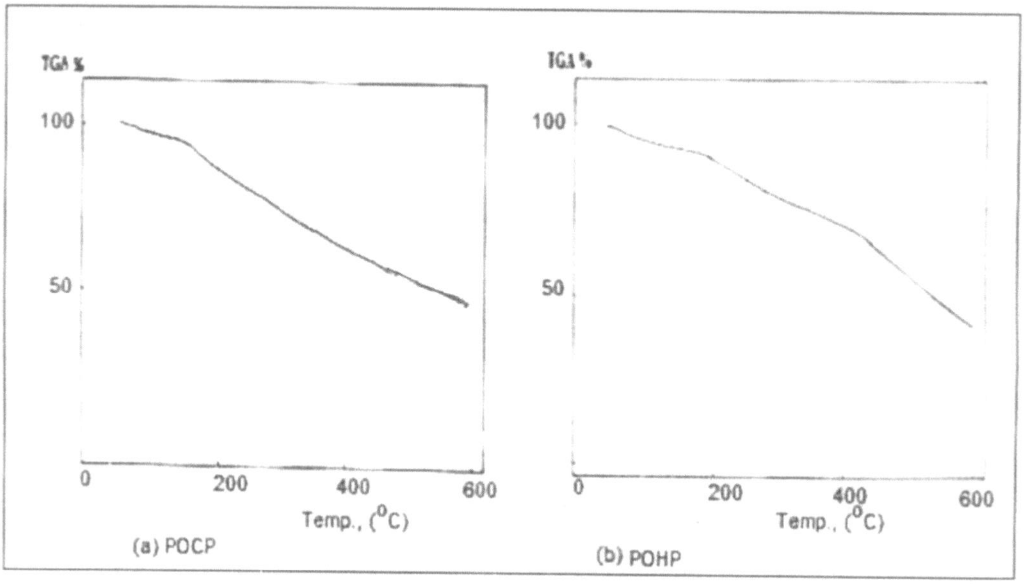

Fig. 13. TGA of POCP and POHP.

The second stage: includes the loss of the dopant species , SO_3 and one OH-radical and H_2O of sulfuric acid in the temperature range between 100 °C and 250 °C , the weight loss for this step was found to be(14.68%) which is in good agreement with the calculated value (14.7%)

The third stage: in the temperature range between 250 °C and 400 °C, includes the benzenoid ring and 5 molecules of OH-.The weight loss for this step was found to be (24.8 %) which is in good agreement with the calculated value(24. 6%).

The fourth step: the remaining part of the polymer decomposition,was found to be 56.17 in the range 400-600 °C and the calculated value is equal to 56.1. c.f. Table 6

Temperature range	Polymer name		The released molecules or Radicals
	POCLP	POHP	
	Calculated / found %	Calculated/ found %	
25-125	5.70 / 5.30	4.60 / 4.35	Two water molecule
100-250	--------	14.70 / 14.68	SO_3+H_2O+hydroxyl radical
120-400	24.80 / 24.04	-------	SO_3+ one molecule of benzenoide
250-400	--------	24.60 / 24.80	One molecule of benzenoide + 5hydroxyl radicals
400-600	69.50 / 70.66	56.10 / 65.17	The remained polymeric chain

Table 6. TGA analysis of POCP and POHP

3.4.2.5 Surface morphology study

In most conditions, a homogeneous, smooth and well-adhering brown POCP or black POHP films were electrodeposited on platinum electrode surfaces. The X-ray diffraction pattern shows that the prepared polymers are amorphous as shown in Figure 14 (a and b).

The surface morphology of the polymers obtained at the optimum conditions was examined by scanning electron microscopy. The SEM micrographs show a smooth lamellar surface feature of POCP and a smooth feature with uniform thickness for POHP (c.f., Fig 14 (c and d)).

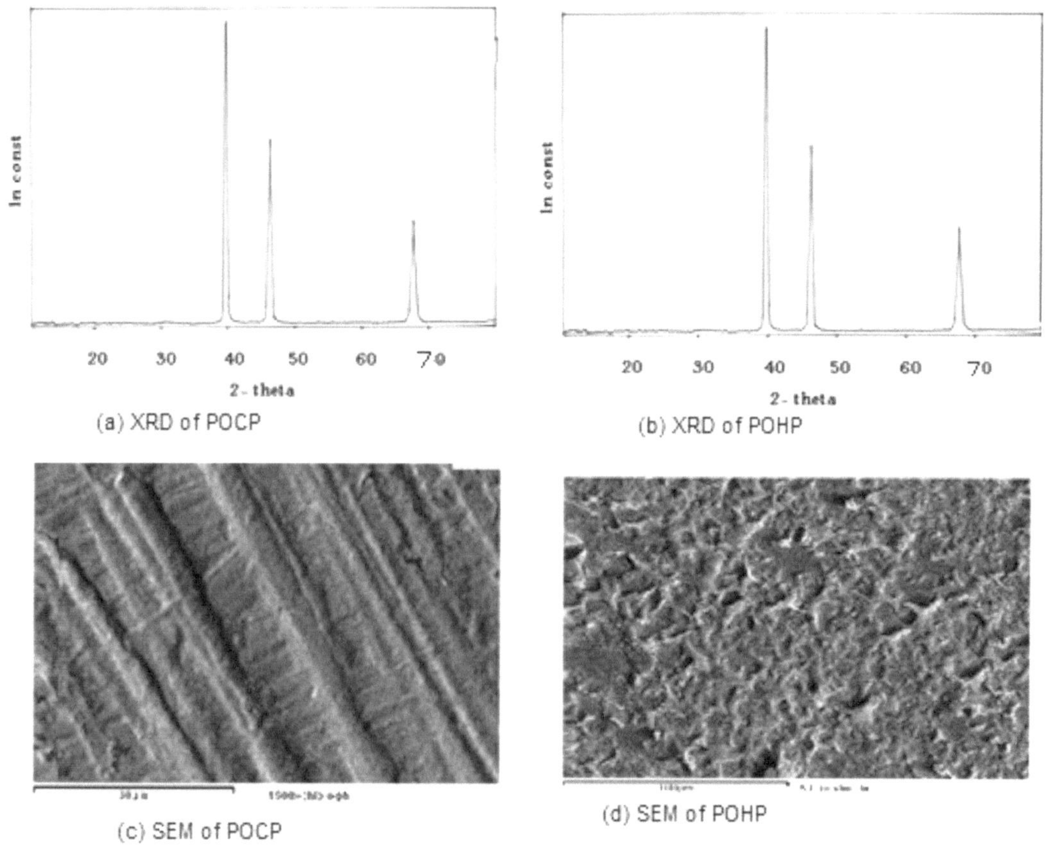

Fig. 14. XRD and SEM of POCP and POHP adhere Pt surfaces.

3.5 Application of the polyphenols as a pH sensor and dye removal.

3.5.1 The pH sensors

Hydrogen ion is ubiquitous species encountered in most chemical reactions. It quantified in terms of pH -the negative logarithm of its activity:

$$pH = - \log [H] \tag{5}$$

The pH sensors are widely used in chemical and biological applications such as environmental monitoring (water quality), blood pH measurements and laboratory pH measurements amongst others. The earliest method of pH measurement was by means of chemical indicators, e.g. litmus paper that changes its colour in accordance to a solution's pH. For example, when litmus is added to a basic solution it turns blue, while when added to an acidic solution the produced colour is red. Since many chemical processes are based on pH, almost all aqua samples have their pH tested at some point. The most common systems for pH sensing are based upon either amperometric or potentiometric devices. The most popular potentiometric approach utilises a glass electrode because of its high selectivity for hydrogen ions in a solution, reliability and straight forward operation. Ion selective membranes, ion selective field effect transistors, two terminal microsensors, fibre optic and fluorescent sensor, metal oxide and conductometric pH-sensing devices have also been developed . However, these types of devices can often suffer from instability or drift and, therefore, require constant re-calibration. Although litmus indicators and other above-mentioned pH sensors are still widely used in numerous areas, considerable research interest is now focused on the development of chemical or biological sensors using functional polymers. Thus, electrosynthesized polymers are considered to be good candidates as pH sensors due to the fact that they are strongly bonded to the electrode surfaces during the Electropolymerization step.

3.5.1.1 POCP modified Pt- electrode as pH sensor

3.5.1.1.1 Potentiometric study of POCP

In recent years, there has been a growing interest in electropolymerized film chemically modified electrodes and their application as potentiometric sensor particularly as P^H sensor. To investigate the effect of the thickness on the potentiometric response. POCP modified electrode of different thickness were prepared. Fig. (15) shows the potentiometric response of POCP film electrode in a wide range of pH (2-11) as a function of thickness ,we observed that this electrode gave a linear response over pH range with potentiometric slope values ranging from 26.6 to 40.72 mV/pH with the difference of the thickness as summarized in Table (7).

	At pH range (2-12)		At pH range (5-9)	
No of Cycles	-slope, (mV/pH)	r^2	-slope, (mV/pH)	r^2
3	40.72	0.96	56.19	0.98
5	37.70	0.99	42.12	0.98
10	31.70	0.96	35.09	0.99
15	26.60	0.93	28.11	0.85

Table 7. the potentiometric response of POCP modified electrode with different thickness at different range of pH.

From Figure 15 and Table 7, it is clear that the calculated potentiometric slope decreased as the polymer film thickness increased. Where the thin film of POCP electrode shows potentiometric slope equal to 40.7mV/pH and the thickest POCP shows decrease in potentiometric slope equal to 26.60 mV/pH, this decrease may be attributed to the diffusion rate of hydrogen ion across this modified electrode.
But when we use more limited pH range (4-9) as shown in Fig. 16 we noticed that the potentiometric response improved where the potentiometric slope was found to be 56.19 mV/pH for the thin POCP and decreased to 28.11 mV/pH for the thickest POCP film, from this result we noticed that this electrode might not an effective pH sensor for more basic or more acidic solutions but it is a good sensor in moderate pH range.

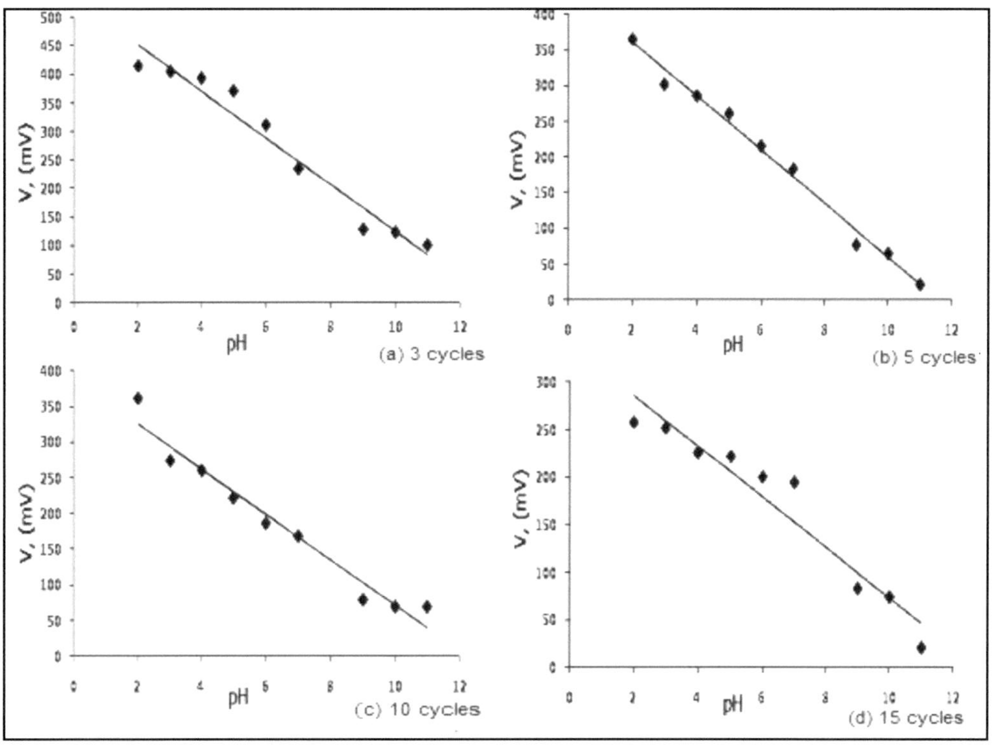

Fig. 15. POCP response at different pH values (2-12), the prepared film after different no of cycles.

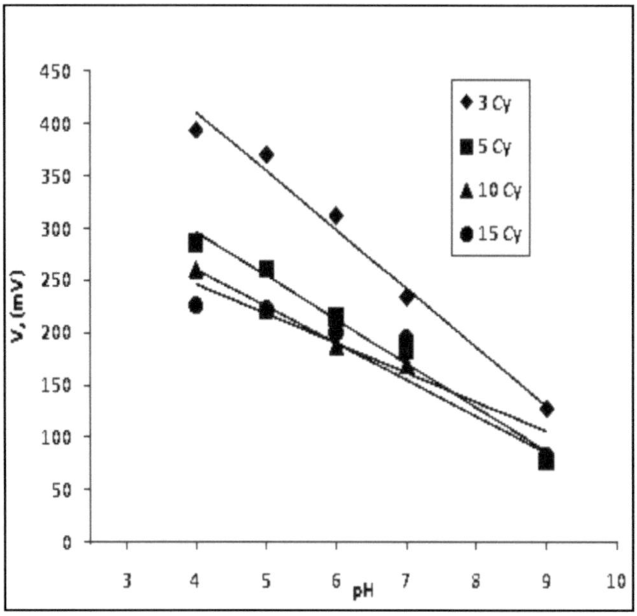

Fig. 16. POCP response at pH range (4-9) with different thicknesses

3.5.1.1.2 Electrode stability

The Potentiometric response of the chemically modified POCP electrode was examined over a period of 9 days in order to test the stability of the electrode. We observed that the chemically modified electrode shows linear behavior from pH (4-10) during 9 days as shown from Table 8 and Fig. 17 and the sensitivity of this coating decrease considerably with time since the slope varies from 40.7 at the first day to25.5mV/pH unit at the ninth day. Consequently polymer coating is not stable over a period of 9 days and the sensor cannot be used for several measurements. However, this electrode can be sensitive for one use sensor due to its sensitivity during the first measurement. It is important to note that the POCP film was found to peel off the supported electrode and it is a possible reason for the change of response slope.

Time, (day)	-slope, (mV/pH)	r^2
1	42.66	0.97
2	34.71	0.94
3	29.2	0.94
4	27.46	0.87
5	24.46	0.89
6	24.51	0.93
7	29.23	0.92
8	31.60	0.90
9	24.51	0.93

Table 8. The slope and r^2 values of POCP at different time intervals.

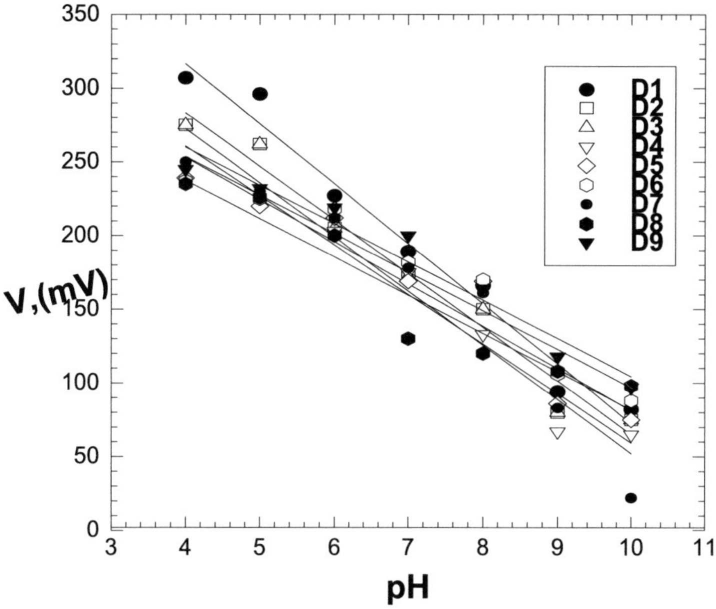

Fig. 17. P OCP response at different days, the prepared film from the first day to the ninth day

3.5.1.1.3 Response mechanism

We have shown that the potentiometric responses to pH changes of the different modified electrodes are linear in the range 4–9. These responses must be mainly attributed to the polymer films rather than the platinum substrate. Possible explanation is the affinity of the numerous hydroxyl groups and Cl atoms to the protons in solutions. The reaction of H^+ onto polymer creates local charge density excess at the electrode surface. Surface reactions seem to take place on the polymer film, essentially protonation and deprotonation of superficial OH groups of the polymers as symbolically described as follow;

$$\text{P (Polymer)} + H^+ \longrightarrow PH^+ \quad (6)$$

When the equilibrium is reached at the polymer/solution interface, we can write the equilibrium expression K of the surface reaction (1) and the equilibrium potential E as:

$$K = \frac{[PH^+]}{([P][H^+])} \quad (7)$$

and

$$E = E_0 + \left(\frac{RT}{F}\right)\ln\left(\frac{[PH^+]}{[P]}\right) = E_0' + \left(\frac{RT}{F}\right)\ln[H^+ \quad (8)$$

According to this mechanism of reaction, we are waiting for a potentiometric response slope of 59 mV/pH unit at 25 °C at all pH values. But our electrodes showed lower response slope. The presence of anionic and cationic responses of the polymer film electrodes, due to the presence of ions (K^+, Na^+, Cl^-,....etc of the buffer solutions) in the different solutions probably caused this difference of response slope.

3.5.2 Dye removal

Water resources are of critical importance to both natural ecosystem and human developments. Increasing environmental pollution from industrial wastewater particularly in developing countries is of major concern. Many industries like dye industries, textile, paper and plastics, use dyes in order to color their products. As a result they generate a considerable amount of colored wastewater. The presence of small amount of dyes (less than 1 ppm) is highly visible and undesirable. Many of these dyes are also toxic and even carcinogenic and pose a serious threat to living organisms. Hence, there is a need to treat wastewaters containing toxic dyes and metals before they are discharged into the water bodies.

Many different treatment methods, including biological treatment, coagulation/flocculation, ozone treatment, chemical oxidation, membrane filtration, ion exchange, photocatalytic degradation and adsorption have been developed for the removal of dyes from wastewaters to decrease their impact in environment. Among these methods, adsorption is a well known separation process and is widely used to remove certain classes of chemical pollutants from waters, especially those that are practically unaffected by conventional biological treatments.

3.5.2.1 POHP as MB dye adsorbent

3.5.2.1 Adsorption kinetics

The adsorption kinetics was conducted to determine the optimum adsorption time for the adsorption of MB dye by POHP. The effect of the contact time on adsorption of MB onto POHP is represented in Fig. 18.

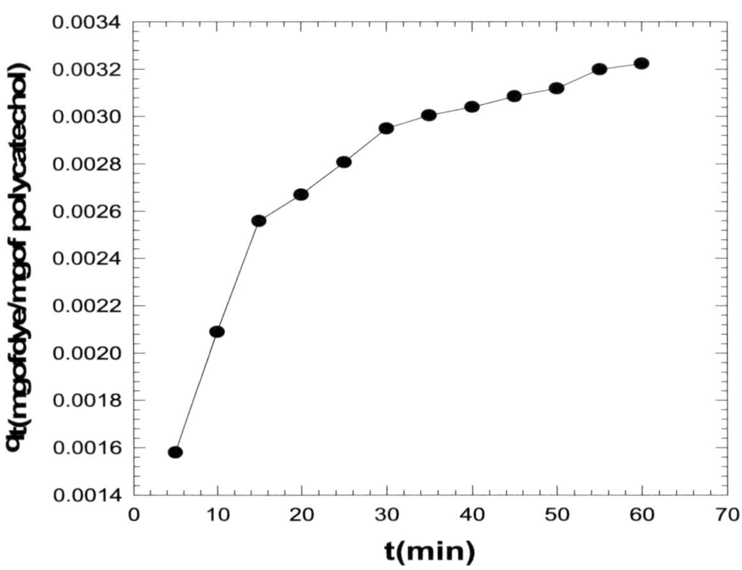

Fig. 18. The effect of contact time on adsorption of MB onto POHP.

The adsorption capacity increases rapidly during the initial adsorption stage and then continues to increase at a relatively slow speed with contact time. The obtained result reveals that, at the beginning, the adsorption mainly occurs on the surface of POHP so the adsorption rate is fast. After the surface adsorption is saturated, the adsorption gradually proceeds into the inner part of POHP via the diffusion of MB dye into the polymer matrix, leading to a lower adsorption rate In order to evaluate the adsorption kinetics of MB dye onto POHP, the pseudo-first-order, pseudo-second-order, intraparticle diffusion and Elovich models were employed to interpret the experimental data, as shown below:
Pseudo-first-order equation:

$$\text{Log}\,(q_e - q_t) = \log q_e - k_1 t/2. \quad (9)$$

Pseudo-second-order equation:

$$t/q_t = 1/k_2 q^2 e + t/ \quad (10)$$

Intraparticle diffusion equation:

$$q_t = kit^{0.5} + \quad (11)$$

Elovich equation:

$$q_t = b + a\ \ln \quad (12)$$

Where q_e and q_t are the amounts of dye adsorbed (mg mg^{-1}) at equilibrium and at time t (min), and t is the adsorption time (min). The other parameters are different kinetics constants, which can be determined by regression of the experimental data. The validities of these four kinetic models are checked and graphically represented in Fig. 19. The corresponding kinetic parameters and the correlation coefficients are summarized in Table 9. Based on linear regression (r^2) values, the adsorption kinetics of MB dye have the following order. pseudo-second order > Elovich > pseudo-first-order > intraparticle diffusion models.

Kinetic models	Parameters		r^2
Pseudo-first-order	k_1(min^{-1}) q_e, (mg mg^{-1})	0.02303 1.81×10^{-3}	0.934
Pseudo-second-order	k_2 (mg mg^{-1} min^{-1}) q_e, cal (mg mg−1)	40.32 0.0036	0.999
Intraparticle diffusion	K_i(mg $mg^{-1}min^{-0.5}$) C(mg mg^{-1})	2.68×10^{-4} 1.31×10-3	0.891
Elovich	a b	6.46×10^{-4} 6.59×10^{-4}	0.972

Table 9. Kinetic parameters for adsorption of MB onto POHP.

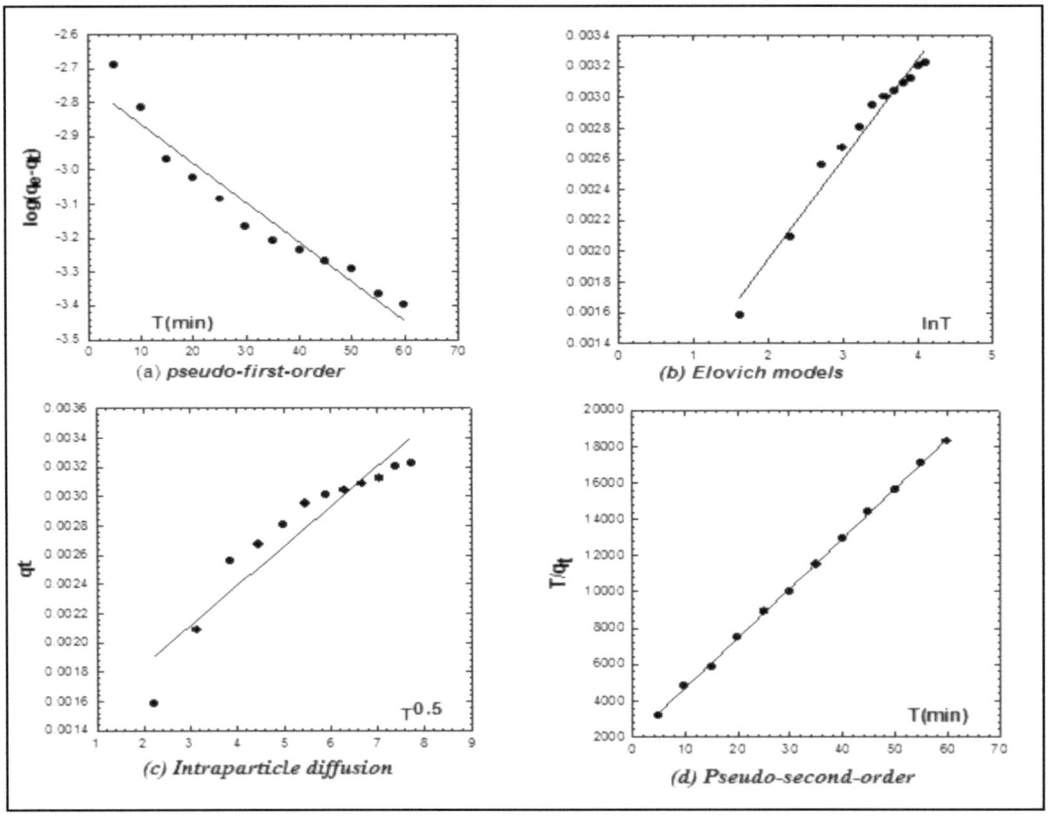

Fig. 19. Adsorption kinetics of MB dye onto POHP.

3.5.2.2 Adsorption isotherm

The adsorption isotherm of MB dye onto POHP is represented in Fig 20. The adsorption capacity of MB dye increases with the increase of dye concentration. This may be attributed to the extent of a driving force of concentration gradients with the increase of dye concentration. Then tends to level off, this is due to the saturation of the sorption site on the adsorbent.

3.5.2.3 Equilibrium isotherm models

The adsorption isotherm is fundamental in describing the specific relation between the concentration of adsorbate and the adsorption capacity of an adsorbent, and it is important for the design of adsorption system. In this study, two important isotherms are selected, that is, Langmuir and Freundlich models, to describe the experimental results of MB dye adsorption on POHP are summarized in Table 10.

The Langmuir isotherm assumes that the adsorption occurs at specific homogeneous sites on the adsorbent and is the most commonly used model for monolayer adsorption process, as represented by Fig. 21 and the following equation:

$$C_e/\ q_e = 1/bqm + C_e/\ gm \qquad (13)$$

Where the constant b is related to the energy of adsorption (Lmg^{-1}), and q_m is the Langmuir monomolecular adsorption capacity ($mg\ mg^{-1}$).

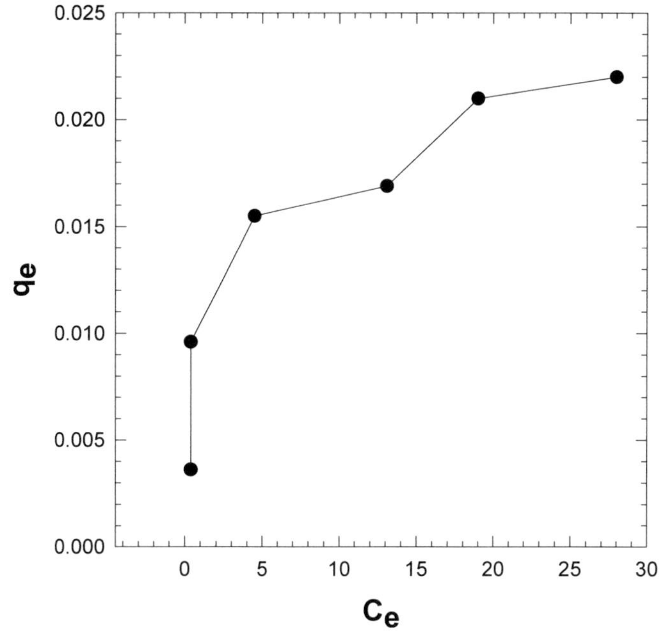

Fig. 20. Adsorption isotherms of MB onto POHP

The essential characteristics of the Langmuir isotherm can be expressed in terms of a dimensionless constant separation factor R_L represented by the equation .

$$R_L = 1/(1 + bC_o) \tag{14}$$

where C_o (mg L^{-1}) is the initial dye concentration. If the value of R_L lies between 0 and 1, the adsorption is favorable.

The Freundlich isotherm is an empirical equation assuming that the adsorption process takes place on heterogeneous surfaces and adsorption capacity is related to the concentration of dye at equilibrium is described by Fig. (21) and the following equation:

$$\log q_e = \log K_f + 1/n \log C_e \tag{15}$$

where K_f (mg mg^{-1}) is roughly an indicator of the adsorption capacity and 1/n is the adsorption intensity.

Sample	Langmuir				Freundlich		
POHP	q_m	b	R_L	r^2	n	Ln K_f	r^2
	0.023	0.34	0.42	0.99	1.74	5.16	0.972

Table 10. Langmuir and freundlich isotherm constants of the adsorption of MB dye on POHP.

It is shown that the applicability of the above adsorption isotherms was compared by judging the correlation coefficients. The results indicate that the Langmuir isotherm fits quite well with the experimental data (r^2= 0.99), whereas, the low correlation coefficients (r^2 > 0.97) show the poor agreements of the Freundlich isotherm with the experimental data. The value of n for Freundlich isotherm is greater than 1, mean while the values of R_L lie between 0 and 1, indicating that MB dye is favorably adsorbed by POHP.

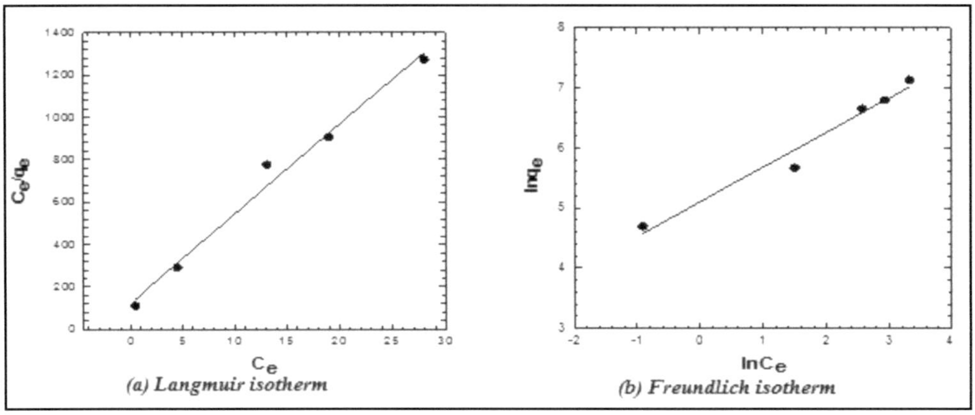

Fig. 21. Langmiur and Frendlich isotherms for adsorption of MB dye on POHP.

4. Conclusions

Cv is useful tool in oxidation of pollutants as phenols. The electropolymerization, by cyclic voltammetry, of phenols on Pt-surfaces is a notoriously complex process which depends on the phenols structure, the potential scan rate, the pH, the temperature and the phenols concentration. Comparing the voltammograms from the different monomer compound solutions, it was demonstrated that the polymer film resulting from both OCP and OHP oxidation leads to the higher surface deactivation degree. This is probably due to the higher adhering properties of POCP and POHP to Pt-surface.

The proposed mechanisms are well confirmed by different tools and agree with that mentioned in literatures.

The oxidation processes are partially controlled processes with stable diffusion coefficients at different scan rates.

The Obtained polymers are amorphous with higher thermal stability with smooth lamellar surface feature for POCP and a smooth feature with uniform thickness for POHP.

POCP modified Pt electrode could be used as one use pH sensor with good response and perfect Nernstian- slope especially at pH range 4-9 but its poor pH sensor at more acidic or basic solution and loose its response by time.

More work must be done to improve both response and stability of the sensor.

POHP is a perfect adsorbent to MB dye from aqueous solution and must be used in purification of waste water.

5. Nomenclatures

CV	Cyclic Voltammetry
OCP	Ortho-chlorophenol

OHP	Ortho Hydroxyphenol
POCP	Poly orthochlorophenol
POHP	Poly orthohydroxy phenol
MB	Methelene blue dye
TGA	Thermal gravemetric analysis
XRD	X-ray diffraction analysis
SEM	Scaninng electron microscopy
UV-vis.	UV-visible spectroscopy
IR	Infra red spectroscopy
^{1}HNMR	Nuclear Magnetic Resonance Spectroscopy
WE	Working electrode
SB	Salt bridge
CE	Counter Electrode
SCE	Saturated Calomel electrode

6. References

Ardakani, M.M., Sadeghiane, A., Moosavizadeh, S., Karimi, M. A., Mashhadizadeh, M. H.(2009). Electrocatalytic Determination of Hydrazine using Glassy Carbon Electrode with Calmagates, Anal. Bioanal. Electrochem. 1(3) : 224-338.

Arslan, G.; Yazici, B & Erbil, M. (2005) . The effect of pH, temperature and concentration on electrooxidation of phenol, Journal of Hazardous Materials B124: 37–43

Comninellis, C. (1994). Electrocatalysis in the electrochemical conversion/combustion of organic pollutants for waste water treatment, Electrochimica Acta, 39 (11-12): 1857-1862.

Ezerskis Z., & Jusys, Z. (2002). Electropolymerization of chlorinated phenols on a Pt electrode in alkaline solution. Part IV: A gas chromatography mass spectrometry study, Journal of Applied Electrochemistry, 32 (5): 543-550.

Fleszar, B. & Jolanta, P. (1985), An attempt to define benzene and phenol electrochemical oxidation mechanism, Electrochimica Acta, 30 (1) : 31-42.

Gattrell, M., & Kirk, D.W.(1993). Study of electrode passivation during aqueous phenol electrolysis, Journal of the Electrochemical Society, 140(4) : 903-911

Gutiérrez, C. (2002).Electrooxidation of 2,4-dichlorophenol and other polychlorinated phenols at a glassy carbon electrode, Electrochim. Acta, 47(15): 2399. Iotov, P.I.& Kalcheva, S.V. (1998). Mechanistic approach to the oxidation of phenol at a platinum/gold electrode in an acid medium , Journal of Electroanalytical Chemistry, 442(1-2) : 19-26. Randles, J. E. B.(1948). A cathode ray polarograph. II. The current voltage curves, Trans. Faraday Soc, 44 : 327-338.

Sayyah, S. M., Abd El-Rehim, S. S., El-Deeb, M. M., Kamal, S. M. & Azooz, R. E., (2010). Electropolymerization of p- Phenylenediamine on Pt-Electrode from Aqueous Acidic Solution: Kinetics, Mechanism, Electrochemical Studies, and Characterization of the Polymer Obtained, J Appl Polym Sci 117: 943–952

Sevick, A. (1948), Oscillographic polarography with periodical triangular voltage, Collect. Czech. Chem. Commun., 13 349-377.

Wang, J., Jiang, M. & Lu, F. (1998), Electrochemical quartz crystal microbalance investigation of surface fouling due to phenol oxidation, Journal of Electroanalytical Chemistry, 444(1, 5) : 127-132

Wang, J., Martinez, T. Daphna, Y. R. & McCormick, L.D. (1991). Scanning tunneling microscopic investigation of surface fouling of glassy carbon surfaces due to phenol oxidation, Journal of Electroanalytical Chemistry and Interfacial Electrochemistry, 313 (1-2): 129-140

Oxidation of Porphyrins in the Presence of Nucleophiles: From the Synthesis of Multisubstituted Porphyrins to the Electropolymerization of the Macrocycles

Delphine Schaming[1], Alain Giraudeau[2], Laurent Ruhlmann[1,2] et al.*
[1]*Laboratoire de Chimie Physique, UMR 8000 CNRS / Université Paris-Sud 11, Faculté des Sciences d'Orsay, bât. 349, 91405 Orsay cedex*
[2]*Laboratoire d'Electrochimie et de Chimie-Physique de Corps Solide, UMR 7177 CNRS / Université de Strasbourg, 4 rue Blaise Pascal, CS 90032, 67081 Strasbourg cedex, France*

1. Introduction

Porphyrins and porphyrins-containing materials have attracted a considerable attention these last decades, not only because of their roles as biological photosensitizers, redox centers and oxygen carriers, but also because of their attractive chemical properties and potential technological applications (Kadish et al., 2000). In particular, molecular engineering and design of controlled spatial assemblies and architectures of porphyrins are fields undergoing wide growth (Griveau & Bedioui, 2011). For example, macromolecular porphyrins-based systems can potentially mimic natural metalloenzyme structures. Indeed, in such systems, proteins are replaced by metalloporphyrins which can mimic the structure and/or the activity of the prosthetic groups of enzymes (Traylor, 1991). Another area of applications of such assembled porphyrins systems consists in the field of nanomaterials, the electronic communication between the macrocycles allowing developments of molecular photonic, electronic or optoelectronic devices (Jurow et al., 2010). Finally, nanocomposite porphyrins-based materials have also been investigated for applications involving energy storage systems, fuel cells and sensors (Di Natale et al., 2010; Ma et al., 2006).

For this purpose, porphyrins-based polymers have received peculiar attention these last decades. For instance, it has been established that a polymeric matrix may provide the best arrangement for a catalytically active center. The immobilized porphyrins appeared also

* Clémence Allain[1], Jian Hao[1], Yun Xia[1], Rana Farha[3], Michel Goldmann[3], Yann Leroux[4] and Philippe Hapiot[4]
[1]*Laboratoire de Chimie Physique, UMR 8000 CNRS / Université Paris-Sud 11, Faculté des Sciences d'Orsay, bât. 349, 91405 Orsay cedex, France*
[3]*Institut des NanoSciences de Paris, UMR 7588 CNRS / Université Paris 6, 4 place Jussieu, boîte courrier 840, 75252 Paris cedex 05, France*
[4]*Sciences Chimiques de Rennes, équipe MaCSE, UMR 6226 CNRS / Université de Rennes 1, campus de Beaulieu, bât. 10C, 35042 Rennes cedex, France*

more stable than biomimetic enzymes in aqueous or organic media, particularly at extreme pH and temperature (Griveau & Bedioui, 2011). Finally, conductive porphyrin polymers allow interesting applications for molecular electronics (Wagner & Lindsey, 1994).
Different strategies for the formation of porphyrin polymers have already been described in the literature. A first common method to polymerize porphyrins consists in the use of bridging ligands which can coordinate the metal center of metalloporphyrins. For instance, such coordination polymers have been obtained by coordination of nitrogenous ligands with ruthenium, osmium or iron porphyrins (Collman et al., 1987; Marvaud & Launay, 1993). Nevertheless, more frequently, covalent polymers have been obtained. Generally, the formation of such polymers relies on the use of porphyrins with polymerizable substituents attached on the ring periphery. Even if classical chemical ways can be used to obtain such polymers (Li et al., 2004), electrochemical route appears as an elegant, attractive and easy strategy to perform polymerization.
Indeed, one of the main advantages of electropolymerization lies in the non-manual addressing of polymers allowing formation of films with a good reproducibility and a controlled thickness. Electropolymerization is also an easy way to functionalize conductive surfaces with a good precision. Moreover, electropolymerization of porphyrins provides densely packed layers that facilitate the electron hopping process between macrocycles. Furthermore, the immobilization of such porphyrins-based materials on an electrode is a convenient way to design electrochemical sensing devices, catalytic electrodes, and to study the enzyme reactions by excluding the requirement of a chemical redox mediator to shuttle electrons between the porphyrins and the electrode (Griveau & Bedioui, 2011).
Electropolymerization of porphyrins has been initially developed by Macor and Spiro, who have polymerized vinyl-substituted porphyrins, the coupling occurring after the formation of vinyl radicals by electrooxidation (Macor & Spiro, 1983). Afterwards, other methods of electropolymerization have been published, for example from hydroxy-, amino-, pyrrole- or thiophene-substituted porphyrins (Bedioui et al., 1995; Bettelheim et al., 1987; Li et al., 2005).
In this context, we have developed an original methodology for the electropolymerization of porphyrins, based on nucleophilic attacks of di-pyridyl compounds directly onto electrooxidized porphyrins. This chapter presents an overview of this new method of electropolymerization.

2. Reactivity of porphyrin π-radical cations and dications with nucleophilic compounds

It is well-known that the oxidation of the π-ring of a porphyrin proceeds *via* two one-electron steps generating the π-radical cation and the dication (Fajer et al., 1970). The reactivity of these porphyrin π-radical cations and dications with nucleophilic compounds has been intensively studied. Dolphin and Felton have observed for the first time the reactivity of oxidized porphyrins with nucleophilic solvents or halides (Dolphin & Felton, 1974). In the case of *β*-octaethylporphyrin derivatives, stable *meso*-substituted porphyrins have been obtained. This is the first example of nucleophilic substitution at the periphery of porphyrin nucleus, allowing afterwards the development of a new and simple convenient synthetic route to obtain varied substituted porphyrins. Indeed, several further works report the nucleophilic attack of nitrogeneous, phosphorous or sulphurous nucleophiles (nitrite ion, pyridine, phosphine, thiocyanate…) onto π-radical cations in *meso*-position of *β*-

octaethylporphyrins or in β-position of *meso*-tetraphenylporphyrins (Barnett & Smith, 1974; Barnett et al., 1976; Rachlewicz & Latos-Grażyński, 1995; Shine et al., 1979). However, it must be noted that the π-radical cations have been in most cases prepared by chemical ways, principally by oxidation of macrocycles with iodine, bromine or peroxide.

Giraudeau *et al.* have for the first time investigated the reactivity of the radical cation of the zinc *meso*-tetraphenylporphyrin (ZnTPP), obtained by a direct electrochemical oxidation (electrolysis), in the presence of pyridine as Lewis base, leading to a macrocycle substituted by a pyridinium in β-position (El Kahef et al., 1986). Afterwards, a similar work has permitted to obtain zinc β-octaethylporphyrin (ZnOEP) with pyridinium *meso*-substituted (Giraudeau et al., 1996, 2001) (Fig. 1.a).

When 4,4'-bipyridine is used instead of pyridine, dimers of porphyrins can also be electrosynthesized (El Baraka et al., 1998; Giraudeau et al., 1996). Indeed, 4,4'-bipyridine presents two accessible nucleophilic sites which can both react with porphyrin rings (Fig. 1.b and c).

More recently, dimers of porphyrins with pyridinium spacer have also been synthesized from macrocycles substituted by a pendant pyridyl group which can react with a ZnOEP radical cation (Schaming et al., 2011a) (Fig. 1.e).

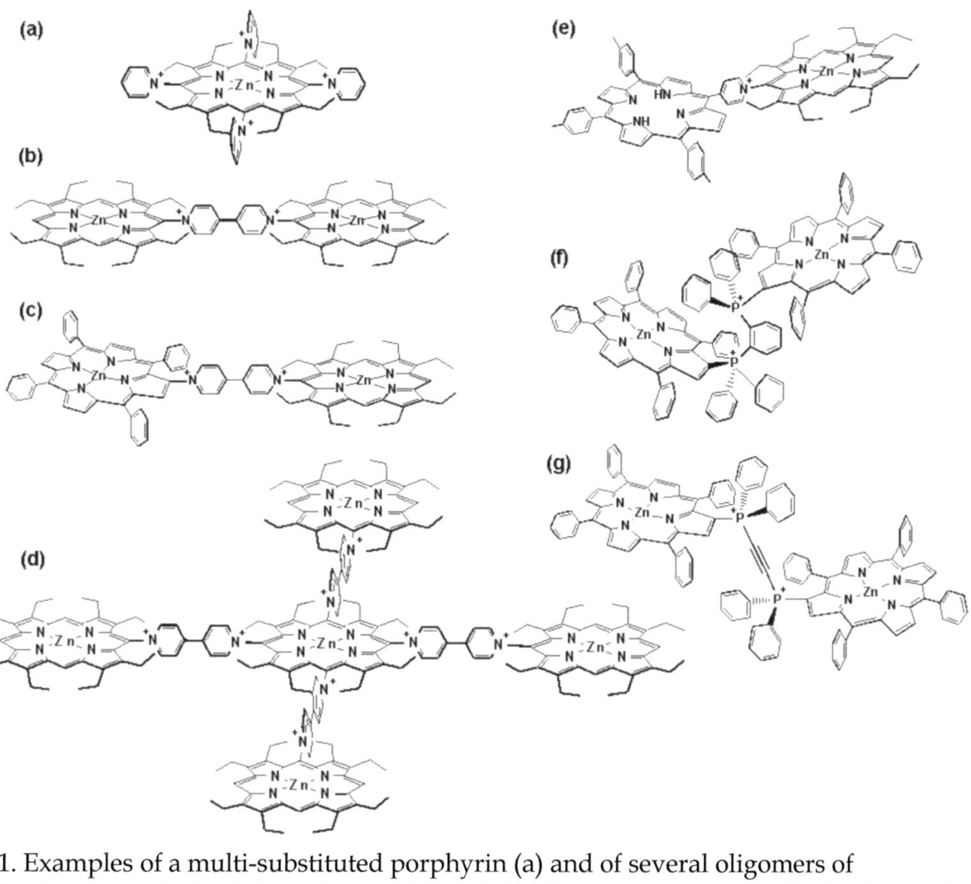

Fig. 1. Examples of a multi-substituted porphyrin (a) and of several oligomers of porphyrins (b-g) obtained from the reactivity of the electrogenerated π-radical cation with Lewis bases such as pyridine-derived species (a-e) or phosphane-derived species (f-g).

To perform all these electrosyntheses, electrolyses at a potential allowing the formation of the radical cation of the porphyrin have been carried out. Nevertheless, an increase of the electrolysis potential can allow multi-substitutions onto macrocycles. Indeed, a judicious control of the applied potential has permitted to obtain di-, tri- and tetra-*meso*-substitutions, leading to the formation of trimers, tetramers and pentamers of porphyrins, respectively (Ruhlmann, 1997; Ruhlmann et al., 1999a) (Fig. 1.d).
Finally, instead of the use of pyridyl-derived compounds as nucleophile, phosphanes have also been used to perform substitutions onto porphyrin rings, allowing the synthesis of dimers or trimers with more flexible di- or tri-phosphonium spacers (Ruhlmann & Giraudeau, 2001; Ruhlmann et al., 2003) (Fig. 1.f and g).
The mechanism of the *meso*-substitutions (onto zinc *β*-octaethylporphyrin (ZnOEP) for instance) has been described as an $E_1C_{Nmeso}E_2C_B$ process (Giraudeau et al., 1996) (fig. 2). Indeed, after formation of the radical cation (step E_1), nucleophilic attack can occur directly onto the *meso*-position (C_{Nmeso}) where initially a proton is present. After a second oxidation (step E_2), this proton can be removed (step C_B), allowing the re-aromatization of the macrocycle. This mechanism will be more precisely described in the case of the electropolymerization process (see part 3.2).

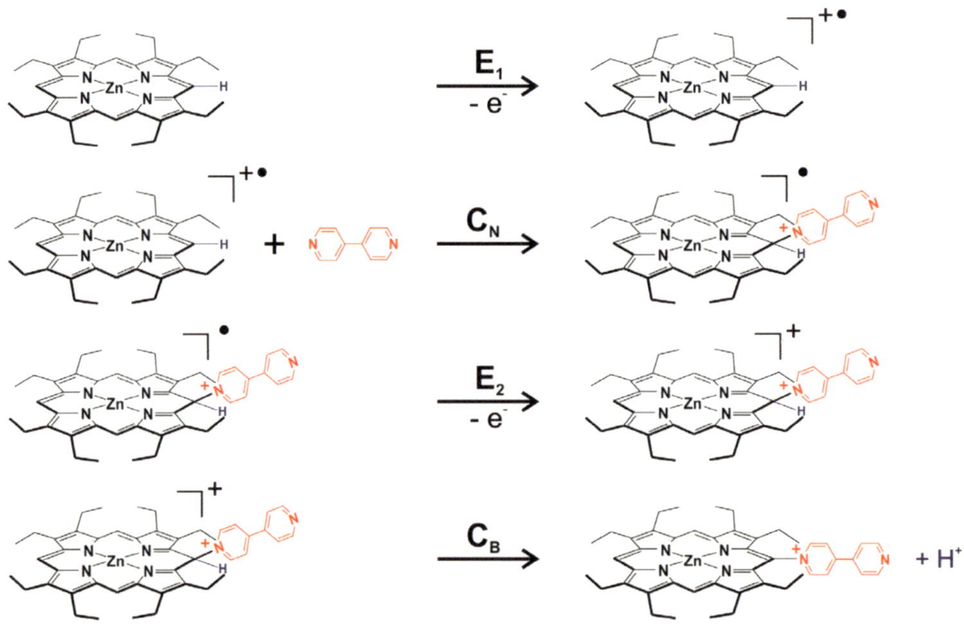

Fig. 2. $E_1C_NE_2C_B$ mechanism explaining the reactivity of a 4,4′-bipyridine (bpy) in *meso*-position of a zinc *β*-octaethylporphyrin (ZnOEP).

In the case of a nucleophilic substitution in *β*-position (allowed when *meso*-positions are already substituted as in the case of the zinc *meso*-tetraphenylporphyrin (ZnTPP)), the mechanism is a little bit different. Indeed, an $E_1C_{Nmeso}E_2C_{N\beta}C_B$ process occurs with two different successive nucleophilic attacks (Rachlewicz et al., 1995). First an attack at a *meso*-position (step C_{Nmeso}), which is more favorable, occurs after the electrogeneration of the radical cation (step E_1). After the second oxidation (step E_2), as no proton can be removed from the substituted *meso*-position, a second nucleophilic attack occurs at a *β*-position of the macrocycle (step $C_{N\beta}$) which leads to the loss of the pyridinium attached at the *meso*-position. Then, the spare proton in β-position can be removed (step C_B) in a last step in order to re-aromatize the macrocycle.

3. Towards an electropolymerization process of porphyrins

3.1 Electrochemical investigation of the electropolymerization

During previous electrolyses, electrodes were systematically coated by a thin colored film, ascribed to polymers. Consequently, a more detailed study of these polymers has been investigated.

Thus, in a first report, we have published the formation of a polymer of porphyrins obtained from the electrogenerated dications of ZnOEP macrocycles substituted by a bipyridinium group in *meso*-position (zinc *meso*-bipyridinium-β-octaethylporphyrin, abbreviated $ZnOEP(bpy)^+$) (Fig. 3) (Ruhlmann et al., 1999b, 2008).

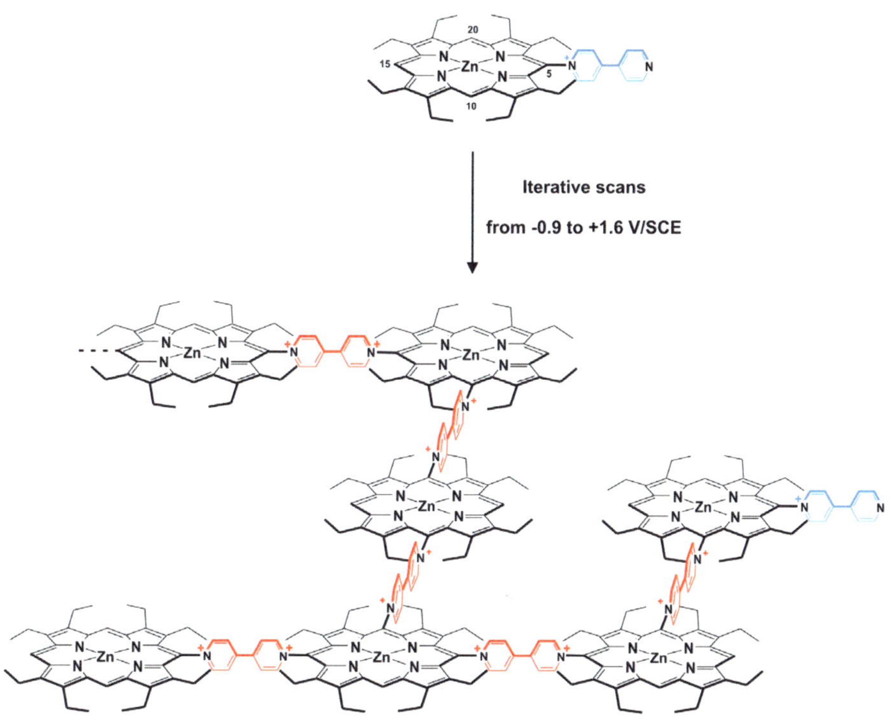

Fig. 3. Electropolymerization scheme of the mono-substituted $ZnOEP(bpy)^+$.

This polymer is obtained with iterative scans by cyclic voltammetry (Fig. 4). During this iterative process, the current increases progressively, showing the formation of a conducting polymer coating the electrode. Moreover, the oxidation waves of the macrocycle become irreversible, suggesting well the reactivity of the dication. Furthermore, these oxidation waves are progressively shifted to higher anodic potential values during electropolymerization. This shift is a consequence of the electron-withdrawing effect of the viologen spacers formed between porphyrins during the polymerization process. These viologen spacers are also responsible of the two reversible reduction waves which appear and grow around 0.00 and –0.65 V/SCE (peaks a and b in Fig. 4). Indeed, the two pyridiniums of the viologen spacers are reducible in two successive steps due to their mutual interaction (Schaming et al., 2011b).

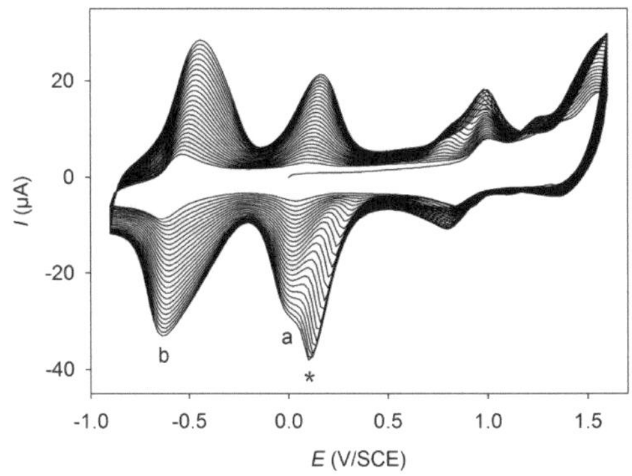

Fig. 4. Cyclic voltammograms recorded during the iterative scans between –0.90 and +1.60 V/SCE of ZnOEP(bpy)$^+$ (0.25 mM) in 1,2-$C_2H_4Cl_2$ and 0.1 M NEt_4PF_6. Working electrode: ITO; S = 1 cm^2; v = 0.2 V s^{-1}.

3.2 Mechanism of the electropolymerization process

The mechanism explaining the formation of polymers is similar to the one proposed for the electrosynthesis of mono-substituted porphyrins, excepted for the first oxidation step. Indeed, for electropolymerization, an $E(EC_NEC_B)_nE$ mechanism can be proposed (Fig. 5), where C_N corresponds to a nucleophilic attack and C_B to an acid-base reaction. Firstly, after formation of the dication (two steps E), the nucleophile (bipyridinium-substituted porphyrin) can react with it in *meso*-position to give an isoporphyrin (step C_N). This result correlates with the previous analysis of the redox behavior of porphyrins in the presence of nucleophiles and in particular the well documented report of Hinman *et al.* (Hinman et al., 1987). This isoporphyrin corresponds in fact to a porphyrin for which aromaticity is broken, because of the simultaneous presence of a pyridinium group and a proton on the same *meso*-carbon. Afterwards, this isoporphyrin can be oxidized again (step E). Finally, this spare proton is lost, leading to the re-aromatization of the macrocycle (step C_B) and allowing the formation of a di-substituted porphyrin. In a similar way, this one can react again, leading gradually to the formation of a polymer. One can also note that the isoporphyrins (which have not lost removable proton) are reduced during the further cathodic scan and conduct to the irreversible wave (peak * in Fig. 4) around +0.10 and +0.40 V/SCE, leading to the regeneration of the porphyrin.

One can notice that if the iterative sweeps are stopped in the anodic part at a potential corresponding to the formation of the radical cation of the porphyrin, no change of the cyclic voltammograms is observed (Schaming et al., 2011b). Moreover, in that case, the oxidation wave of the macrocycle remains reversible, showing that the electrogenerated radical cation does not react further. Thus, these results show that no polymerization occurs when the radical cation is formed. While the formation of the radical cation was sufficient to perform mono-substitutions onto macrocycles, the impossibility to carry out electropolymerization in this case can be explained by a kinetic problem. Indeed, the nucleophilic attack is certainly faster onto the dication than onto the radical cation. Consequently, during

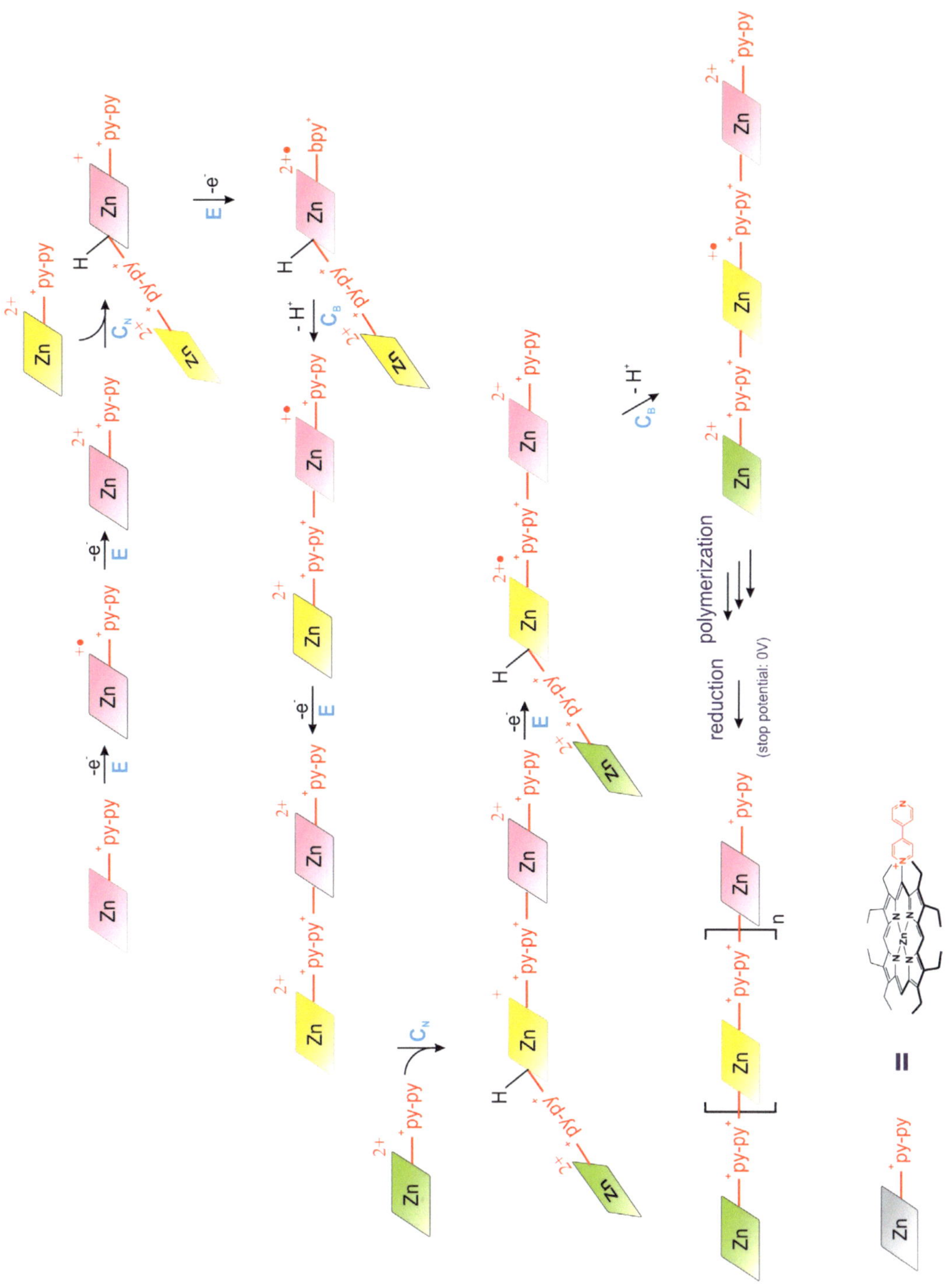

Fig. 5. E(EC$_N$EC$_B$)$_n$E mechanism proposed for the electropolymerization of the mono-substituted ZnOEP(bpy)$^+$.

electropolymerization, the rate of the sweeps is too fast to perform nucleophilic attacks onto the radical cation, the reverse scan in the cathodic part (leading to its reduction) being too fast. Nevertheless, if an electrolysis is carried out at a potential corresponding to the formation of the radical cation, the electrode is coated, showing also the formation of oligomers and/or polymers. However, this polymeric film formation may also be due to the disproportionation of the radical cation (slow kinetic) leading to the non-oxidized ZnOEP and the porphyrin dication which can be attacked more rapidly by the nucleophile. Nevertheless, the advantage to perform electropolymerization by iterative scans instead of electrolysis at a fixed potential is double: the protons released during the polymerization can be reduced during the cathodic scans, and the iterative scans allow a better homogeneity of the polymeric film deposited onto the electrode, therefore enhancing its conductivity. Another point which can explain the need to apply a higher potential for electropolymerization concerns the degree of substitution of macrocycles. Indeed, into the polymers, porphyrins are at least substituted twice by positively charged groups. Moreover, when the chain of polymer grows, the quantity of positively charged groups increases. Consequently, porphyrins are more and more difficult to oxidize, needing an increase of the oxidative potential to perform the polymerization. This explanation is also supported by the fact that a higher potential has always been required to perform electrosyntheses of multi-substituted porphyrins by electrolysis. As a matter of fact, it can also be noted that the higher the limit potential during the anodic sweeps is, the longer the polymeric chains could be obtained. Otherwise, if the limit potential is too low, only small oligomers can be obtained.

3.3 Control of the geometry of the polymer

The use of ZnOEP(bpy)$^+$ as monomer allows multi-substitutions onto macrocycles because three *meso*-positions remain free (positions 10, 15 and 20, Fig. 3). Thus, nucleophilic attacks can occur on each of them, and consequently that can lead to the formation of zig-zag polymers and eventually of hyper-branched polymers. In order to permit a better control of the geometry of the obtained polymers, porphyrins substituted by two protecting groups as chlorides (zinc 5-bipyridinium-10,20-dichloro-*β*-octaethylporphyrin, abbreviated ZnOEP(Cl)$_2$(bpy)$^+$) can be used (Ruhlmann et al., 1999b). In this case, linear wires of polymers are obtained because only one *meso*-position remains free. Transmission electron micrographs of this polymer show long wires having length of several micrometers and diameter of about 20 Å (Fig. 6). This is in agreement with the 19 Å molecular width of a ZnOEP molecule.

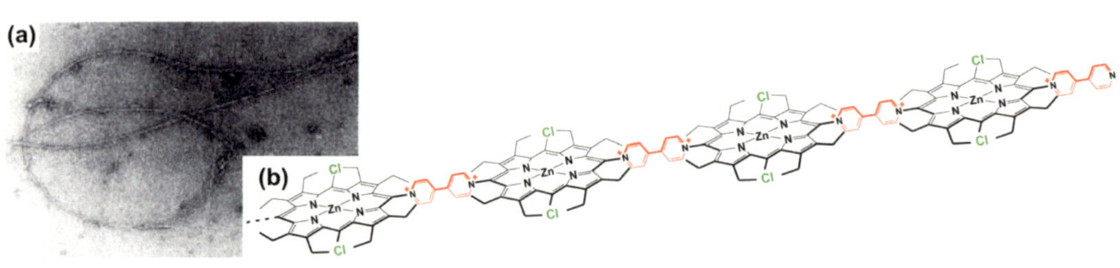

Fig. 6. (a) Transmission electron micrograph of a linear fiber obtained by electropolymerization of ZnOEP(Cl)$_2$(bpy)$^+$. (b) Scheme of the linear chain obtained.

4. A new process of Easy Polymerization Of Porphyrins (EPOP process)

4.1 A first example

The way of electropolymerization of porphyrins presented above requires a first step consisting in the synthesis of the starting monomeric subunit (ZnOEP(bpy)$^+$). The synthesis of this monomer can be performed by the electrochemical process described before, (reaction of 4,4′-bipyridine with the electrogenerated radical cation of the ZnOEP, see part 2). Even if this synthesis appears easy, the further purification of the obtained compound is more difficult, due to the possibility of multi-substitutions onto the macrocycle. In order to avoid the synthesis of this monomeric subunit, we have recently proposed an alternative method consisting in the direct use of commercial and non-substituted ZnOEP, with the presence of free 4,4′-bipyridine (bpy) in the solution (Fig. 7) (Giraudeau et al., 2010).

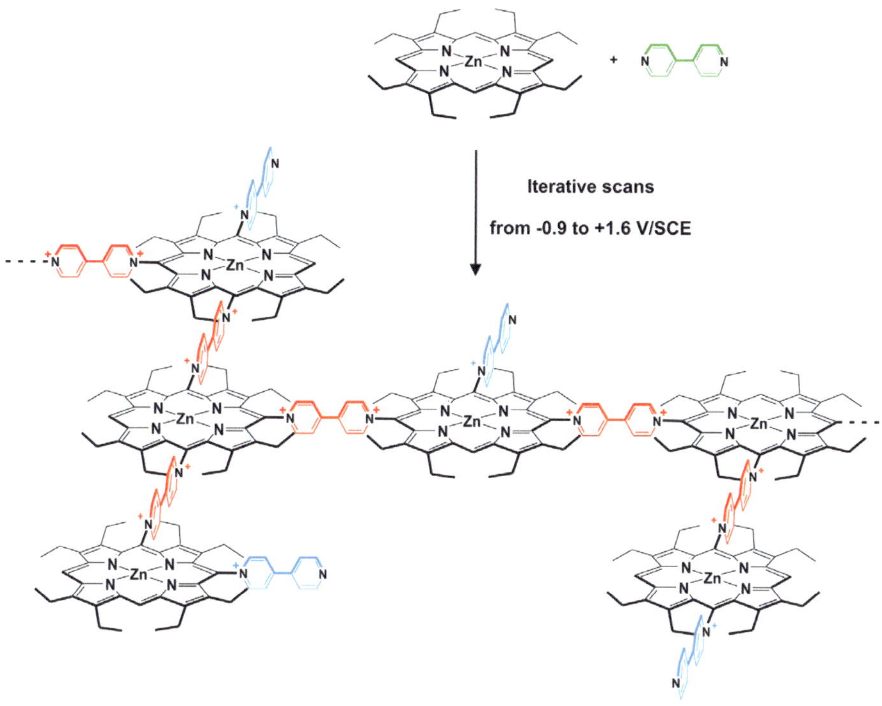

Fig. 7. Electropolymerization scheme of the non-substituted ZnOEP with free bpy.

As previously, electropolymerization can be performed with iterative scans by cyclic voltammetry (Fig. 8). The current increases again during the iterative sweeps. Compared to the cyclic voltammograms obtained in the previous case (Fig. 4), an additional reversible wave is observed at -0.30 V/SCE (peak c in Fig. 8). This additional wave can be attributed to the reduction of the bipyridinium substituents (Schaming et al., 2011b). Indeed, in this case, many bipyridinium groups can be grafted on the macrocycles, without further polymeric chains growth, because of the too important hindrance between the macrocycles. Consequently, the polymer obtained from non-substituted ZnOEP and free bpy can have many bipyridinium groups substituted onto the porphyrins (represented in blue in Fig. 7), leading to the appearance of this new reversible wave corresponding to the reduction of these bipyridinium substituents. On the contrary, when ZnOEP(bpy)$^+$ is used to carry out the electropolymerization, each chain of polymer can contain only one bipyridinium substituent (at the end of the chain, represented in blue in Fig. 3). In this case, the number of bipyridinium substituents is negligible compared to the number of viologen spacers (present between each macrocycle and represented in red in Fig. 3). As a matter of fact, the signal corresponding to the reduction of these bipyridinium substituents is not observed for the polymer obtained from the substituted monomer $ZnOEP(bpy)^+$.

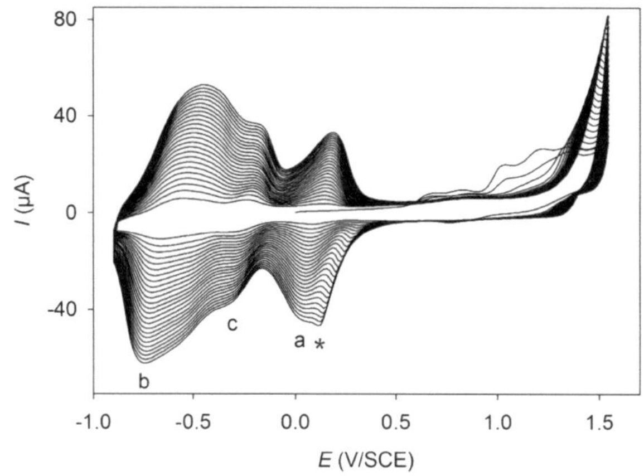

Fig. 8. Cyclic voltammograms recorded during the iterative scans between -0.90 and +1.60 V/SCE of ZnOEP (0.25 mM) in the presence of bpy (0.25 mM) in 1,2-$C_2H_4Cl_2$ and 0.1 M NEt_4PF_6. Working electrode: ITO; S = 1 cm^2; v = 0.2 V s^{-1}.

4.2 Extension to the use of other spacers

As previously said, this second method of electropolymerization allows to avoid the synthesis of the monomeric substituted porphyrin ($ZnOEP(bpy)^+$ for example). Thus, this new way of electropolymerization more conveniently allows the use of the commercial free base porphyrin (H_2OEP) or metalloporphyrins (MOEP) with different central metals (M = Co^{II}, Mg^{II}, Ni^{II}, $Ru^{II}(CO)$...) (Giraudeau et al., 2010). For all of them, an electropolymerization process has been observed, even when porphyrins with oxidable central metal (CoOEP and NiOEP) have been used. Consequently, further applications could be envisaged. For example, polymer with CoOEP could be promising for dioxygen reduction, cobalt porphyrins being known to catalyze this reaction (Chen et al., 2010; Collman et al., 1980).

Moreover, it is also possible to modulate easily the nature of the bridging spacers between the porphyrin macrocycles. Indeed, instead of using free bpy, this process of electropolymerization can be extended to other spacers by varying the nature of the nucleophile, since each compound having two pendant pyridyl groups can be used as spacer. The spacers can be selected for their specific chemical and structural properties: rigid or not; long or short; electron conducting or not; electroreducible or not; with conjugated π bonds (aromatic/alkene/alkyne chains) or successive σ bonds (alkyl chains)... For instance, electropolymerization has been successfully carried out with different nucleophilic compounds (named Py-R-Py) as 1,2-bis(4-pyridyl)ethane (bpe), trans-1,2-bis(4-pyridyl)ethylene (tbpe), but also with reducible spacers as 4,4'-azopyridine (azpy) and 3,6-bis(4-pyridyl)-s-tetrazine (tzpy) (Fig. 9) (Giraudeau et al., 2010; Schaming et al., 2011b).

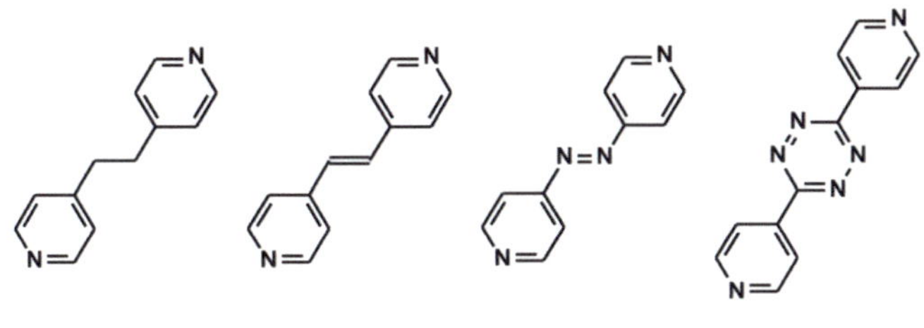

Fig. 9. Several Lewis bases (Py-R-Py) used as spacers: 1,2-bis(4-pyridyl)ethane (bpe), trans-1,2-bis(4-pyridyl)ethylene (tbpe), 4,4′-azopyridine (azpy) and 3,6-bis(4-pyridyl)-*s*-tetrazine (tzpy).

5. Studies of the polymers

When the electropolymerization process is finished (stopped after a pre-defined number of scans: 25 scans for all studies described in this paragraph, unless otherwise indicated), the electrode is removed from the electrochemical cell. Then, the electrode is systematically coated with a brown thin film corresponding to the polymer. In order to study in more details these polymers, it is necessary to previously wash the electrode with CH_3CN or water in order to remove the supporting electrolyte (NEt_4PF_6).

5.1 UV-visible absorption and fluorescence spectroscopies

Firstly, the polymers can be characterized by UV-visible absorption spectroscopy. When an ITO electrode is used to perform the electropolymerization, the spectrum can be recorded directly onto this optically transparent electrode. Whatever the method of electropolymerization, and whatever the spacer presented before, the spectra are similar (blue spectrum, Fig. 10.a). They consist in a large Soret band whose maximum is red-shifted compared to the ZnOEP monomer. Similarly, Q bands are also red-shifted and larger. That can be attributed to the intra- and intermolecular interactions between the porphyrins subunits (Giraudeau et al., 2010; Ruhlmann et al., 2008). The red-shift of the Soret and Q bands can also result from the electron-withdrawing effect of the positively charged pyridinium groups on the macrocycles (Giraudeau et al., 1996, 2010).

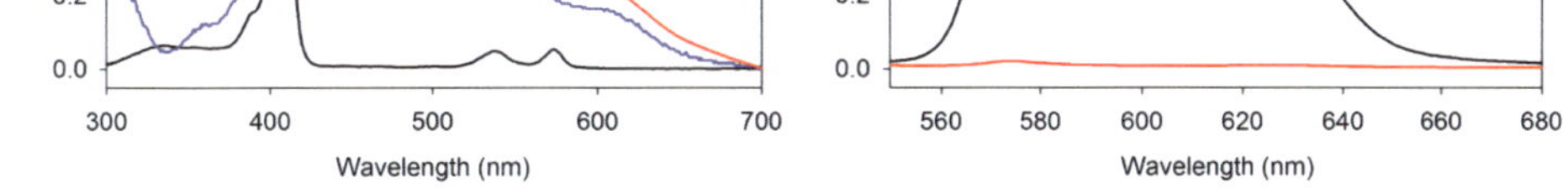

Fig. 10. (a) Normalized UV-visible absorption spectra of ZnOEP in DMF (—), of an ITO electrode modified with the polymer obtained from ZnOEP and bpy after 25 iterative scans (—) and of this same polymer in solution in DMF (—). (b) Luminescence spectra of ZnOEP (—) and of the polymer obtained from ZnOEP and bpy (—) in DMF. λ_{exc} = 420 nm.

Moreover, the polymers can be removed from the electrode by dissolution in dimethylformamide (DMF). That allows to record the spectra of the polymers in solution (red spectrum, Fig. 10.a). Comparing the spectra obtained onto ITO electrodes, the red-shift of the Soret band for spectra recorded in solution appears smaller. Such an evolution can be explained by a decrease of the interactions between the macrocycles when the polymers are in solution. Indeed, in solution, polymeric chains are partially unfolded.
When polymers are in solution, it is also possible to study their luminescence properties: all the investigated polymers show a total quenching (Fig. 10.b).

5.2 Atomic force microscopy (AFM)

The morphology of the polymers can be observed directly onto ITO electrodes by atomic force microscopy (AFM).
Depending on the solvent used to wash the electrode, the morphologies of the deposited polymers appear different (Ruhlmann et al., 2008). Indeed, when water is used to wash the polymer obtained from the mono-substituted porphyrin $ZnOEP(bpy)^{+}$, this one appears in the form of coils (diameter of *ca.* 50 nm) quasi-packed in the same direction (Fig. 11.a). This orientation effect might be induced by a self-alignment of the polycationic coils in the applied electric field during the electropolymerization. Moreover, after changing of the position of the ITO electrode, the coils cross over each, as shown in Fig. 11.a. That demonstrates the direct influence of the applied electric field onto the polymer arrangement during the electropolymerization process. Nevertheless, when CH_3CN is used instead of water to wash the polymer, a drastic reorganization of these tightly packed coils is observed. Indeed, the alignment of the polymers disappears and a homogeneous dispersion of the coils is observed (Fig. 11.b).

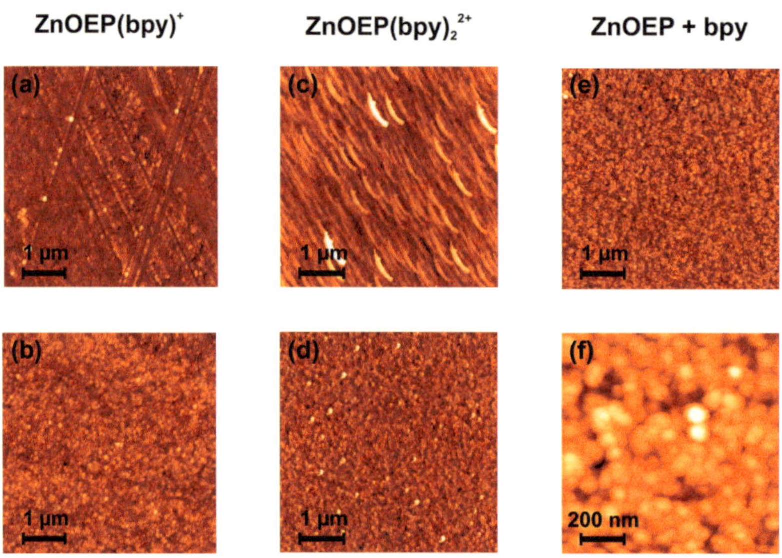

Fig. 11. Atomic force micrographs of different polymers obtained onto ITO electrodes after 25 iterative scans: (a) and (b) polymer obtained from $ZnOEP(bpy)^{+}$, after washing with (a) water and (b) CH_3CN; (c) and (d) polymer obtained from $ZnOEP(bpy)_2^{2+}$, after washing with (c) water and (d) CH_3CN; (e) and (f) polymer obtained from ZnOEP + bpy, after washing with (e) water and (f) CH_3CN (zoom).

It can be noticed that similar electropolymerization has also been performed from the ZnOEP macrocycles substituted by two bipyridinium groups in *cis*-position (zinc 5,10-dibipyridinium-*β*-octaethylporphyrin, abbreviated 5,10-ZnOEP$(bpy)_2^{2+}$) (Ruhlmann et al., 2008). In that case, a regular arrangement of the coils aggregated in the form of "peanuts" (width of *ca.* 100 nm and length of *ca.* 900 nm), all oriented in the same direction, is observed when the washing is performed with water (Fig. 11.c). This enhancement of the self-alignment of the coils induced by the applied electric field could be explained by the increase of the number of positive charges in this case (because of the presence of two bipyridinium substituents onto each macrocycle) which would induce stronger coulombic repulsions between the different subunits. However a treatment with CH_3CN induces also a homogeneous dispersion of the coils (Fig. 11.d) as previously observed for the polymer obtained from the mono-substituted ZnOEP$(bpy)^+$.

In the case of the polymer obtained from the non-substituted ZnOEP with free bpy, tightly packed coils are also observed, but without specific alignment (Giraudeau et al., 2010), whatever the solvent used to wash the electrode (Fig. 11.e and f). That can be tentatively explained by the presence of free bpy which could lead to a disorganization of the film. Moreover, the free bpy can also axially coordinate the Zn central metal of the macrocycles (Giraudeau et al., 2010) which increases the distance between the macrocycles and consequently could lead to a decrease of the electrostatic effect between the macrocycles. As a result, effect of the applied electric field should be lesser onto the electropolymerization process, the orientation effect disappearing in this case.

5.3 Scanning Electrochemical Microscopy (SECM)

Scanning electrochemical microscopy (SECM) in feedback mode has also been used to investigate electronic and permeation properties of the polymeric films obtained from the mono-substituted ZnOEP$(bpy)^+$ and the non-substituted ZnOEP in presence of free bpy, in order to compare these two different electropolymerization ways allowing the formation of similar polymers (Leroux et al., 2010). Briefly, the SECM principle is based on the interaction of the polymeric film (deposited onto ITO electrode) under investigation with a redox probe (the mediator) that is electrogenerated at a microelectrode. This interaction is followed through the analysis of the current flowing at the microelectrode while it approaches the substrate. Depending on the nature of the mediator, the study of either the permeability or the reactivity is possible.

Firstly, ferrocene (Fc) and tetrathiafulvalene (TTF), which work in oxidation, have been chosen as mediator (Fig. 12.b). According to their redox potentials, their oxidized forms (Fc^+ and $TTF^{\bullet+}$, respectively) cannot react with the polymeric films (their redox potentials do not permit the oxidation of the porphyrin macrocycles) (Fig. 12.a). Thus, they are good candidates to probe the permeability of the polymeric films, because they can rapidly exchange electrons with the non-coated ITO substrate (positive feedback). Whatever the polymer studied as substrate onto the ITO electrode, a negative feedback is observed. These negative feedback characters are less important in the case of the polymer obtained from the non-substituted ZnOEP, showing that this polymer is more permeable than the one obtained from the mono-substituted ZnOEP$(bpy)^+$. That can be explained by an increase of the distance between the macrocycles due to the formation of bipyridine-bridged zinc porphyrins (bpy axially ligated to Zn) mentioned before (see part 5.2).

Secondly, tetracyanoquinodimethane (TCNQ) and 4-nitrobenzonitrile (4NB), which work in reduction, have been chosen as mediator (Fig. 12.b). Their redox potentials are sufficiently negative to permit to the reduced forms ($TCNQ^{\bullet-}$ and $4NB^{\bullet-}$, respectively) to reduce the viologen spacers of the polymers (Fig. 12.a). Thus, they are good candidates to probe the reactivity of the polymeric films, because their electron exchanges with the non-coated ITO substrate are negligible (negative feedback). Nevertheless, a more important positive feedback is observed when the ITO electrodes are coated with a polymeric film, showing well the predicted reactivity of the polymers with these two mediators. But the positive feedback characters are lower in the case of the polymer obtained from the non-substituted ZnOEP. That can be explained by a lower conductivity of this polymer compared to the other one, because of an increase of the distance between the redox active centers also due to the formation of bipyridine-bridged zinc porphyrins.

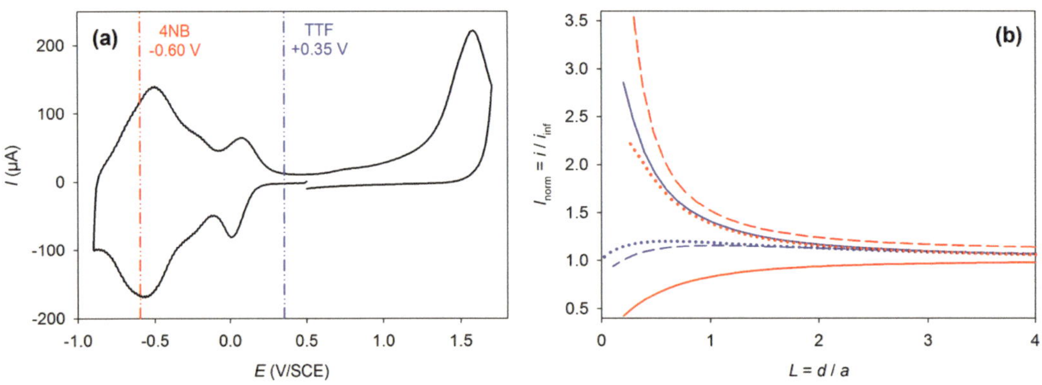

Fig. 12. (a) Cyclic voltammogram of the polymer obtained from ZnOEP and bpy after 25 iterative scans onto ITO electrode in CH_3CN and 0.1 M NBu_4PF_6 (v = 0.2 V s^{-1}), with indications of the redox potential values of several mediators used. (b) SECM approach curves, in CH_3CN and 0.1 M NBu_4PF_6 on non-coated ITO electrodes (full lines), ITO electrodes coated with the polymer obtained from $ZnOEP(bpy)^+$ after 25 iterative scans (dashed lines) and ITO electrodes coated with the polymer obtained from ZnOEP and bpy after 25 iterative scans (dotted lines), with TTF (—) and 4NB (—) used as redox mediators. L is the normalized distance and I_{norm} is the normalized current, where i corresponds to the current at a platinum microelectrode (with a radius a equal to 5 µm) localized at a distance d from the substrate and i_{inf} corresponds to the steady-state current when the microelectrode is at an infinite distance from the polymeric substrate.

5.4 Control of the polymeric film thickness

It is also interesting to note that electropolymerization allows easily to control the thickness of the polymeric film deposited onto the electrode (Schaming et al., 2011b).

As represented Fig. 13, plot of the absorbance of the polymer at the maximum of the Soret band as a function of the number n of iterative scans shows a linear increase. The thicknesses of the films determined from AFM confirm this point since a linear increase of the thickness with the value of n is observed (Fig. 13, inset).

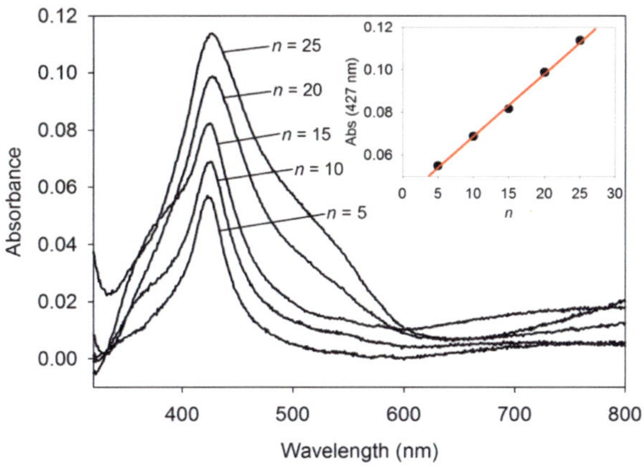

Fig. 13. UV-visible absorption spectra of ITO electrodes modified with the polymer obtained from ZnOEP and tzpy after different numbers *n* of iterative scans. Inset: absorbance at 427 nm *vs*. *n*.

6. Formation of original copolymers from the EPOP (Easy Polymerization Of Porphyrins) process

As previously explained, all compounds having two pendant pyridyl groups can play the role of spacers between macrocycles. Consequently, more original organic or inorganic compounds, with specific interesting properties, can be used, if two pendant pyridyl groups are present. Thus, this novel easy way of electropolymerization of porphyrins appears as a promising approach to elaborate new functional materials.

6.1 Porphyrin–porphyrin copolymers

6.1.1 Electropolymerization

In a first example, porphyrins having two pendant pyridyl groups have been used in order to obtain copolymers of porphyrins containing two different types of macrocycles. (Xia et al., 2012)

Fig. 14.b illustrates the cyclic voltammograms obtained in the case of the use of the 5,15-dipyridyl-10,20-diphenyl free base porphyrin (5,15-$H_2Py_2Ph_2P$) in the presence of zinc 5,15-dichloro-β-octaethylporphyrin (5,15-ZnOEP$(Cl)_2$). These two macrocycles have been chosen in order to allow the formation of a linear copolymer as represented in Fig. 14.a.

As expected, the current increases progressively during the iterative scans. The new reduction processes appearing around -0.40 and -0.60 V/SCE (peaks *) can be attributed to the reduction of the pyridinium spacers.

One can underline that even if the 5,15-$H_2Py_2Ph_2P$ porphyrin is oxidized during the iterative scans, no substitution in β-positions of this macrocycle by a pyridyl group of another 5,15-$H_2Py_2Ph_2P$ macrocycle could occur. Indeed, the kinetic of a nucleophilic substitution in β-position is too slow compared to the kinetic of a substitution in *meso*-position (see part 2). Consequently, while mono-substitutions are possible in β-position of porphyrins (Giraudeau et al., 1996), electropolymerization is very difficult and even impossible, because the rate of the iterative sweeps is too fast to let enough time for the β-substitutions in such experimental conditions.

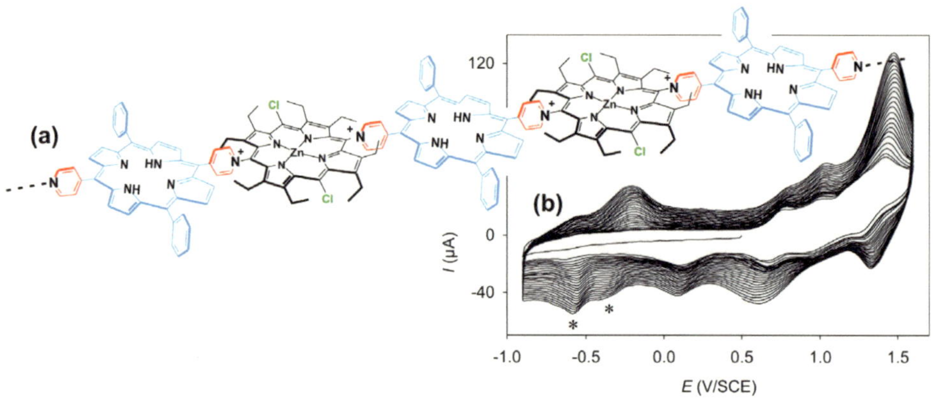

Fig. 14. (a) Scheme of the linear bis-porphyrin copolymer obtained. (b) Cyclic voltammograms recorded during the iterative scans between -0.90 and +1.60 V/SCE of 5,15-ZnOEP(Cl)$_2$ (0.25 mM) in the presence of 5,15-H$_2$Py$_2$Ph$_2$P (0.75 mM) in 1,2-C$_2$H$_4$Cl$_2$/CH$_3$CN (4:1) and 0.1 M NBu$_4$PF$_6$. Working electrode: ITO; S = 1 cm^2; v = 0.2 V s^{-1}.

6.1.2 Characterization

The characterization of this copolymer has also been performed by UV-visible absorption spectroscopy and by atomic force microscopy. No important change in comparison with the previous polymers has been observed. Indeed, its spectrum is red-shifted and larger compared to the ones of the porphyrins alone (Fig. 15.a). Furthermore, the copolymer appears as tightly packed coils without alignment (Fig. 15.b).

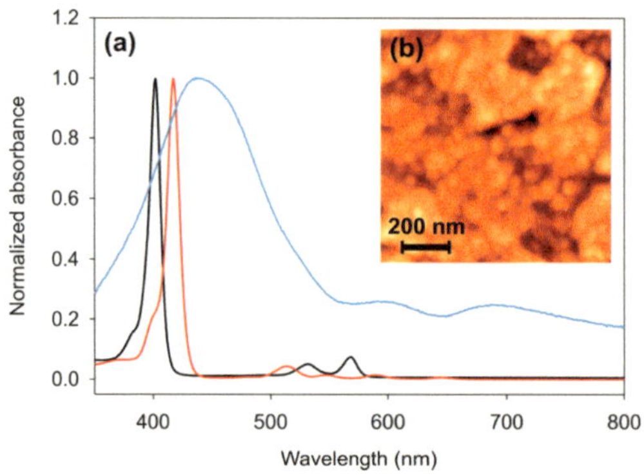

Fig. 15. (a) Normalized UV-visible absorption spectra of ZnOEP in 1,2-C$_2$H$_4$Cl$_2$ (—), of 5,15-H$_2$Py$_2$Ph$_2$P in 1,2-C$_2$H$_4$Cl$_2$ (—) and of an ITO electrode modified with the copolymer obtained from ZnOEP and 5,15-H$_2$Py$_2$Ph$_2$P after 25 iterative scans (—). (b) Atomic force micrograph of the same copolymer after washing with water.

6.2 Porphyrin–polyoxometalate copolymers

Polyoxometalates (POMs) are well-defined metal-oxygen cluster anions constituted of early metal elements in their highest oxidation state with a wide variety of structures and properties (Jeannin, 1998; Katsoulis, 1998). They have particularly attractive catalytic, electrocatalytic and photocatalytic applications. For instance, POMs are known to photocatalyze the reduction of noble or heavy metal cations (Costa-Coquelard et al., 2008; Troupis et al., 2001, 2006).

Nevertheless, POMs are efficient only under UV light, which is a drawback for environmental applications. Indeed, it seems preferable to use solar light, principally in the visible domain. The development of photosensitized systems could overcome this issue. For instance, {POM–porphyrin} systems appear interesting for photocatalytic applications in the visible domain (Schaming et al., 2010a, 2011c; Schaming, 2010c). In these hybrid organic-inorganic systems, porphyrins act as photosensitizers able under visible illumination to transfer electrons to POMs, which are known to be good electron acceptors. Thus, the excitation of the porphyrins leads to their oxidation and to the simultaneous reduction of the POMs. Then, the reduced POMs are able to transfer electrons to M^{n+} cations to give M^0.

Using an Anderson-type POM substituted by two pendant pyridyl groups ($[MnMo_6O_{18}\{(OCH_2)_3CNHCO(4\text{-}C_5H_4N)\}_2]^{3-}$, abbreviated py–POM–py), we have recently obtained a porphyrin–POM copolymer by the method of electropolymerization presented before (Fig. 16) (Schaming et al., 2010b).

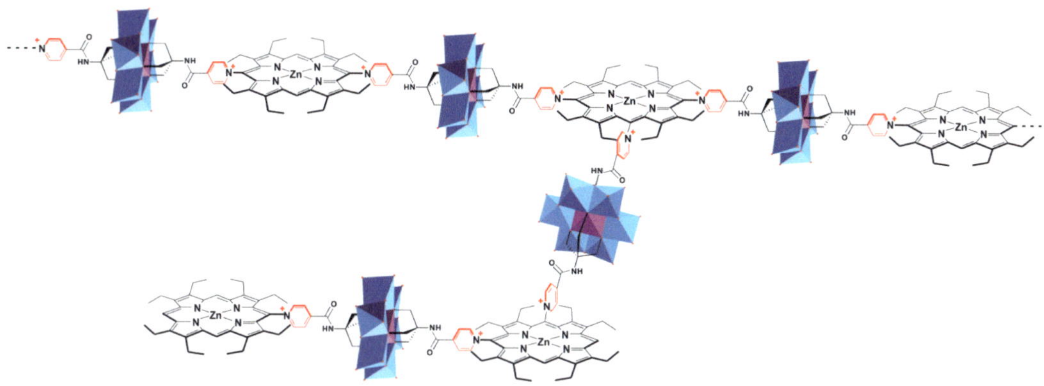

Fig. 16. Scheme of the porphyrin–POM copolymer.

6.2.1 Electropolymerization

Fig. 17.a shows the cyclic voltammograms obtained in the case of the use of the non-substituted ZnOEP macrocycle in the presence of the py–POM–py compound. As previously, the current increases progressively during the iterative scans. As already described, the new peak appearing around -1.00 V/SCE (peak *) can be attributed to the reduction of the pyridinium spacers. Nevertheless, one can also notice the appearance of a new peak around +0.20 V/SCE (peak •) during the anodic scans. This one could be assigned to the oxidation of the adsorbed H_2 formed upon the reduction of the protons during the cathodic scans, these protons being formed during the nucleophilic substitution onto the porphyrins. To explain the presence of this additional anodic peak (not observed previously), one can suggest that it is due to the presence of POMs which can be easily protonated and consequently the released protons are not dispersed in the solution but remain close to the electrode. As a matter of fact, they can be easier and in bigger quantity reduced during the cathodic scans, and the H_2 formed can be further re-oxidized.

In order to confirm the assignment of this wave, the electropolymerization process has also been carried out with iterative scans performed only in the anodic part (scans stopped at 0 V/SCE), in order to avoid the reduction of the protons. As expected, the signal assigned to the oxidation of the adsorbed H_2 disappears (Fig. 17.b). Surprisingly, in this case, the current decreases during the iterative scans. Thus, the copolymer obtained by this way seems less conductive. This decrease in the conductivity can be tentatively explained by the fact that when the electropolymerization is performed by oxidative and reductive scans, the film should be more organized and consequently more conductive than when performed only by oxidative scans.

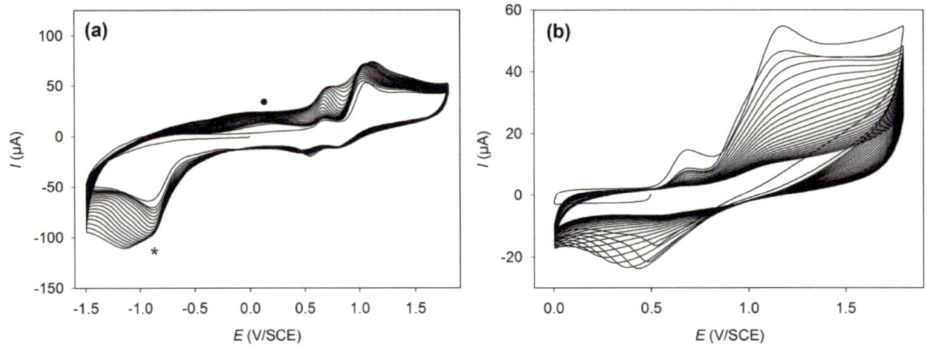

Fig. 17. Cyclic voltammograms recorded during the iterative scans (a) between -1.50 and +1.80 V/SCE and (b) between 0 and +1.80 V/SCE of ZnOEP (0.25 mM) in the presence of py-POM-py (0.25 mM) in 1,2-$C_2H_4Cl_2$/CH_3CN (7:3) and 0.1 M NBu_4PF_6. Working electrode: ITO; $S = 1$ cm^2; $v = 0.2$ V s^{-1}.

6.2.2 Characterization

Compared to the previous polymers, the UV-visible absorption spectrum of the porphyrin-POM copolymer appears less large and more structured (Fig. 18.a). That can be attributed to the presence of POM between each macrocycle, which avoids interactions between adjacent porphyrins. Concerning its morphology, this copolymer appears again as little coils (Fig. 18.b).

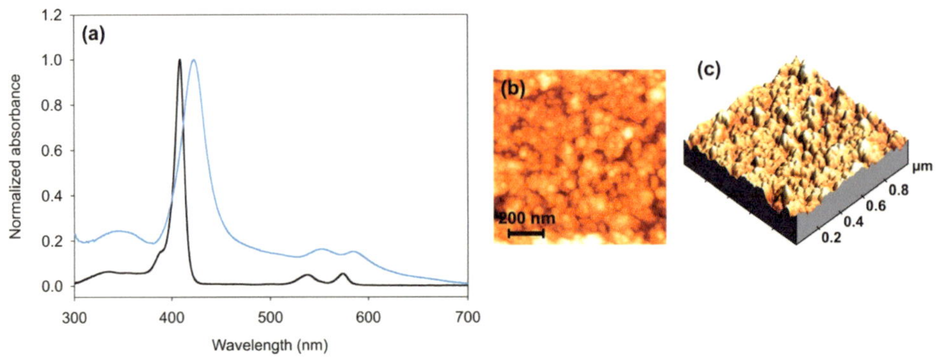

Fig. 18. (a) Normalized UV-visible absorption spectra of ZnOEP in DMF (—) and of an ITO electrode modified with the copolymer obtained from ZnOEP and py-POM-py after 25 iterative scans performed between -1.50 and +1.80 V/SCE (—). (b) 2D and (c) 3D atomic force micrographs of the same copolymer.

6.2.3 Use of di-substituted ZnOEP porphyrins

Different di-substituted ZnOEP macrocycles have also been used in order to control the geometry of the copolymers. As previously explained, 5,15-ZnOEP(Cl)$_2$ allows the formation of linear polymers. But it can be noticed that the zinc 5,15-dipyridinium-β-octaethylporphyrin (5,15-ZnOEP(py)$_2^{2+}$) can also be used in the same purpose. It is interesting to note that in this case the morphology of the obtained copolymer is quite different. Indeed, the coils appear more agglomerated (Fig. 19.b), while in the case of the use of 5,15-ZnOEP$(Cl)_2$, non-agglomerated coils are observed (Fig. 19.a).

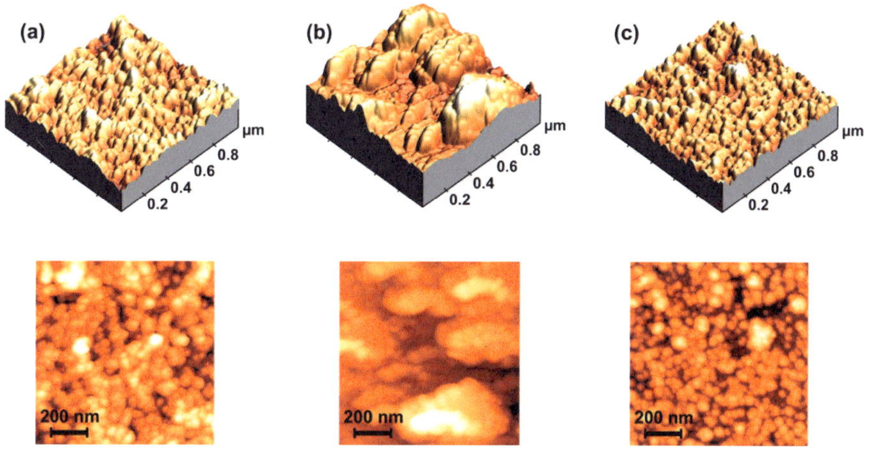

Fig. 19. 2D and 3D atomic force micrographs of different porphyrin–POM copolymers obtained onto ITO electrodes after 25 iterative scans: (a) polymer obtained from 5,15-ZnOEP$(Cl)_2$ by iterative scans between -1.30 and +1.80 V/SCE; (b) polymer obtained from 5,15-ZnOEP$(py)_2^{2+}$ by iterative scans between -1.50 and +1.80 V/SCE and (c) polymer obtained from 5,15-ZnOEP$(py)_2^{2+}$ by iterative scans between 0 and +1.80 V/SCE.

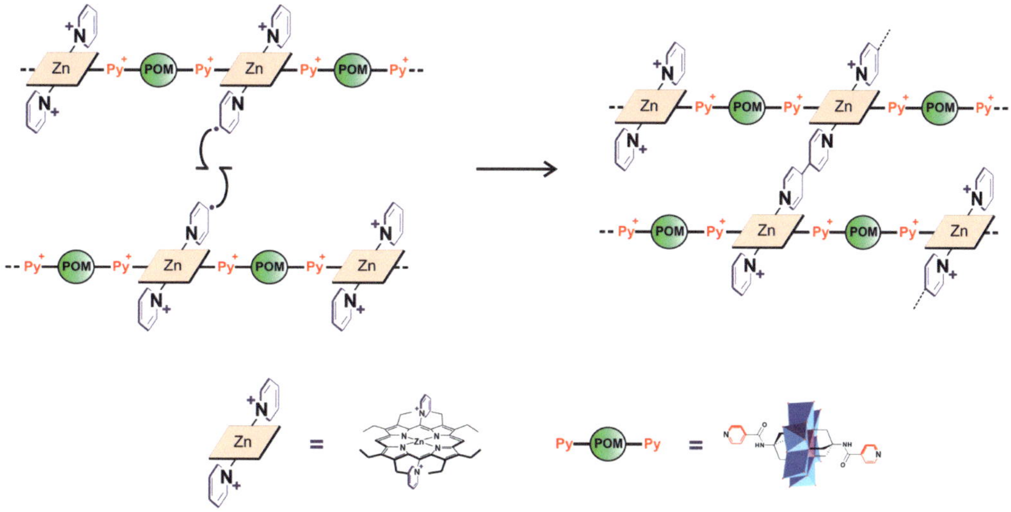

Fig. 20. Scheme of the coupling in 4-position of the electrogenerated pyridyl radicals leading to their oligomerization.

This agglomeration of the coils when 5,15-ZnOEP$(py)_2^{2+}$ is used can be explained by the chemical reactivity of the pyridyl radicals obtained during the cathodic scans by reduction of the pyridiniums used as protecting groups. Indeed, these electrogenerated pyridyl radicals can react each other to give oligomeric species resulting from the coupling at the 4-position of the radicals (Fig. 20). This coupling has already been described in several works (Brisach-Wittmeyer et al., 2005; Carelli et al., 1998, 2002; Karakostas et al. 2010; Schaming et al., 2009). To check this explanation, the electropolymerization with 5,15-ZnOEP(py)$_2^{2+}$ has also been performed by iterative scans limited to the anodic part (cyclic voltammograms performed between 0 and +1.80 V/SCE) in order to avoid the reduction of the pyridinium substituents. As expected, the coils appear non-agglomerated in this case (Fig. 19.c).

6.2.4 Photocatalytic tests

As previously said, this porphyrin–POM copolymer has been prepared in order to use it as photocatalyst for the reduction of metal cations. For this purpose, we have chosen silver cations as a model system to perform photocatalytic tests (Schaming et al., 2010b).

To carry out this experiment, the copolymer has previously been removed from the electrode by dissolution in DMF, and then, a drop of the copolymer solution has been deposited on a slide of quartz and the solvent evaporated in air. Then, the covered slide has been plunged in an aqueous solution of Ag_2SO_4 (80 µM) containing 0.13 M of propan-2-ol used as sacrificial electron donor. Finally, the sample has been exposed during 8 hours to visible light. The reaction has been followed by UV-visible absorption spectroscopy (Fig. 21.a): during light irradiation, a very large plasmon band appeared progressively, suggesting the formation of silver aggregates. Transmission electron micrographs and electron diffraction patterns have confirmed this result: silver triangular nanosheets were principally obtained (Fig. 21.b and c). The mechanism explaining the reaction has been discussed in detail in our previous works (Schaming et al., 2010a, 2011c).

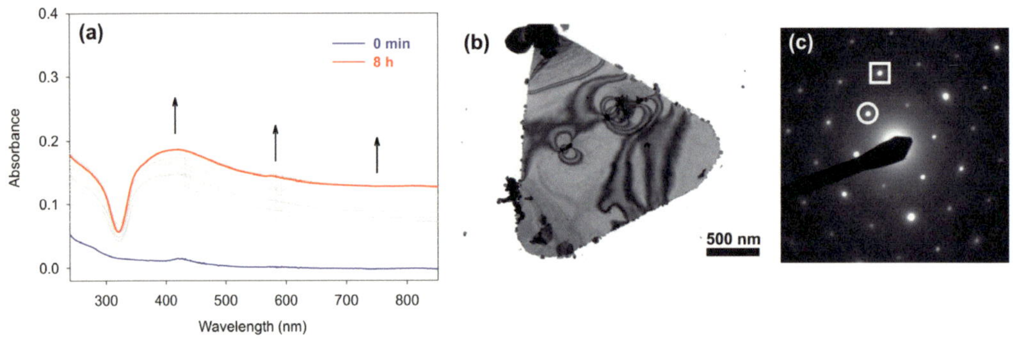

Fig. 21. (a) Change in the UV-visible absorption spectrum of a porphyrin–POM copolymer deposited on quartz in aerated aqueous solution containing 80 µM of Ag_2SO_4 and 0.13 M of propan-2-ol under visible illumination. (b) Transmission electron micrograph of a silver triangular nanosheet obtained. (c) Selected-area electron diffraction pattern of the previous silver nanostructure. The first spots (circled) correspond to the formally forbidden ⅓{422} reflections and the second spots (squared) correspond to the {220} reflections.

For this study, silver has been chosen as a model system for metal reduction, but this copolymer could be used in water depollution for reduction or recovery of valuable or toxic metals that could also be combined with degradation of organic pollutants (used instead of sacrificial donor) as already proposed (Troupis et al., 2006).

6.2.5 Electrostatic {porphyrin–POM} systems

It can be noticed that hybrid {porphyrin–POM} systems can be obtained for similar application purposes by plunging an ITO electrode coated with a cationic polymer of porphyrins, as described in the first parts of this chapter, into an aqueous solution of POMs. The aim of this process is to replace the counteranions of the polycationic polymers (PF_6^- further to the electropolymerization) by POMs which are large polyanions. Thus, electrostatic interactions between the cationic porphyrin polymers and the anionic POMs can occur, leading to a hybrid supramolecular assembly. For instance, the dipping of an ITO electrode coated with a polymer obtained from 5,10-ZnOEP(bpy)$_2^{2+}$ in an aqueous solution of $SiW_{12}O_{40}^{4-}$ for 10 hours has been studied (Hao et al., 2008). Atomic force microscopy studies of the deposit show strong modifications to its morphology (Fig. 22). Indeed, the initial regular arrangement of the coils aggregated in the form of "peanuts" (Fig. 11.c) has disappeared and regular cloudy assemblies which are still oriented in the same direction are now observed. Such supramolecular assemblies are the consequence of an agglomeration of "peanuts" by the POMs.

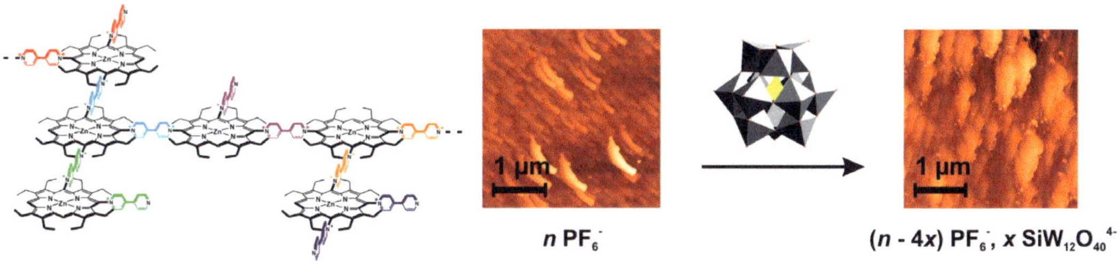

Fig. 22. Atomic force micrographs of an ITO electrode coated with the polymer obtained from 5,10-ZnOEP(bpy)$_2^{2+}$ after plunging for 10 hours in water (left) and in a solution of $SiW_{12}O_{40}^{4-}$ (1 mM) (right).

7. Conclusion

In this chapter, we have shown that electropolymerization of porphyrins can be easily performed by nucleophilic attacks of di-pyridyl compounds directly onto electrogenerated dications of octaethylporphyrins, by an ECEC-type mechanism.
The first tests have been performed with porphyrins substituted with bipyridinium(s). According to the degree of substitution of the monomer (which is linked to its charge), different morphologies of the polymers can be observed, which seem to be induced by the electric field imposed during the electropolymerization process.
We have further shown that it is also possible to use non-substituted β-octaethylporphyrins in the presence of free Lewis bases having two nucleophilic sites. This second way of electropolymerization allows to avoid the somewhat complicated synthesis of the monomeric substituted porphyrin, and consequently allows to modulate easily the nature of the bridging spacers between the macrocycles. Thus, original organic or inorganic compounds, with specific interesting properties, can be used, on condition that two pyridyl groups have beforehand been grafted. Consequently, this new easy way of electropolymerization of porphyrins appears as a promising approach to elaborate new functional materials, for example with catalytic properties if polyoxometalates are used as spacers. Other applications can be envisaged according to the spacer chosen.
Finally, it is interesting to note that as all the electropolymerization methods, it is easy to control the thickness of the polymers, according to the number of scans performed, which is also a great advantage compared to the classical chemical methods of polymerization.

8. References

Barnett, G.H. & Smith K.M. (1974). Reactions of some metalloporphyrin and metallochlorin π-cation radicals with nitrite. *Chem. Commun.*, pp. 772-773

Barnett, G.H.; Evans, B. & Smith K.M.; Besecke, S. & Fuhrhop, J.-H.(1976). Synthesis of meso-pyridinium porphyrin salts. *Tetrahedron Lett.*, Vol. 44, pp. 4009-4012

Bedioui, F.; Devynck, J. & Bied-Charreton, C. (1995). Immobilization of metalloporphyrins in electropolymerized films: design and applications. *Acc. Chem. Res.*, Vol. 28, pp. 30-36

Bettelheim, A.; White, B.A.; Raybuck, S.A. & Murray R.W. (1987). Electrochemical polymerization of amino-, pyrrole-, and hydroxy-substituted tetraphenylporphyrins. *Inorg. Chem.*, Vol. 26, pp. 1009-1017

Brisach-Wittmeyer, A.; Lobstein, S., Gross, M. & Giraudeau A. (2005). Electrochemical reduction of 1-(*meso*-tetraphenylporphyrin)-pyridinium cations. *J. Electroanal. Chem.*, Vol. 576, pp. 129-137

Carelli, V.; Liberatore, F.; Casini, A.; Tortorella, S.; Scipione L. & Di Rienzo, B. (1998). On the regio- and stereoselectivity of pyridinyl radical dimerization. *New. J. Chem.*, pp. 999-1004

Carelli, V.; Liberatore, F.; Tortorella, S.; Di Rienzo, B. & Scipione L. (2002). Structure of the dimers arising from one-electron electrochemical reduction of pyridinium salts 3,5-disubstituted with electron-withdrawing groups. *Perkin Trans.*, pp. 542-547

Chen, W.; Akhigbe, J.; Brückner, C.; Li, C.M.& Lei Y. (2010). Electrocatalytic four-electron reduction of dioxygen by electrochemically deposited poly{[*meso*-tetrakis(2-thienyl)porphyrinato]cobalt(II)}. *J. Phys. Chem. C*, Vol. 114, pp. 8633-8638

Collman, J.P.; Denisevich, P.; Konai, Y.; Marrocco, M.; Koval, C. & Anson, F.C. (1980). Electrode catalysis of the four-electron reduction of oxygen to water by dicobalt face-to-face porphyrins. *J. Am. Chem. Soc.*, Vol. 102, pp. 6027-6036

Collman, J.P.; McDevitt, J.T.; Leidner, C.R.; Yee, G.T.; Torrance, J.B. & Little, W.A. (1987). Synthetic, electrochemical, optical, and conductivity studies of coordination polymers of iron, ruthenium, and osmium octaethylporphyrin. *J. Am. Chem. Soc.*, Vol. 109, pp. 4606-4614

Costa-Coquelard, C.; Schaming, D.; Lampre, I. & Ruhlmann, L. (2008). Photocatalytic reduction of Ag_2SO_4 by the Dawson anion α-$[P_2W_{18}O_{62}]^{6-}$ and tetracobalt sandwich complexes. *Appl. Catal. B: Environ.*, Vol. 84, pp. 835-842

Di Natale, C.; Monti, D. & Paolesse, R. (2010). Chemical sensitivity of porphyrin assemblies. *Mat. Today*, Vol. 13, pp. 46-52

Dolphin, D. & Felton R.H. (1974). The biochemical significance of porphyrin π cation radicals. *Acc. Chem. Res.*, Vol. 7, pp. 26-32

El Baraka, M.; Janot, J.M.; Ruhlmann, L.; Giraudeau, A.; Deumié, M. & Seta P. (1998). Photoinduced energy and electron transfers in the porphyrin triad (zinc octaethylporphyrin-4,4′ bipyridinium-tetraphenylporphyrin)$^{2+}$, 2 ClO_4^-). *J. Photochem. Photobiol. A: Chem.*, Vol. 113, pp. 163-169

El Kahef, L.; El Meray, M.; Gross, M. & Giraudeau A. (1986). Electrochemical synthesis of β-pyridinium zinc tetraphenylporphyrin. *Chem. Commun.*, 621-622

Fajer, J.; Borg, D.C.; Forman, D.; Dolphin, D. & Felton, R.H. (1970). π-cation radicals and dications of metalloporphyrins. *J. Am. Chem. Soc.*, Vol. 92, pp. 3451-3459

Giraudeau, A.; Ruhlmann, L.; El Kahef, L. & Gross, M. (1996). Electrosynthesis and characterization of symmetrical and unsymmetrical linear porphyrin dimers and their precursor monomers. *J. Am. Chem. Soc.*, Vol. 118, pp. 2969-2979.

Giraudeau, A.; Lobstein, S.; Ruhlmann, L.; Melamed, D.; Barkigia, K.M. & Fajer J. (2001). Electrosynthesis, electrochemistry, and crystal structure of the tetracationic Zn-meso-tetrapyridiniumyl-β-octaethylporphyrin. *J. Porphyrins Phthalocyanines*, Vol. 5, pp. 793-797

Giraudeau, A.; Schaming, D.; Hao, J.; Fahra, R.; Goldmann, M. & Ruhlmann, L. (2010). A simple way for the electropolymerization of porphyrins. *J. Electroanal. Chem.*, Vol. 638, pp. 70-75

Griveau S. & Bedioui, F. (2011). Electropolymerized thin films of metalloporphyrins for electrocatalysis and electroanalysis, In: The Porphyrin Handbook, Vol. 12, Applications. Kadish, K.M.; Smith, K.M. & Guilard, R. (Ed.), Academic Press Ed.

Hao, J.; Giraudeau, A.; Ping, Z. & Ruhlmann, L. (2008). Supramolecular assemblies obtained by large counteranion incorporation in a well-oriented polycationic copolymer. *Langmuir*, Vol. 24, pp. 1600-1603

Hinman, A.S.; Pavelich, B.J.; Kondo, A.E. & Pons S. (1987). Oxidative voltammetry of some tetraphenylporphyrins in the presence of nucleophiles leading to isoporphyrins. *J. Electroanal. Chem.*, Vol. 234, pp. 145-162

Jeannin, Y.P. (1998). The nomenclature of polyoxometalates: how to connect a name and a structure. *Chem. Rev.*, Vol. 98, pp. 51-76

Jurow, M.; Schuckman, A.E., Batteas, J.D. & Drain, C.M. (2010). Porphyrins as molecular electronic components of functional devices. *Coord. Chem. Rev.*, Vol. 254, pp. 2297-2310

Kadish, K.M.; Smith, K.M. & Guilard, R. (Ed.) (2000). The Porphyrin Handbook, Vol. 6, Applications: past, present and future. Academic Press Ed.

Karakostas, N.; Schaming, D.; Sorgues, S.; Lobstein S.; Gisselbrecht, J.-P.; Giraudeau, A.; Lampre, I. & Ruhlmann, L. (2010). Photophysical, electro- and spectroelectrochemical properties of the nonplanar porphyrin [ZnOEP(Py)$^{4+}$,4Cl^{-}] in aqueous media. *J. Photochem. Photobiol. A: Chem.*, Vol. 213, pp. 52-60

Katsoulis, D.E. (1998). A survey of applications of polyoxometalates *Chem. Rev.*, Vol. 98, pp. 359-387

Leroux, Y.; Schaming, D.; Ruhlmann, L. & Hapiot, P. (2010). SECM investigations of immobilized porphyrins films. *Langmuir*, Vol. 26, pp. 14983-14989

Li, G.; Wang, T.; Schulz, A.; Bhosale, S.; Lauer, M.; Espindola, P.; Heinze, J. & Fuhrhop, J.-H. (2004). Porphyrin-acetylene-thiophene polymer wires. *Chem. Commun.*, pp. 552-553

Li, G.; Bhosale, S.; Tao, S.; Guo, R.; Bhosale, S.; Li, F.; Zhang, Y.; Wang, T. & Fuhrhop, J.-H. (2005). Very stable, highly electroactive polymers of zinc(II)-5,15-bisthienylphenyl porphyrin exhibiting charge-trapping effects. *Polymer*, Vol. 46, pp. 5299-5307

Ma, Z.-F.; Xie, X.-Y.; Ma, X.-X.; Zhang, D.-Y.; Ren, Q.; Heβ-Mohr, N. & Schmidt, V. M. (2006). Electrochemical characteristics and performance of CoTMPP/BP oxygen reduction electrocatalysts for PEM fuel cell. *Electrochem. Commun.*, Vol. 8, pp. 389-394

Macor, K.A. & Spiro, T.G. (1983). Porphyrin electrode films prepared by electrooxidation of metalloprotoporphyrins. *J. Am. Chem. Soc.*, Vol. 105, pp. 5601-5607

Marvaud, V. & Launay, J.P. (1993). Control of intramolecular electron transfer by protonation: oligomers of ruthenium porphyrins bridged by 4,4′-azopyridine. *Inorg. Chem.*, Vol. 32, pp. 1376-1382

Rachlewicz, K. & Latos-Grażyński, L. (1995). Novel reactions of iron(III) tetraphenylporphyrin π-cation radicals with pyridine. *Inorg. Chem.*, Vol. 34, pp. 718-727

Ruhlmann, L. (1997). Couplage anodique de porphyrines : nouvelle méthodologie pour l'obtention de multiporphyrines. PhD Thesis, Université Louis Pasteur, Strasbourg, France

Ruhlmann, L.; Lobstein, S.; Gross, M. & Giraudeau, A. (1999 a). An electrosynthetic path toward pentaporphyrins. *J. Org. Chem.*, Vol. 64, pp. 1352-1355

Ruhlmann, L.; Schulz, A.; Giraudeau, A.; Messerschmidt, C. & Fuhrhop, J.-H. (1999 b). A polycationic zinc-5,15-dichlorooctaethylporphyrinate-viologen wire. *J. Am. Chem. Soc.*, Vol. 121, pp. 6664-6667

Ruhlmann, L. & Giraudeau A. (2001). A first series of dimeric porphyrins electrochemically linked with diphosphonium bridges. *Eur. J. Inorg. Chem.*, pp. 659-668

Ruhlmann, L.; Gross, M. & Giraudeau A. (2003). Bisporphyrins with bischlorin features obtained by direct anodic coupling of porphyrins. *Chem. Eur. J.*, Vol. 9, pp. 5085-5096

Ruhlmann, L.; Hao, J.; Ping, Z. & Giraudeau A. (2008). Self-oriented polycationic copolymers obtained from bipyridinium meso-substituted-octaethylporphyrins. *J. Electroanal. Chem.*, Vol. 621, pp. 22-30

Schaming, D.; Giraudeau, A.; Lobstein, S.; Farha, R.; Goldmann, M; Gisselbrecht, J.-P. & Ruhlmann, L. (2009). Electrochemical behavior of the tetracationic porphyrins (py)ZnOEP(py)$_4^{4+}$4PF_6^- and ZnOEP(py)$_4^{4+}$4Cl^-. *J. Electroanal. Chem.*, Vol. 635, pp. 20-28

Schaming, D.; Costa-Coquelard, C.; Sorgues, S.; Ruhlmann, L. & Lampre, I. (2010 a). Photocatalytic reduction of Ag_2SO_4 by electrostatic complexes formed by tetracationic zinc porphyrins and tetracobalt Dawson-derived sandwich polyanions. *Appl. Catal. A: Gen.*, Vol. 373, pp. 160-167

Schaming, D.; Allain, C.; Farha, R.; Goldmann, M.; Lobstein, S.; Giraudeau, A.; Hasenknopf, B. & Ruhlmann, L. (2010 b). Synthesis and photocatalytic properties of mixed polyoxometalate–porphyrin copolymers obtained from Anderson-type polyoxomolybdates. *Langmuir*, Vol. 26, pp. 5101-5109

Schaming, D. (2010 c). Assemblages hybrides porphyrines-polyoxométallates : étude électrochimique, photochimique et photocatalytique. PhD Thesis, Université Paris-Sud 11, Orsay, France

Schaming, D.; Marggi-Poullain, S.; Ahmed, I.; Farha, R.; Goldmann, M.; Gisselbrecht, J.-P. & Ruhlmann, L. (2011 a). Electrosynthesis and electrochemical properties of porphyrins dimers with pyridinium as bridging spacer. *New. J. Chem.*, Vol. 35, pp. 2534-2543

Schaming, D.; Ahmed, I.; Hao, J.; Alain-Rizzo, V.; Farha, R.; Goldmann, M.; Xu, H.; Giraudeau, A.; Audebert, P. & Ruhlmann. L. (2011 b). Easy methods for the electropolymerization of porphyrins based on the oxidation of the macrocycles. *Electrochim. Acta*, Vol. 56, pp. 10454-10463

Schaming, D.; Farha, R.; Xu, H.; Goldmann, M. & Ruhlmann, L. (2011 c). Formation and photocatalytic properties of nanocomposite films containing both tetracobalt Dawson-derived sandwich polyanions and tetracationic porphyrins. *Langmuir*, Vol. 27, pp. 132-143

Shine, H.J.; Padilla, A.G. & Wu, S.-M. (1979). Ion radicals. 45. Reactions of zinc tetra phenylporphyrin cation radical perchlorate with nucleophiles. *J. Org. Chem.*, Vol. 44, 4069-4075

Traylor, T.G. (1991). Kinetics and mechanism studies in biomimetic chemistry: metalloenzyme model systems. *Pure & Appl. Chem.*, Vol. 63, pp. 265-274

Troupis, A.; Hiskia, A. & Papaconstantinou, E. (2001). Reduction and recovery of metals from aqueous solutions with polyoxometallates. *New J. Chem. Lett.*, Vol. 25, pp. 361-363

Troupis, A.; Gkika, E.; Hiskia, A. & Papaconstantinou, E. (2006). Photocatalytic reduction of metals using polyoxometallates: recovery of metals or synthesis of metal nanoparticles. *C. R. Chim.*, Vol. 9, pp. 851-857

Wagner, R.W. & Lindsey, J.S. (1994). A molecular photonic wire. *J. Am. Chem. Soc.*, Vol. 116, pp. 9759-9760

Xia, Y.; Schaming, D.; Farha, R.; Goldmann, M. & Ruhlmann, L. (2012). Bis-porphyrin copolymers covalently linked by pyridinium spacers obtained by electropolymerization from β-octaethylporphyrins and pyridyl-substituted porphyrins. *New J. Chem.*, DOI: 10.1039/c1nj20790c

4

Electrochemical Polymerization of Aniline

Milica M. Gvozdenović[1,*], Branimir Z. Jugović[2], Jasmina S. Stevanović[3], Tomislav Lj. Trišović[2] and Branimir N. Grgur[1]
[1]*Department of Physical Chemistry and Electrochemistry, Faculty of Technology and Metallurgy, University of Belgrade, Belgrade,*
[2]*Institute of Technical Sciences, Serbian Academy of Science and Arts,*
[3]*Center for Electrochemistry, Institute of Technology and Metallurgy, Belgrade*
Serbia

1. Introduction

From its revolutionary discovery, up to day electroconducting polymers attracts attention of researchers along scientific community. Among numerous known electroconducting polymers, polyaniline and its derivates are probably the most investigated. Such popularity, in both theoretical and practical aspects, is a consequence of its unique properties: existence of various oxidation states, electrical and optical activity, low cost monomer, red/ox reversibility, environmental stability etc. (Inzelt, 2008; Horvat-Radošević & Kvastek, 2006; Gospodinova & Terlemezyan, 1998). These diverse and important features seem to be promising in vide area of practical applications in: rechargeable power sources (Gvozdenović et al., 2011; Jugović et al., 2006; Jugović et al., 2009), sensors (Bezbradica et al., 2011; Dhaoui, 2008; Grummt et al. 1997, Gvozdenović et al., 2011; Mu & Xue, 1996), magnetic shielding, electrochemical capacitors (Xu et al., 2009), electrochromic devices (Kobayashi et al. as cited in Mu, 2004; Wallace et al., 2009), corrosion protection etc. (Biallozor & Kupniewska, 2005; Camalet et al., 1996, 1998, 2000a, 2000b; Gvozdenović & Grgur, 2009; Gvozdenović et al., 2011; Kraljić et al., 2003; Popović & Grgur, 2004; Popović et al. 2005, Grgur et al., 2006; Özylimaz et al., 2006). Polyaniline is commonly obtained by chemical or electrochemical oxidative polymerization of aniline (Elkais et al., 2011; Jugović et al., 2009; Lapkowski, 1990; Stejskal & Gilbert, 2002; Steiskal at al., 2010) although photochemically initiated polymerization (Kim et al., 2001 as cited in Wallace et al., 2009; Teshima et al., 1998) and enzyme catalyzed polymerization were also reported (Bhadra et al., 2009; Nagarajan et al., 2000).

Polyanilne obtained by electrochemical polymerization is usually deposited on the electrode, however electrochydrodinamic route was also developed resulting in polyaniline colloids of specific functionalities (Wallace et al., 2009). Electrochemical polymerization of aniline is routinely carried out in strongly acidic aqueous electrolytes, through generally accepted mechanism which involves formation of anilinium radical cation by aniline oxidation on the electrode (Hussain & Kumar, 2003). Electrochemical polymerization of aniline is proved to be auto-catalyzed (Mu & Kan, 1995; Mu et al., 1997). The experimental conditions, such as: electrode material, electrolyte composition, dopant anions, pH of the

* Corresponding Author

electrolyte etc., all have strong influence on the nature of the polymerization process (Camalet et al., 2000; Ćirić Marjanović et al., 2006; Córdova et al., 1994; Duić & Mandić, 1992; Giz et al., 2000; Gvozdenović & Grgur, 2009; Inzlet, 2008; Lippe & Holze, 1992; Mu & Kan, 1998; Mun & Kan, 1998; Nunziante & Pistoia, 1989; Okamoto & Kotaka, 1998a, 1998b, 1999; Popović & Grgur, 2004; Pron et al., 1993; Pron & Rannou, 2010; Wallace et al., 2009). The low pH is almost always needed for preparation of the conductive polyaniline in the form of emeraldine salt, since it is evidenced that at higher pH, the deposited film is consisted of low chain oligomeric material (Stejskal et al., 2010). The doping anion incorporated into polymer usually determines the morphology, conductivity, rate of the polyaniline growth during electrochemical polymerization, and has influence on degradation process (Córdova et al., 1994; Lippe & Holze, 1992; Mandić et al., 1997; Okamoto & Kotaka, 1999, Pron & Rannou, 2010). The electrochemical polymerization of aniline is practically always carried out in aqueous electrolytes, although polymerization in organic solvents such is acetonitrile was also reported. Recently it was observed that ionic liquids electrolytes might be used for successful preparation of conductive polyaniline (Heinze et al., 2010; Wallace et al., 2009). Finally, electrochemical polymerization and co-polymerization of numerous substituted aniline derivates, resulted in polymer materials with properties different from the parent polymer, were also investigated (Karyakin et al. 1994; Kumar, 2000; Mattoso & Bulhões, 1992; Mu, 2011; Zhang, 2006)

Since there is a still growing interest for the research in the field of conducting polymers with polyaniline as the most representative, the aim of this paper is to review the extremely rich literature attempting to describe all important aspects of electrochemical polymerization of aniline.

2. Polyaniline

Polyaniline is probably the eldest known electroconducting polymer, since it was used for textile coloring one century ago (Sayed & Dinesan, 1991; Wallace et al., 2009).The great interest in research of polyaniline is connected to discovery of its conductivity in the form of emeraldine salt and existence of different oxidation forms (Inzelt, 2008; MacDiarmid et al., as cited in Wallace et al., 2009; Syed & Dinesan, 1991).

2.1 Different oxidation states of polyaniline

Unlike other know electroconducting polymers, polyaniline can exist, depending on degree of oxidation, in different forms, known as: leuoemeraldine, emeraldine and perningraniline. Leucoemeraldine, eg. leucoemeraldine base, refers to fully reduced form; emeraldine, eg. emeraldine base, is half-oxidized, while perningraniline, eg. perningraniline base, is completely oxidized form of polyaniline. The only conducting form of polyaniline is emeraldin salt, obtained by doping or protonation of emeraldine base (Fedorko et al., 2010; MacDiarmid et al., 1987, Pron & Rannou, 2002).

The unique feature of mentioned polyaniline forms is ease of its mutual conversions by both chemical and electrochemical reactions as it can be seen in Fig.1. (Gospodinova & Terlemezyan, 1998; Kang et al., 1998; Stejskal et al.,1996).

Apart from the changes in oxidation levels, all the transitions among polyaniline forms are manifested by color and conductivity changes (Stejskal et al., 1996). The conducting protonated emeraldine in the form of green emeraldine salt, obtained as a product of electrochemical polymerization of aniline in acidic electrolytes, can be easily transformed by further oxidation to fully oxidized dark blue perningraniline salt, which can be treated by alkali to form violet perningraniline. Emeraldine salt can also be reduced to transparent leucoemeraldine, or can be transformed by alkali to blue non conducting emeraldine. The two blue forms of polyanilne, perningranilne salt and emeraldine have different shades of blue (Stejskal et al., 1996). Both reduction of emeraldine salt to leucoemeraldine and oxidation to perningraniline states are followed by decrease in conductivity (Stejskal et al., 2010).

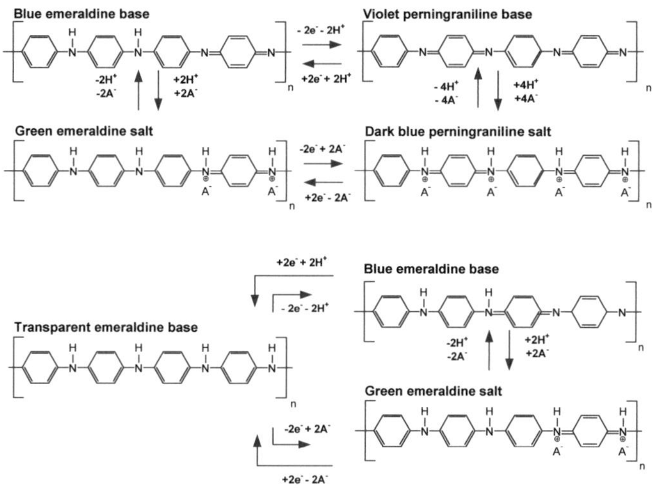

Fig. 1. Different forms of polyaniline

2.2 Polyaniline conductivity

The mechanism of polyaniline conductivity differs from other electroconducting polymers, owing to the fact that nitrogen atom is involved in the formation of radical cation, unlike most of the electroconducting polymers whose radical cation is formed at carbon. On the other hand, nitrogen is also involved in the conjugated double bonds system. Therefore, electrical conductivity of polyaniline is dependent both on the oxidation and protonation degrees (Fedorko et al., 2010; Genies at al., 1990; Pron & Rannou, 2002; Wallace et al., 2009).
As mentioned before, polyaniline is characterized by existence of various oxidation forms. Polyaniline in the form of emeraldine base can be doped (protonated) to conducting form of emeraldine salt. Emeraldine base, half oxidized form, is consisted of equal amount of amine (-NH-) and imine (=NH-) sites. Imine sites are subjected to protonation to form bipolarone or dicatione (emeraldine salt form). Bipolarone is further dissociated by injection of two electrons both from electron pairs of two imine nitrogen, into quinodiimine ring, and the third double bond of benzenoid ring is formed (Stejskal, 2010).
Unpared electrons at nitrogen atoms are cation radicals, but essentially they represent polarons. The polaron lattice, responible for high conductivity of polyaniline in the form of emeraldine salt is formed by redistribution of polarons along polymer chain, according to shematic representation given in Fig. 2. (Wallace et al., 2009)
Although both bipolaron and polaron theoretical models of emeraldine salt conductivity were proposed (Angelopulus et al., as cited in Wallace et al., 2009; Tanaka et al., 1990), it was lately confirmed that, beside from the fact that few of spineless bipolarons exist in

polyaniline, formation of polarons as charge carriers explained high conductivity of polyaniline (Mu et al., 1998; Patil et al., 2002). As mentioned, unique property of polyaniline is conductivity dependence on the doping (proton) level (Chiang et al., 1986; Wallace et al., 2009). The maximal conductivity of polyaniline is achieved at doping degree of 50%, which corresponds to polyaniline in the form of emeraldine salt (Tanaka et al., 1989). For higher doping degrees some of the amine sites are protonated, while lower doping degrees means that some of the imine sites were left unprotonated (Wallace et al., 2009), explaining why, in the light of the polaron conductivity model, reduction of emeraldine salt to leucoemeraldine and oxidation to perningraniline states decrease the conductivity. The order of magnitude for conductivity varies from 10^{-2} S cm^{-1}, for undoped emeraldine, up to 10^{3} S cm^{-1} for doped emeraldine salt (Inzelt, 2008; Wallace et al., 2009).

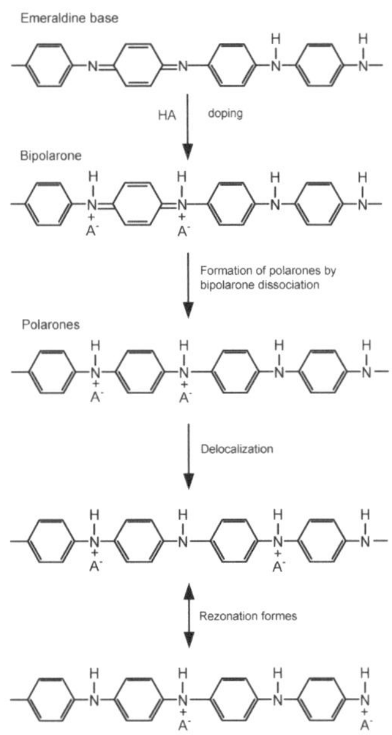

Fig. 2. Schematic presentation of polyaniline conductivity

Beside the fact that doping degree has the pronounced effect on the conductivity, various other factors such as: moisture amount (Kahol et al., 1997; Patil et al., 2002), morphology (Monkman & Adams, 1991; Zhou et al., 2007), temperature (Probst & Holze, 1995) etc. were also found to had influence on the polyaniline conductivity.

3. Mechanism and kinetics of the electrochemical polymerization of aniline

Generally, electroconducting polymers are obtained by either chemical or electrochemical oxidative polymerization, although reductive polymerization was also reported (Heinze et al., 2010; Inzlet et al., 2000; Yamamoto & Okida, 1999;Yamamoto, 2003, as cited in Inzelt, 2008). Chemical polymerization is used when large quantity of polymer is requested. Electrochemical polymerization is favorable, since in the most cases the polymer is directly deposited on the electrode facilitating analysis. On the other hand, electrochemical polymerization is especially useful if polymer film electrode is needed. By proper design of the electrochemical experiment, polymer thickness and conductivity can be easily controlled. It is believed that electrochemical polymerization is consisted of three different steps, in first, oxidation of the monomer at anode lead to formation of soluble oligomers in the diffusion layer, in the second, deposition of oligomers occurs through nucleation and growth process, and finally, the third step is responsible for chain propagation by solid state polymerization (Heinze et al., 2010). Unfortunately, a general mechanism of electrochemical polymerization could not be established, since it was evidenced that various factors had influence. However, it was observed that first step of the electrochemical polymerization was formation of reactive cation radicals (Heinze et al., 2010; Kankare, 1998). The next step, strongly dependent on the experimental conditions, is believed to be essential for the polymer growth (Inzelt, 2008). The knowledge on the kinetics of the nucleation and growth process during electrochemical synthesis of electroconducting polymers is also of great interest, since it would be useful in control of the morphology, density, crystallinity etc. of the desired polymer.

3.1 Mechanism

Both the mechanism and the kinetics of the electrochemical polymerization of aniline were extensively investigated (Andrade et al. 1998; Arsov et al. 1998; Carlos et al. 1997; Hussain & Kumar, 2003; Inzelt et al., 2000, Bade et al. 1992; Lapkowski, 1990; Malinauskas et Holze, 1998; Mandić et al., 1997; Mu & Kan, 1995; Mu et al., 1997). Electrochemical, similarly to chemical, polymerization of aniline is carried out only in acidic electrolyte, since higher pH leads to formation of short conjugation oligomeric material, with different nature (Wallace et al., 2009).
As stated before, it is generally accepted that the first step of the polymerization process of aniline involves formation of aniline cation radicals, by anodic oxidation on the electrode surface, which is considered to be the rate-determining step (Zotti et al., 1987, 1988). The existence of aniline radical cation was experimentally confirmed, by introducing molecules, (resorcinol, hydroquinone, benzoquinine etc.), capable of retarding or even stopping the reaction, which evidenced a radical mechanism (Mu et al., 1997). The oxidation of the aniline monomer is an irreversible process, occurring at higher positive potentials than redox potential of the polyaniline (Inzelt, 2008).
The following step is dependent on numerous factors such as: electrolyte composition, deposition current density, or potential scan rate, nature and state of the anode material, temperature etc. (Inzelt, 2008). There is a request for relatively high concentration of radical cations near the electrode surface. Radical cations can be involved, depending on reactivity, in different reactions. If it is quite stable, it may diffuse into the solution and react to form soluble products of low molecular weights. On the other hand, if is very unstable, it can react rapidly with anion or the solvent, in the vicinity of the electrode and form soluble products with low molecular weights (Park & Joong, 2005). In favorable case, coupling of the anilinium radicals would occur, followed by elimination of two protons and rearomatization leading to formation of dimer (lately oligomer). The aniline dimer, or oligomer, is further oxidized on the anode together with aniline. The chain propagation is achieved by coupling radical cations of the oligomer with anilinium radical cation. Finally, the counter anion originating from the acid, normally present in the electrolyte, dopes the polymer, meeting the request of electroneutrality. The mentioned mechanism of aniline electrochemical polymerization is schematically presented in Fig. 3 (Hussain & Kumar 2003, Wallace et al. 2009).

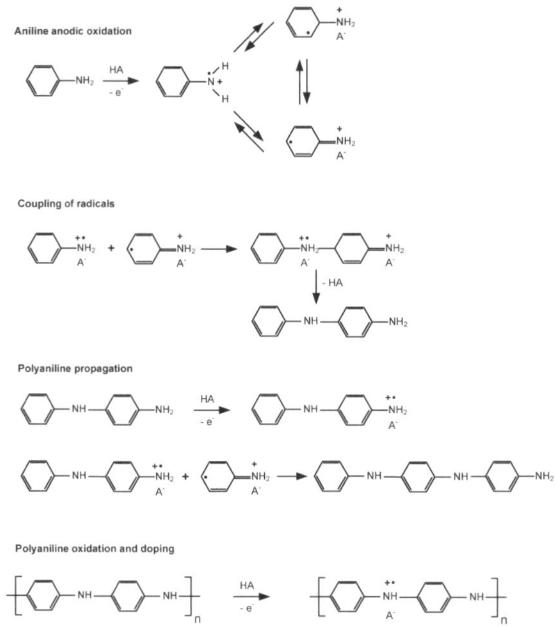

Fig. 3. Schematic presentation of mechanism of electrochemical polymerization of aniline.

It was evidenced that electrochemical polymerization of aniline is an autocatalytic process (Inzelt, 2008; Mu & Kan, 1996; Mu et al.,1997; Stilwell & Park, 1996; Wallace et al., 2009). It was observed that current increased over time, at constant potentials higher than 0.80 V and that anodic peak potentials shifted to more negative values upon increasing cycle number (Mu et al., 1997). Generally, it means that the more polymer formed on the anode, the higher the rate of the electrochemical polymerization.

3.2 Nucleation and growth of polyaniline

Explanation of the kinetics of the nucleation and growth process during electrochemical synthesis of electroconducting polymers is usually relaying on the metal deposition theory (Heinze et al., 2010; Kankare, 1998). According to the theory, two kinds of nucleation process exist, instantaneous and progressive, with three types of growth refering to: one- (1D), two- (2D), and three- (3D) dimensional processes. Instantaneous nucleation implies constant number of nuclei, growing without the further formation of nuclei. In the case of progressive nucleation, nuclei are constantly generated. 1D growth implies growth in only one direction, e.g. perpendicular to the electrode surface. In the 2D growth, the nuclei has preference to grow parallel to the electrode surface, while the 3D growth is characterized by the similar rates for these processes perpendicular and parallel to the electrode are quite similar (Heinze et al., 2010).

It was shown, mostly base on potentiostatic experiments, that several stages of polyaniline growth during electrochemical polymerization of aniline were involved, proceeding through different mechanism (Mandić et al.,1997). Cyclic voltammetry studies indicated that polyaniline growth was strongly dependent on type and concentration of anion in the electrolyte (Zotti et al.,1988). Studies on the early stages of the polyaniline growth indicate progressive nucleation of the polyaniline film, with 2D or 3D growth mechanism or 3D instantaneous nucleation, depending on electrolyte concentration and composition (Bade et al., 1992; Córdova et al., 1994; Mandić et al., 1997). The mass transfer controlled early stage of the polyaniline growth leads to formation of a compact layer (Inzlet 2010). In the case of perchloric acid, depending on monomer concentration, the nucleation process proceeds from progressive, at lower, to instantaneous nucleation at higher concentration (Mandić et al., 1997). At advanced stage, characterized by exponential current increase, 1D growth was assumed, resulted in continual branching and formation of the open structure (Cruz & Ticianelli, 1997)

3.3 Factors affecting electrochemical polymerization of aniline

Electrochemical synthesis of electroconducting polymers is strongly dependent on numerous parameters involved, such as: nature of the doping anion (affecting morphology, order of the polymer rate growth, nature and the composition of the solvent (to nucleofilic solvent would react with cation radicals formed by monomer oxidation on the anode) electrode material (depending on its surface energy controls the ease of the desired polymer deposition), temperature of the electrochemical polymerization etc. (Heinze et al., 2010; Inzelt, 2008; Pron & Rannou, 2002; Pruneanu et al., 1998; Wallce et al., 2009).

3.3.1 Doping anions

Electrochemical polymerization of aniline, as mentioned before, is practically always carried out in strong acidic aqueous electrolytes. Doping anions incorporated in polyaniline originate from the acid, and represent its conjugated base. The dopant anions are inserted during electrochemical polymerization fulfilling the request of electroneutrality, and therefore their concentrations are on the stoichiometric levels, for its reasonable that their presence have strong influence on, polyanilne morphology, conductivity, and electrochemical activity and the polymerization process itself (Arsov et al., 1998; Cordova et al,. 1994; Dhaoui et al., 2008; Koziel, 1993, 1995; Lapkowski, 1990, 1993; Lippe & Holze, 1992; Okamoto & Kotaka, 1998; Pron et al., 1992; Pron & Rannou, 2002).

It was experimentally confirmed that polyanilne obtained in the presence of so called "large dopant anions", originated from hydrochloride acid, sulfuric acid, nitric acid, *p*-toluensulfonic acid, and sulfosalicylic acid promoted formation of more swollen and open structured film, while the presence of "small ions" such is ClO_4^- or BF_4^-, resulted in formation of a more compact structure (Nunziante & Pistoria, 1989; Pruneanu et al., 1998; Zotti et al., 1988). The order of the polyaniline growth was also proved to increase with the size of the dopant anion (Inzelt et al., 2000). It was shown that addition of polyelectrolytes in polymerization electrolyte resulted in insertion of these molecules as dopants (Hyodo & Nozaki 1988, as cited in Wallace et al. 2009). It was also possible to obtain optically active polyaniline by electrochemical polymerization in the presence of (+) or (-) camphorsulfonic acid, leading to insertion of chiral dopants (Majidi et al., 2009).

3.3.2 Electrolyte composition

As mentioned previously, electrochemical polymerization of aniline is usually performed in aqueous electrolytes. There is limited number of studies referred to electrochemical polymerization in non-aqueous solvents (Genies & Lapkowski, 1987; Lapkowski, 1990; Pandey & Singh, 2002; Şahin et al., 2003). In the early studies acetronitrlie was mostly used as solvent, for example Watambe et al. (Miras et al. 1991; Watambe et al. 1989, as cited in

Wallace et al., 2009) made first electrochemical polymerization of aniline in acetonitrile solution containing lithium perchlorate, the resulted polymer exhibited similar redox properties as "ordinary" prepared polyaniline.
Lately, apart from acetonitrile, the studies involved use of: dichlormethane, nitrobenzene with various electrolytes such as: sodium tetraphenylborate, tetraethilammonium tetrafluoroborate and tetraetilammoniumperchlorate and resulted films showed different microstructures and electrochemical activity (Pandey & Singh, 2002). Other studies referred to electrochemical polymerization of aniline and fluoro- and chloro- substituted anilines and their coopolymers with aniline in acetonitrile containing tetrabutilammonium perchlorate and perchloric acid. The obtained polymers exhibited similar electrochemical and UV-behavior to "ordinary" polyaniline, but their conductivities were remarkably lower, explained by the steric effects of the substituent (Şahin et al. 2003). Various alkyl substituted anilines were electrochemically polymerized in both acetonytrile and dimethiylsulfoxide, their conductivities were also very low suggested that there were not in the typical state of the emeraldine salt (Yano et al., 2004). Successful electrochemical co-polymerization of aniline and pyrrole was also carried out in acetonitrile.
Finally, electrochemical polymerization of aniline was performed in various ionic liquids (Heinze et al., 2010; Innis et al., 2004; Li et al., 2005; Mu, 2007). For example, using IR and NMR spectroscopy Mu showed that the ionic liquid, namely 1-ethyl-3-methylimidazolium ethyl sulfate was incorporated in polyaniline during electrochemical polymerization. It was also observed that resulted polyaniline had exhibited good electrochemical activity in solutions with pH 12, and also considerably wider window of the detectable color changes at higher pH values, this effect was explained by the fact that used ionic liquid possessed high buffer capacity, which improved the redox activity and the electrochemical activity in broader pH range (Mu, 2007). Apart from the strong influence of the solvent, the presence of other components in the electrolyte solution, used for electrochemical polymerization of aniline, also had influence, primarily, on the morphology of the deposit (Inzelt, 2008). It was shown that presence of alcohols in the electrolyte would lead to polyaniline in the form of nanofibres agglomerated into interconnected network, FTIR spectra of the resulted polymer revealed strong interactions between alcohol and polyaniline molecules (Zhou et al., 2008).

3.3.3 Electrode material

Electrochemical polymerization of aniline is easily performed at so called inert electrodes, such as: platinum, gold, various graphite, carbons or indium-tin-oxide glasses, according to previously described mechanism. But the fact that relatively high electrode potential is required for oxidation of aniline, restricts the usage of other materials. The electrochemical polymerization of aniline on active metals is usually considered for application in corrosion protection (Biallozor & Kupniewska, 2005; Tallman et al., 2002). The problem connected to electrochemical polymerization onto active metals is either dissolution, or formation of non-conducting passive layer, on the potentials necessary for oxidation of aniline.
In the case of iron and steel the potential at which polymerization starts is in the region of active dissolution, leading to lost of the metal and contamination of the electrolyte, therefore it necessary to find a suitable electrolyte that would enable strong passivation of the metal without suppressing further electrochemical polymerization. The most common electrolyte used to electrochemical deposition of polyaniline on steel and aluminum is oxalic acid (Camalet et al., 1996, 2000a, 2000b; Martyak et al., 2002). The use of oxalic acid permitted

formation of passive layer consisted of iron oxalate, on which aniline polymerize. It was also showed that *p*-toluen sulfonic acid can be used for electrochemical polymerization of aniline, the deposition occurred after passivation. The passive film, in contrast to oxalic acid, was consisted mainly of iron oxide (Camalet et al., 1998). Other approach involves the pretreatment of the steel surface by polypyrrole, which can be easily formed electrochemically on iron and steel, with low extent of the metal dissolution, after this treatment aniline is easily electrochemically polymerized (Lacroix et al., 2000). The problem with electrochemical polymerization of aniline on aluminum is occurrence of two simultaneous processes electrochemical polymerization and passivation of the electrode by very stable protective oxide (Biallozor & Kupniewska, 2005). The studies of electrochemical polymerization on aluminum and its alloys involved pretreatment of the metal, and further polymerization (Huerta-Vilca et al., 2005; Wang & Tan, 2006). Similarly to steel, oxalic acid and *p*-toluen sulfonic acid electrolytes were used to grow polyaniline on aluminum (Conroy & Breslin, 2005; Karpagam et al., 2008). It was shown that sodium benzoate could be used to electrochemically polymerize aniline, without need for pretreatment, on steel, copper and aluminum (Gvozdenović & Grgur, 2009; Gvozdenović et al. 2011; Popović & Grgur, 2004). Electrochemical polymerization of substituted anilines on various metals and alloys were also carried out (Chaudhari & Patil 2007; Chaudhari et al., 2009; Pawar et al., 2007).

3.4 Experimental performance

Generally, experimental set-up for electrochemical synthesis of electroconducting polymers in laboratory conditions is simple. It involves, in majority of cases, standard three-electrode electrochemical cell, although in some cases of galvanostatic polymerization, two electrode cell can be used (Wallace et al., 2009). The polymer obtained by this procedure is deposited directly on the electrode. Novel experimental set-up, enabling electrochemical generation of polyaniline colloids, using flow-through electrochemical cell, was also reported (Aboutanos et al., 1999; Innis et al., 1998). In this novel electrochemical cell, anode was separated from two and cathodes by ion exchange membrane. The anodic and cathodic electrolytes were passed through electrode compartments at specified flow, while polymerization was achieved at constant potential.

The most common experimental techniques used for electrochemical polymerization of aniline are: cyclic voltammetry (potentiodynamic), galvanostatic and potetntiostatic techniques. Polymerization using cyclic voltammetry is characterized by cyclic regular change of the electrode potential and the deposited polymer is, throughout the experiment, changing between its non-conducting and conducting (doped) state, followed by exchange of the electrolyte through polymer (Heinze et al., 2010). At the end of polymerization the obtained polymer is in its non-conducting form, moreover, cyclic voltammetry favors formation of disordered chains and open structure (Heinze et al., 2010) As stated before, relatively high potential is required for electrochemical oxidation of aniline monomer, therefore at first 2 - 10 cycles, the upper potential limit is high, but owing to autocatalytic nature of aniline electropolymerization, the upper potential limit can be decreased to avoid degradation resulted from over oxidation of perningranilin form of polyaniline (Inzelt, 2008). Recently, it was shown that cyclic voltammetry can even be useful for formation of nanostructure polyaniline . It was shown that different sized polyaniline nanofibres were electrochemicaly polymerized, by different scan rates, in the presence of ferrocensulfonic acid (Mu & Yang, 2008).

Galvanostatic polymerization, owing to current control, enables reaction to proceed at constant rate. Galvanostatic synthesis permits estimation of the polymer mass deposited on the electrode (Kankare, 1998). On the other hand, galvanostatic polymerization leads to formation of polyaniline in its conductive form.
Electrochemical polymerization of aniline on graphite electrode from hydrochloride acid electrolyte, obtained by cyclic voltammetry (numbers on the figure refers to cycle number) given in Fig. 4., while in the insert of the Fig.4, hronopotnetiometric curve of galvanostatic polymerization from the same electrolyte is shown (Gvozdenović et al., 2011; Jugović, PhD thesis, 2009; Jugović et al., 2009)
Electrochemical polymerization of aniline proceeds together with insertion of chloride anions (dopant) from the electrolyte, according to:

$$(PANI)_n + nyCl^- \rightarrow [PANI^{y+} (Cl^-)]_n + nye^-$$

Where y refers to doping degree, ration between the number of charges in the polymer and the number of monomer units (Kankare, 1998).
As seen on cyclic voltammograms in Fig.4., doping of chloride ions started at potential of ~ -0.1 V (SCE), the first well defined anodic peak, situated at potential of 0.2 V (SCE) indicate transition of leucoemeraldine form of polyaniline to emeraldine salt, followed by the changes of y between 0 and 0.5.

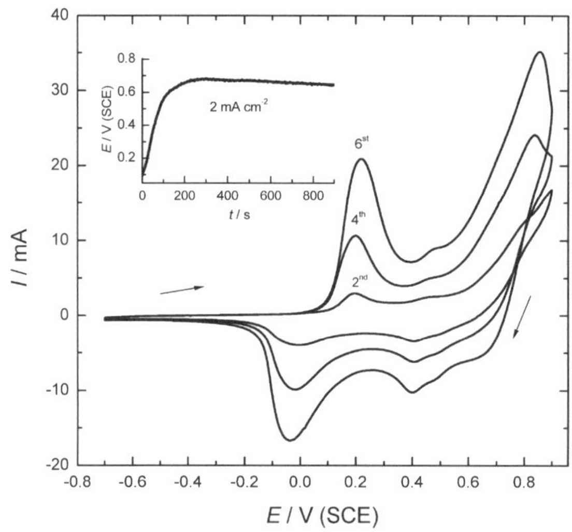

Fig. 4. Cyclic voltammograms of electrochemical polymerization of aniline on graphite electrode from aqueous solution of 1.0 mol dm^{-3} HCl and 0.25 mol dm^{-3} aniline, at scan rate of 20 mV s^{-1}. Insert: Chronopotentiometric curves of aniline polymerization at constant current density of 2.0 mA cm^{-2}.

Second anodic peak, occurred at potenial of ~ 0.5 V (SCE) denotes formation of fully doped perningraniline salt (y = 1).
The potentiostatic technique of electrochemical polymerization is characterized by pronounced changes in the current i.e. polymerization rate, and similarly to galvanostatic polymerization obtained polymer is in its doped form (Heinze et al., 2010). It was observed

that potentiostatic method could be useful in obtaining polyaniline nanowiers (Gupta & Miura, 2007). Modified potentiostatic techniques were also reported.
Some of the researches used pulse potentiostatic technique to obtained polyaniline electrochemically (Tang et al., 2000; Tsakova et al., 1993; Zhou et al., 2007). The potentiostatic pulse technique implies application of periodic cathodic and anodic pulses, with important parameters, lower (cathodic) and upper (anodic) limit potentials with additional cathodic and anodic pulses duration, during given time. It was observed (Zhou et al., 2007) that mentioned parameters had strong influence on the morphology of polyaniline, thus on its electrochemical activity.

3.5 Electrochemical co-polymerization of aniline and aniline derivates

Owing to its conductivity and redox activity, polyaniline is considered for practical application in various fields. Unfortunately, beside its unique properties, application of polyaniline in biochemical systems is limited as a consequence of the lost of activity at pH above 4 (Karyakin et al., 1994; Malinauskas, 1999; Mu 2011). This problem might be overcome by introduction of so- called pH functional groups into polyaniline chain (Mu, 2011). This could be achieved either by sulfonation (Wei et al., 1996) or by co-polymerization, which is more efficient way to alter the properties of parent polymer. Electrochemical polymerization of aniline and aniline derivates with pH functional groups, sulpho, carboxyl or hydroxyl was reported. It was observed that self-doped polyanilines, obtained by electrochemical co-polymerization of aniline with: *o*-aminobenzoic acid, *m*-aminobenzoic acid, or *m* - aminobenzensulfonic acid had exhibited redox activity at high pH (Karyakin et al., 1996).
Apart from aniline and aniline derivate, electrolyte solution also contains acid necessary for protonation of nitrogen atom. The obtained co-polymers are often called self-doped polyanilines, since the introduced negatively charged functional groups plays role of an intermolecular dopant which is able to compensate the charge on positively charged nitrogen atoms of the polymer. The presence of intermolecular anion alters properties of "ordinary" polyaniline, and has influence on the polymerization process as well. It was shown that upper switching potential limit had important influence on self-doping, the limit of 0.9 V was proven to be optimal, and while in the case of un- substituted aniline, upper limit was lower.
The problem related to electrochemical activity of self-doped polyaniline is its rapid lost. Recently, it was showed that electrochemical polymerization of aniline and 5-aminosalycylic acid, which nears two acidic functional groups, had lead to co-polymer with satisfactory redox activity (Mu, 2011).

4. Conclusion

Although polyaniline is among the first know electroconductive polymers, the interest in this field of study still exist, since its diverse and unique properties can be useful in various practical applications. Electrochemical polymerization of aniline and aniline derivates were intensively investigated. Various factors such as: electrode material, dopant anions, electrolyte composition, monomer type, pH etc. were proven to exhibit influence in the electropolymerization process and properties of the desired polymer. The electrochemical synthesis of polyaniline, similar to chemical, is practically always performed in strong acidic electrolytes, according to radical mechanism, and the polymer is deposited on anode, permitting easy way for further analysis. It was also proved that, using inconvenient experimental setup, polyaniline colloids can also be obtained electrochemically. Finally, studies of electrochemical co-polymerization of aniline and its derivates with acidic functional groups were also performed, leading to so called self-doped polyanilines, with benefit of being electrochemically active even in high pH solutions.

5. Acknowledgment

This work was supported by the Ministry of Education and Science, Republic of Serbia, under Contract No. 172046

6. References

Aboutanos V.; Barisci J.; Kane-Maguire L. & Wallace G. (1999). Electrochemical preparation of chiral polyaniline nanocomposites. *Synthetic Metals*, Vol. 106, No. 2, (October 1999), pp. 89-95, ISSN0379-6779

Andrade G.; Aguirre M. & Biaggi S.(1998). Influence of the first potential scan on the morphology and electrical properties of potentiodynamically grown polyaniline films. *Electrochimica Acta*, Vol. 44, No. 4, (September 1998), pp. 633-642, ISSN0013-4686

Arsov Lj.; Plieth W. & Koβmehl. (1998). Electrochemical and Raman spectroscopic study of polyaniline; influence of the potential on the degradation of polyaniline. *Journal of Solid State Electrochemistry,* Vol. 2, No. 5, (August 1998), pp. 355-361, ISSN1432-8488

Bade K.; Tsakova V. & Schultze J. (1992). Nucleation, growth and branching of polyaniline from microelectrode experiments. *Electrochimica Acta,* Vol. 37, No. 12, (September 1992), pp. 2255-2261, ISSN0013-4686

Bernard M.-C.; Cordoba de Torresi S. & Hugot-Le Goff A.(1999). In situ Raman study of sulfonate-doped polyaniline. *Electrochimica Acta*, Vol. 44, No. 12, (January 1999), pp. 1989-1997, ISSN0013-4686

BezbradicaD.; Jugović B.; Gvozdenović M.; Jakovetić S. & Knežević-Jugović Z. (2011). Electrochemically synthesized polyaniline as support for lipase immobilization. *Journal of Molecular Catalysis B: Enzymatic*, Vol. 70, No. 1-2, (June 2011), pp. 55-60, ISSN1381-1177

Bhadani, S.; Gupta M. & Sen Gupta S. Cyclyc Voltammetry and Conductivity Investigations of Polyaniline. (1993), *Journal of Applied Polymer Science*, Vol. 49, No. 3 (July 1993), pp. 397-403, ISSN0021-8995

Bhadra S.; Khastgir D.; Singha N. & Lee J. (2009). Progress in preparation, processing and applications of polyaniline. *Progress in Polymer Science*, Vol. 34, No. 8, (August 2009), pp. 783-810, ISSN0079-6700

Biallozor S. & Kupniewska A. (2005). Conducting polymers electrodeposited on active metals. *Synthetic Metals*, Vol. 155, No. 3, (December 2005), pp. 443-449, ISSN0379-6779

Camalet J.; Lacroix J.; Aeiyach S. & Lacaze P.(1998). Characterization of polyaniline films electrodeposited on mild steel in aquous *p*-toluensulfonic acid solution. *Journal of Electroanalytical Chemistry,*Vol. 445, No.1-2 (March 1998) pp. 117-129, ISSN1572-6657

Camalet J.; Lacroix J.; Aeiyach S.; Chane-Ching K. & Lacaze P. Electrodeposition of protective polyaniline films on mild steel. *Journal of Electroanalytical Chemistry*, Vol. 416, No. 1-2, (November 1996), pp. 179-182, ISSN1572-6657

Camalet J.; Lacroix J.; Aeiyach S.; Chane-Ching K.; Petit Jean J.; Chauveau E. & Lacaze P. (2000). Aniline electropolymerization on mild steel and zinc in a two-step process. *Journal of Electroanalitical Chemistry*,Vol. 481, No.1 (January 2000) pp. 76-81 ISSN1572-6657

Camalet J.; Lacroix J.; Ngyen T.; Aeiyach S.; Pham M.; Petit Jean J. & Lacaze P.-C. (2000). Aniline electropolymerization on platinum and mild steel from neutral aqueous media. *Journal of Electroanalytical Chemistry*, Vol. 485, No.1 (May 2000) pp. 13-20, ISSN1572-6657

Chaudhari S. & Patil P. (2007). Corrosion protective poly(*o*-ethoxyaniline) coatings on copper. *Electrochimica Acta*, Vol. 53, No. 2, (December 2007), pp. 927-933, ISSN0013-4686

Chaudhari S.; Gaikwad A. & Patil P.(2009). Poly(*o*-anisidine) coatings on brass: Synthesis, characterization and corrosion protection. *Current Applied Physics*, Vol. 9, No. 1, (January 2009), pp. 206-218, ISSN1567-1739

Chiang J. & MacDiarmid A. (1986). Polyaniline: Protonic acid doping of the emeraldine form to the metallic regime. *Synthetic Metals*, Vol. 13, No. 1-3, (January 1986), pp. 193-205, ISSN0379-6779

Ćiric-Marjanović G.; Marjanović N., Popović M.; Panić V.& Mišković-Stanković V. (2006). Anilinium 5-sulfosalicylate electropolymerization on mild steel from an aqueous solution of sodium 5-sulfosalicylate/disodium 5-sulfosalicylate. *Russian Journal of Electrochemistry*, Vol. 42, No. 12, (December 2006), pp. 1358-1364, ISSN1023-1935

Conroy K. & Breslin C.(2003).The electrochemical deposition of polyaniline at pure aluminium: electrochemical activity and corrosion protection properties. *Electrochimica Acta*, Vol. 48, No. 6, (February 2003), pp. 721-732, ISSN0013-4686

Córdova R.; del Valle M.; Arratia A.; Gómez H. Schrelber R. (1994). Effects of anions on the nucleation and growth mechanism of polyaniline. *Journal of Electroanalytical Chemistry*, Vol. 377, No. 1-2, (October 1994), pp. 75-83, ISSN1572-6657

Cruz C. & Ticianelli E. (1997).Electrochemical and ellipsometric studies of polyaniline films grown under cycling conditions. *Journal of Electroanalytical Chemistry*, Vol. 428, No. 1-2, (May 1997), pp. 185-192, ISSN1572-6657

Dalmolin C.; Canobre S.; Biaggio S.; Rocha-Filho R.; Bocchi N. Electropolymerization of polyaniline on high surface area carbon substrates. *Journal of Electroanalytical Chemistry*, Vol. 578, No. 1, (April 2005), pp. 9-15, ISSN1572-6657

Dhaoui W.; Bouzitoun M.; Zarrouk H.; Ouada H. & Pron A. (2008). Electrochemical sensor for nitrite determination based on thin films of sulfamic acid doped polyaniline deposited on Si/SiO_2structures in electrolyte/insulator/semiconductor (E.I.S.) configuration. *Synthetic Metals*, Vol. 158, No.17-18, (October 2008), pp. 722-726, ISSN0379-6779

Duić Lj. & Mandić Z. (1992). Counter-ion and pH effect on the electrochemical synthesis of polyaniline. *Journal of Electroanalytical Chemistry*, Vol. 335, No. 1-2, (September 1992), pp. 207-221 , ISSN1572-6657

Elkais A.; Gvozdenović M.; Jugović B.; Stevanović J.; Nikolić N.; Grgur B. (2011). Electrochemical synthesis and characterization of polyaniline thin film and polyaniline powder, *Progress in Organic Coatings*, Vol. 71, No. 1, (May 2011) pp. 32-35, ISSN0300-9440

Fedorko P.; Trznadel M.; Pron A.; Djurado D.; Planès J. & Travers J. (2010). New analytical approach to the insulator-metal transition in conductive polyaniline. *Synthetic Metals*, Vol. 160, No. 15-16, (August 2010), pp. 1668-1671, ISSN0379-6779

Ferreira V.; Cascalheira A. & Abrantes L. (2008). Electrochemical preparation and characterisation of Poly(Luminol–Aniline) films. *Thin Solid Films*, Vol. 516, No. 12, (April 2008), pp. 3996-4001 ISSN0040-6090

Genies E.; Boyle A.; Lapkowski M. & Tsintavis. (1990). Polyaniline: A historical survey. *Synthetic Metals*, Vol. 36, No. 2, (June 1990), pp. 139-182, ISSN0379-6779

Giz M.; de Albuqurque Maranhão S. & Torresi R. (2000). AFM morphological study of electropolymerized polyaniline films modified by sufracant and large anions. *Electrochemistry Communications*, Vol.2, No. 6, (June 2000), pp. 377-381, ISSN1388-2481

Gospodinova N. & Terlemezyan L. Conducting polymers prepared by oxidative polymerization: polyaniline. (1998). *Progress in Polymer Science*, Vol. 23, No. 8, (December 1998), pp. 1443-1484, ISSN0079-6700

Grgur B.; Gvozdenović M.; Mišković-Stanković V. & Kačarević-Popović Z. (2006). Corrosion behavior and thermal stability of electrodeposited PANI/epoxy coating system on mild steel in sodium chloride solution. *Progress in Organic Coatings*, Vol. 56, No. 2-3, (July 2006), pp. 214-219, ISSN0300-9440

Grummt U.; Pron A.; Zagorska M. & Lefrant S. (1997).Polyaniline based optical pH sensor. *Analytica Chimica Acta*, Vol. 357, No. 3, (December 1997), pp. 253-259, ISSN0003-2670

Gupta V. & Miura N. (2005).Large-area network of polyaniline nanowires prepared by potentiostatic deposition process. *Electrochemistry Communications*, Vol. 7, No. 10, (October 2005), pp. 995-999, ISSN1388-2481

Gvozdenović M. & Grgur B. (2009). Electrochemical polymerization and initial corrosion properties of polyaniline-benzoate film on aluminum. *Progress in Organic Coatings*, Vol. 65, No. 3, (July 2009), pp. 401-404, ISSN0300-9440

Gvozdenović M.; Jugović B.; Stevanović J.; Grgur B.; Trišović T. & Jugović Z. (2011). Electrochemical synthesis and corrosion behavior of polyaniline-benzoate coating on copper, *Synthetic Metals*, Vol. 161, No. 13-14, July 2011, pp 1313-1318, ISSN0379-6779

Gvozdenović M.; Jugović B.; Bezbradica D.; Antov M. Knežević-Jugović Z. & Grgur B. (2011). Electrochemical determination of glucose using polyaniline electrode modified by glucose oxidase. *Food Chemistry*, Vol. 124, No.1, (January 2011), pp. 396-400, ISSN0308-8146

Gvozdenović M.; Jugović B.; Trišović T.; Stevanović J. & Grgur B. (2011). Electrochemical characterization of polyaniline electrode in ammonium citrate containing electrolyte. *Materials Chemistry and Physics*, Vol. 125, No. 3, (February 2011), pp. 601-605, ISSN0254-0584

Heinze J.; Frontana-Uribe B. & Ludwigs S. (2010). Electrochemistry of Conducting Polymers-Persistent Models and New Concepts. *Chemical Reviews*, Vol.110, No.8, (June 2010), pp. 4724-4771 ISSN0009-2665

Horvat-Radošević V. & Kvastek K. Role of Pt-probe pseudo-reference electrode in impedance measurements of Pt and polyaniline (PANI) modified Pt electrodes. (2006). *Journal of Electroanalytical Chemistry,* Vol. 591, No. 2, (June 2006), pp. 217-222, ISSN1572-6657

Huerta-Vilca D.; de Moraes S. & de Jesus Motheo A. (2005). Aspects of polyaniline electrodeposition on aluminium. *Journal of Solid State electrochemistry,* Vol. 9, No. 6 (June 2005), pp. 416-420, ISSN1432-8488

Hussain A. & Kumar A. (2003). Electrochemical synthesis and characterization of chloride doped polyaniline. *Bulletin of Material Science,* Vol. 26, No. 3, (April 2003), pp. 329-334 ISSN0250-4707

Innis P.; J. Mazurkiewicz J.; Nguyen T.; Wallace G. & MacFarlane D. (2004). Enhanced electrochemical stability of polyaniline in ionic liquids. *Current Applied Physics,* Vol. 4, No. 2-4, (April 2004), pp. 389-393, ISSN1567-1739

Innis P.; Norris I.; Kane-Maguire L. & Wallace G.(1998).Electrochemical Formation of Chiral Polyaniline Colloids Codoped with (+)- or (−)-10-Camphorsulfonic Acid and Polystyrene Sulfonate. *Macromolecules,* Volume 31, No. 19, (September 1998), pp 6521–6528 ISSN0024-9297

Inzelt G.; Pineri M.; Schultze J. & Vorotyntsev M. Electron and proton conducting polymers: recent developments and prospects. *Electrochimica Acta,* Vol. 45, No. 15-16, (May 2000), pp. 2403-2421 ISSN0013-4686

Inzelt G.(2008). *Conducting Polymers – A New Era in Electrochemistry,* Springer-Verlag, ISBN 978-3-540-75929-4, Berlin, Heidelberg

Jugović B.; Trišović T.; Stevanović J.; Maksimović M. & Grgur B. (2006). Novel electrolyte for zinc–polyaniline batteries, *Journal of Power Sources,* Vol. 160, No. 2, (October 2006), pp. 1447-1450 ISSN0378-7753

Jugović B.; Gvozdenović M.; Stevanović J.; Trišović T. & Grgur B. (2009). Characterization of electrochemically synthesized PANI on graphite electrode for potential use in electrochemical power sources . *Materials Chemistry and Physics,* Vol. 114, No. 2-3, (April 2009), pp. 939-942, ISSN0254-0584

Kahol P.; Dyakonov A.; & McCormick B. (1997). An electron-spin-resonance study of polymer interactions with moisture in polyaniline and its derivatives. *Synthetic Metals,* Vol. 89, No. 1, (July 1997), pp. 17-28, ISSN0379-6779

Kang E; Neoh K. & Tan K. Polyaniline: A polymer with many interesting intrinsic redox states , *Progress in Polymer Science,* Vol.23, No. 2, (1998), pp. 277-324, ISSN0079-6700

Kankare J. (1998). Electronically Conducting Polymers: Basic Methods of Synthesis and Characterization, In: *Electrical and Optical Polymer Systems: Fundamentals: Methods, and Applications,* Ed. by: Wise D.; Wnek G.; Trantolo D.; Cooper J. & Gresser D. , pp. 167-199, Marcel Dekker, ISBN 0824701186, New York

Karpagam V.;Sathiyanarayanan S. & Venkatachari G. (2008).Studies on corrosion protection of Al2024 T6 alloy by electropolymerized polyaniline coating *Current Applied Physics,* Vol. 8, No. 1, (January 2008), pp. 93-98, ISSN1567-1739

Karyakin A.; Strakhova A. & Yatsimirsky A. (1994). Self-doped polyanilines electrochemically active in neutral and basic aqueous solutions.: Electropolymerization of substituted anilines. *Journal of Electroanalytical Chemistry,* Vol. 371, No. 1-2, (June 1994), pp. 259-265, ISSN1572-6657

Koziel K. & Lapkowski M.(1993). Studies on the influence of the synthesis parameters on the doping process of polyaniline. *Synthetic Metals*, Vol. 55, No. 2-3, (March 1993), pp. 1011-1016, ISSN0379-6779

Koziel K.; Lapkowski M. & Lefrant S. (1995). Spectroelectrochemistry of polyaniline at low concentrations of doping anions. *Synthetic Metals*, Vol. 69, No. 1-3, (March 1995), pp. 137-138, ISSN0379-6779

Kraljić M.; Mandić Z. & Duić Lj. (2003). Inhibition of steel corrosion by polyaniline coatings. *Corrosion Science*, Vol. 45, No. 1, (January 2003), pp. 181-198, ISSN0010-938X

Kumar D.(2000). Synthesis and characterization of poly(aniline-co-o-toluidine) copolymer *Synthetic Metals*, Vol. 114, No. 3, (September 2000), pp. 369-372 ISSN0379-6779

Lacroix J.; Camalet J.; Aeiyach S.; Chane-Ching K.; Petitjean J.; Chauveau E. & Lacaze P. (2000). Aniline electropolymerization on mild steel and zinc in a two-step process. *Journal of Electroanalytical Chemistry*, Vol. 481, No.1, (January 2000), pp. 76-81, ISSN1572-6657

Lapkowski M. (1990). Electrochemical synthesis of linear polyaniline in aqouos solutions. *Synthetic Metals*, Vol. 35, No. 1-2, (February-March 1990), pp. 169-182 ISSN0379-6779

Li M.; Ma C.; Liu B. & Jin Z. (2005). A novel electrolyte 1-ethylimidazolium trifluoroacetate used for electropolymerization of aniline. *Electrochemistry Communications*, Vol. 7, No. 2, (February 2005), pp. 209-212, ISSN1388-2481

Macdiarmid A.; Chiang J.; Richter A. & Epstein A. (1987). Polyaniline: a new concept in conducting polymers. *Synthetic Metals*, Vol. 18, N. 1-3, (February 1987), pp. 285-290, ISSN0379-6779

Majidi M.; Kane-Maguire L. & Wallace G. (1994).Enantioselective electropolymerization of aniline in the presence of (+)- or (−)-camphorsulfonate ion: a facile route to conducting polymers with preferred one-screw-sense helicity. *Polymer*, Vol. 35, No. 14, (July 1994), pp. 3113-3115, ISSN0032-3861

Malinauskas A. & Holze R. (1997). Suppression of the "first cycle effect" in self-doped polyaniline. *Electrochimica Acta*, Vol. 43, No. 5-6, (November 1997), pp. 515-520, ISSN0013-4686

Malinauskas A. (1999). Electrocatalysis at conducting polymers. *Synthetic Metals*, Vol. 107, No. 2, (November 1999), pp. 75-83 ISSN0379-6779

Mandić Z.; Duić Lj. & Kovačiček F. (1997). The influence of counter-ions on nucleation and growth of electrochemically synthetized polyaniline film. *Electrochimica Acta*, Vol. 42, No. 9, (1997), pp. 1389-1402, ISSN0013-4686

Martyak N.;McAndrew P.; McCaskie J.& Dijon J. (2002). Electrochemical polymerization of aniline from an oxalic acid medium. *Progress in Organic Coatings*, Vol. 45, No. 1, (September 2002), pp. 23-32, ISSN0300-9440

Mattoso L. & Bulhões L. (1992). Synthesis and characterization of poly(*o*-anisidine) films. *Synthetic Metals*, Vol. 52, No. 2, (October 1992), pp. 171-181, ISSN0379-6779

Mažeikienė R.; Niaura G. & Malinauskas A. (2003). Voltammetric study of the redox processes of self-doped sulfonated polyaniline. *Synthetic Metals*, Vol. 139, No. 1, (August 2003), pp. 89-94, ISSN0379-6779

Miras M.; C. Barbero C. & Haas O. (1991). Preparation of polyaniline by electrochemical polymerization of aniline in acetonitrile solution. *Synthetic Metals*, Vol. 43, No. 1-2, (June 1991), pp. 3081-3084, ISSN0379-6779

Monkman A. & Adams P. (1991). Optical and electronic properties of stretch-oriented solution-cast polyaniline films. *Synthetic Metals*, Vol. 40, No.1, (March 1991), pp. 87-96, ISSN0379-6779

Mu S. & Kan J. (1996). Evidence for the autocatalitic polymerization of aniline, *Electrochimica Acta*, Vol.4, No.10, (November 1996), pp. 1593-1599, ISSN0013-4686

Mu S. & Kan J. The effect of salts on the electrochemical polymerization of aniline. *Synthetic Metals*, Vol. 92, No. 2, (January 1998), pp. 149-155, ISSN0379-6779

Mu S. & Xue H. (1996). film. *Sensors and Actuators B: Chemical*, Vol. 31, No. 3, (March 1996), pp.155-160 ISSN0925-4005

Mu S. & Yang Y. (2008). Spectral Characteristics of Polyaniline Nanostructures Synthesized by Using Cyclyc Voltammetry. *Journal of Physical Chemistry B*. Vol. 112, No. 37, (September 2008), pp. 11558-11563 ISSN1520-6106

Mu S. &Kan J. (1998). The effect of salts on the electrochemical polymerization of aniline. *Synthetic Metals*, Vol. 92, No. 2 (January 1998), pp.149-155, ISSN0379-6779

Mu S. (2004). Electrochemical copolymerization of aniline and *o*-aminophenol. *Synthetic Metals*, Vol.143, No. 3, (June 2004), pp. 259-268, ISSN0379-6779

Mu S. (2007). Pronounced effect of the ionic liquid on the electrochromic property of the polyaniline film: Color changes in the wide wavelength range. *Electrochimica Acta*, Vol. 52, No. 28, (November 2007), pp. 7827-7834, ISSN0013-4686

Mu S. (2011). Synthesis of poly(aniline-*co*-5-aminosalicylic acid) and its properties. *Synthetic Metals*, Vol. 161, No. 13-14, (July 2011), pp. 1306-1312, ISSN0379-6779

Mu S. (2011). Synthesis of poly(aniline-*co*-5-aminosalicylic acid) and its properties. *Synthetic Metals*, Vol. 161, No. 13-14, (July 2011), pp. 1306-1312, ISSN0379-6779

Mu S.; Chen C. & Wan J. (1997). The kinetic behavior for the electrochemical polymerization of aniline in aqueous solution, *Synthetic Metals*, Vol.88, (April 1997), pp. 249-254, ISSN0379-6779

Mu S.; Kan J.; Lu J. & Zhuang L. (1998). Interconversion of polarons and bipolarons of polyaniline during the electrochemical polymerization of aniline. *Journal of Electroanalytical Chemistry*, Vol. 446, No. 1-2, (April1998), pp.107- 112, ISSN1572-6657

Nagarajan R.; Tripathy S; Kumar J.; Bruno F. & Samuelson L. (2000). An Enzymatically Synthesized Conducting Molecular Complex of Polyaniline and Poly(vinylphosphonic acid). *Macromolecules*, Vol. 33, No. 262, (November 2000), pp. 9542–9547 ISSN0024-9297

Nunziante P. & Pistoia G. (1989). Factors affecting the growth of thick polyaniline films by the cyclic voltmmetry technique. *Electrochimica Acta*, Vol. 34, No. 2, (February 1989), pp. 223-228, ISSN0013-4686

Okamoto H. Okamoto M. & Kotaka T. (1998). Structure and properties of polyaniline films prepared via electrochemical polymerization. II: Structure and properties of polyaniline films prepared via electrochemical polymerization. *Polymer*, Vol. 39, No. 18, (1998), pp. 4359-4367 ISSN0032-3861

Okamoto H. & Kotaka T. (1998). Structure and properties of polyaniline films prepared via electrochemical polymerization. I: Effect of pH in electrochemical polymerization media on the primary structure and acid dissociation constant of product polyaniline films. *Polymer*, Vol. 39, No. 18, pp. 4349-4358, ISSN0032-3861

Okamoto H. & Kotaka T.(1998). Structure and properties of polyaniline films prepared via electrochemical polymerization. III: Effect of counter ions in electrochemical polymerization media on the structure responses of the product polyaniline films. *Polymer,* Vol. 40, No. 2, (1989), pp. 407-417, ISSN0032-3861

Özyılmaz A.; Erbil M. & Yazıcı B. (2006). The electrochemical synthesis of polyaniline on stainless steel and its corrosion performance. *Current Applied Physics,* Vol. 6, No. 1, (January 2006), pp. 1-9, ISSN1567-1739

Pandey P. & Singh G. (2002). Electrochemical Polymerization of Aniline in Proton-Free Nonaqueous Media, Dependence of Microstructure and Electrochemical Properties of Polyaniline on Solvent and Dopant. *Journal of Electrochemical Society,* Vol. 149, No. 4, (April 2002) pp. D51-D56, ISSN0013-4651

Park S. & Joong H. (2005). Recent Advances in Electrochemical Studies of π-Conjugated Polymers. *Buillten of the Korean Chemical Society,* Vol. 26, No. 5, (May 2005), pp. 697-706 ISSN0253-2964

Patil R.; Harima Y.; Yamashita K.; Komaguchi K.; Itagaki Y. & Shiotani M. (2002). Charge carriers in polyaniline film: a correlation between mobility and in-situ ESR meausurements. *Journal of Electroanalytical Chemistry,* Vol. 518, No. 1, (January 2002), pp. 13-19, ISSN1572-6657

Pawar P.; Gaikwad A. & Patil P. (2007). Corrosion protection aspects of electrochemically synthesized poly(o-anisidine-co-o-toluidine) coatings on copper. *Electrochimica Acta,* Vol. 52, No. 19, (May 2007), pp. 5958-5967, ISSN0013-4686

Popović & Grgur B. (2004). Electrochemical synthesis and corrosion behavior of thin polyaniline-benzoate film on mild steel. *Synthetic Metals,* Vol. 143, No. 2, (June 2004), pp. 191-195, ISSN0379-6779

Popović M.; Grgur B. & Mišković-Stanković V. (2005). Corrosion studies on electrochemically deposited PANI and PANI/epoxy coatings on mild steel in acid sulfate solution . *Progress in Organic Coatings,* Vol. 52, No. 4, (April 2005), pp. 359-365, ISSN0300-9440

Probst M. & Holze R. (1995). Time- and temperature-dependent changes of the *in situ* conductivity of polyaniline and polyindoline. *Electrochimica Acta,* Vol. 40, No. 2, (February 1995), pp. 213-219, ISSN0013-4686

Proń A.; Laska J.; Österholm J-E. & Smith P. (1993). Processable conducting polymers obtained via protonation of polyaniline with phosphoric acid esters. *Polymer,* Vol. 34, No. 20, (October 1993), pp. 4235-4240 ISSN0032-3861

Pron A. & Rannou P. (2002). Processible conjugated polymers: from organic semiconductors to organic metals and superconductors. *Progress in Polymer Science,* Vol. 27, No. 1, (February 2002), pp. 135-190, ISSN0079-6700

Pruneanu S.; Csahók E.; Kertész V. & Inzelt G. (1998). Electrochemical quartz crystal microbalance study of the influence of the solution composition on the behaviour of poly(aniline) electrodes. *Electrochimica Acta,* Vol 43. No. 16-17, (May 1998), pp. 2305-2323, ISSN0013-4686

Şahin Y.; Perçin P.; Şahin M. & Özkan G. (2003). Electrochemical preparation of poly (2-bromoaniline) and poly (aniline-*co*-2-bromoaniline) in acetonitrile. *Journal of Applied Polymer Science,* Vol. 90, No. 9, (November 2003), pages 2460–2468, ISSN0021-8995

Spinks G.; Dominis A. Wallace G. & Tallman D (2002). Electroactive conducting polymers for corrosion control. Part 2. Ferrous metals. *Journal of Solid State Electrochemistry,* Vol. 6, No. 2, (February 2002), pp. 85-100, ISSN1432-8488

Stejskal J. & Gilbert R. (2002). Polyaniline. Preparation of a conducting polymer (IUPAC Technical Report). *Pure and Applied Chemistry,* Vol. 74, No. 5, (2002), pp. 857-867 doi:10.1351/pac200274050857

Stejskal J.; Kratochvíl & Jenkins A. (1996). The formation of polyaniline and nature of its structures. *Polymer,* Vol. 37, No. 2, pp. 367-369, (February 1996), ISSN0032-3861

Stejskal J.; Sapurina I. & Trchová M. (2010). Polyaniline nanostructures and the role of aniline oligomers in their formation. *Progress in Polymer Science,* Vol. 35, No. 12, (December 2010), pp. 1420-1481, ISSN0079-6700

Syed A. & Dinesan M. (1991). Polyaniline—A novel polymeric material, *Talanta,* Vol. 38, No. 8, (August 1991), pp. 815-837 ISSN0039-9140

Tallman D.; Spinks G.; Dominis A. & Wallace G. (2002). Electroactive conducting polymers for corrosion control. Part1. General introduction and review of non-ferrous metals. *Journal of Solid State Electrochemistry,* Vol. 6, No. 2, (February 2000), pp. 73-84, ISSN1432-8488

Tanaka J.; Mashita N.; Mizoguchi K. & Kume K. (1989). Molecular and electronic structures of doped polyaniline. *Synthetic Metals,* Vol. 29, No. 1, (March 1989), pp. 175-184, ISSN0379-6779

Tanaka K.; Wang S. & Yamabe T. (1990). Will bipolarons be formed in heavily oxidized polyaniline? *Synthetic Metals,* Vol. 36, No.1, (May 1990), pp. 129-135, ISSN0379-6779

Tang H.; Kitani A. & Shiotani M. (1996). Factors of anions on electrochemical formation and overoxidation of polyaniline. *Electrochimica Acta,* Vol. 41, No. 9, (June 1996) pp. 1561-1567 ISSN0013-4686

Tang Z.; Liu S.; Wang Z.; Dong S. & Wang E. (2000).Electrochemical synthesis of polyaniline nanoparticles . *Electrochemistry Communications,* Vol. 2, No.1, (January 2000), pp. 32-35, ISSN1388-2481

Tsakova V.; Milchev A. & Schultze J. (1993).Growth of polyaniline films under pulse potentiostatic conditions. *Journal of Electroanalytical Chemistry,* Vol. 346, No. 1-2, (March 1993), pp. 85-97, ISSN1572-6657

Wallace G.; Spinks G.; Kane-Maguire L. & Teasdale P. (2009). *Conductive Electroactive Polymers,* CRC Press, Taylor & Francis Group, ISBN 978-1-4200-6709-5, Boca Raton

Wang T.& Tan Y. (2006). Understanding electrodeposition of polyaniline coatings for corrosion prevention applications using the wire beam electrode method. *Corrosion Science,* Vol. 48, No. 8, (August 2006), pp. 2274-2290, ISSN0010-938X

Xu G.; Wang W.; Qu X.; Yin Y.; Chu L.; He B.; Wu H.; Fang L.; Bao Y. & Liang L. (2009). Electrochemical properties of polyaniline in *p*-toluene sulfonic acid solution. *European Polymer Journal,* Vol. 45, No. 9, (September 2009), pp. 2701-2707, ISSN0014-3057

Yano J.; Ota Y. & Kitani A. (2004). Electrochemical preparation of conductive poly(*N*-alkylaniline)s with long *N*-alkyl chains using appropriate dopant anions and organic solvents. *Materials Letters,* Vol. 58, No. 12-13, (May 2004), pp. 1934-1937 ISSN0167-577X

Zhang J.; Shan D. & Shaolin M. (2006). Electrochemical copolymerization of aniline with *m*-aminophenol and novel electrical properties of the copolymer in the wide pH range. *Electrochimica Acta*, Vol. 51, No. 20, (May 2006), pp. 4262-4270, ISSN0013-4686

Zhou H.; Wen J.; Ning X.; Fu C.; Chen J. & Kuang Y. (2007). Electrosynthesis of polyaniline films on titanium by pulse potentiostatic method. *Synthetic Metals*, Vol. 157, No. 2-3, (February 2007), pp. 98-103, ISSN0379-6779

Zhou S.; Wu T.; Kan J. Effect of methanol on morphology of polyaniline.(2007). *European Polymer Journal*, Vol. 43, No. 2, (February 2007), pp. 395-402, ISSN0014-3057

Zotti G; Cattarin S. & Comiss N. (1987). Electrodeposition of polythiophene, polypyrrole and polyaniline by the cyclic potential sweep method. *Journal of Electroanalytical Chemistry and Interfacial Electrochemistry*, Vol. 235, No. 1-2, (October 1987), pp. 259-273, ISSN1572-6657

Zotti G; Cattarin S. & Comiss N. (1988). Cyclic potential sweep electropolymerization of aniline: The role of anions in the polymerization mechanism. *Journal of Electroanalytical Chemistry and Interfacial Electrochemistry*, Vol. 239, No. 1-2, (January 1988). pp. 387-396, ISSN1572-6657

5

Electrochemical Preparation and Properties of Novel Conducting Polymers Derived from 5-Amino-2naphtalensulfonic Acid, Luminol and from Mixtures of Them

Luz María Torres-Rodríguez, María Irene López-Cázares, Antonio Montes-Rojas, Olivia Berenice Ramos-Guzmán and Israel Luis Luna-Zavala
Laboratorio de Electroquímica/ Facultad de Ciencias Químicas/ Universidad Autónoma de San Luis Potosí, S. L. P México

1. Introduction

Polyanilines contain anioniogenic functional groups are denominated self-doped polyanilines. These polymers possess properties different from those of polyaniline (PANI) as suppressed or no need on anion doping during oxidation or reduction processes, solubility in aqueous base, extended redox activity for neutral and basic solution, making them promising applications such as biosensors as a result of physiological pH values, and rechargeable batteries due to the fact that self doping polyanilines are capable of storing more specific energy than PANI as a function of self-doping (Malinauskas, 2004). Therefore it is important the development of these types of polymers, the electrochemical oxidation of 5-amino-2-naphtalensulfonic (ANS) and the luminol can be done a self-doped homopolymer, however the electropolymerizaton of these compounds has not been studied. The chemical structures of these monomers are similar in three aspects (Figure 1): (a) both ANS and luminol contain into their chemical structure the aniline; (b) It is difficult obtain a film of homopolymers to these monomers, because the electroxidation in milium acid of luminol produce a dimmer (De Robertis et al., 2008; Ferreira et al., 2008) in fact the solubility of luminol is bass in this milieu, in these conditions is favored the formation of dimmers; on the other hand no deposition of a polymer onto the electrode has been observed when others amino naphtalensulfonic acid has been electrochemically oxidate, probably for the reason that oligomers are very soluble (Mažeikienė & Malinauskas, 2004). (c) A self-doped polymer can be obtained by copolymerization with aniline and whoever of these monomers (De Robertis et al., 2008; Ferreira et al., 2008, Mažeikienė & Malinauskas, 2004).
So large of our knowledge there are no reports of the homopolymerization of ANS and luminol, the importance of the synthesis of each films consist in the case of ANS in obtain information of the effect of the separation of group sulfonic to aniline in the ion exchange of film, because the charge compensation has been evaluated principally for ring substituted

anilines (Barbero et al., 1994; Cano-Márquez et al., 2007; Mello et al.,2000; Varela et al.; 2001). In the case of homopolymer of luminol should be studied as influence the chemical nature of film in the proprieties obtain and in the activity electrocatalitic; in fact these proprieties are only reported for the dimmer of luminol (Chen & Lin, 2002). On other hand as the polymer obtained to luminol not is water soluble, in consequence the copolymerization of these two monomers can be result of a deposition a self doping polymer film. The goal of this study is the electrosynthes of three novel conducting polymers: polyANS, polyluminol (only has been reported dimmers of luminol) and copolymer, obtained to ANS, luminol and both respectivally, the evaluation of proprieties of films obtained as well as: activity electrocatalytic to ascorbic acid, electrochemical activity in pH neutral and study the charge compensation using the electrochemical quartz crystal microbalance (EQCM) during the redox process only for polyANS.

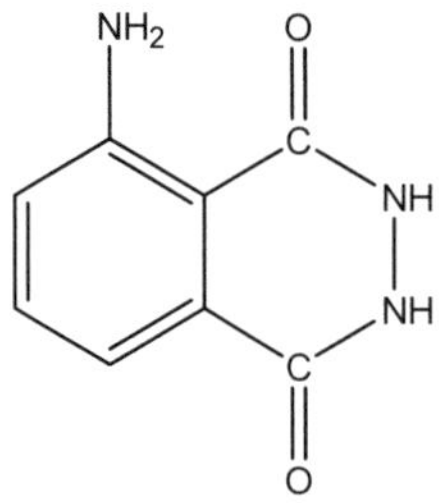

Luminol (3-aminophtalhydrazide)

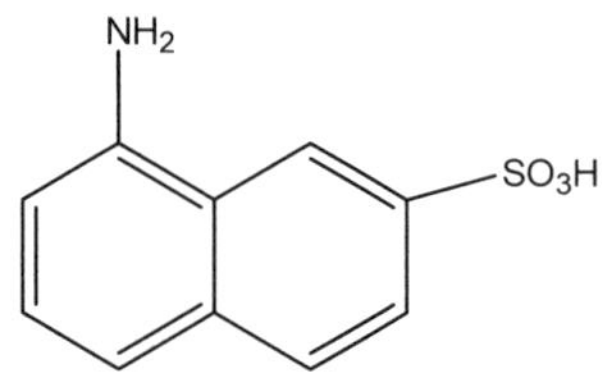

5-Amino-2-naphtalenesulfonic acid

Fig. 1. Chemical structure of momomers used in the present study.

2. Experimental

2.1 Materials

The commercial chemicals: luminol (Fluka), ANS (Aldrich), graphite powder (sigma), nujol (Alfa- Aesar), HCl (Fermont), $HClO_4$ (Fermont), HNO_3 (Caledón), H_2SO_4 (Fermont) and camphorsulfonic acid (Aldrich) were of analytical reagents grade and were used without further purification except the aniline (Sigma) which was distilled before use at stored to low temperature in the dark. The aqueous solutions were prepared using deionized water, and the solutions were deoxygenated by purging with nitrogen gas. After this, a nitrogen atmosphere was kept over the solution during each run.

2.2 EQCM measurements

EQCM measurements were conducted using a PAR 273A (Princeton Applied Research) potentiostat-galvanostat coupled to an electrochemical quartz crystal microbalance Seiko model QCA922, both controlled by WinEchem V. 1.5 installed in a personal computer. The quartz crystal resonator was mounted in a home-made acrylic cell (Figure 2). This cell is characterized by the commercial holder of quartz and it is united to made in home solution container, so the sample container also hold the quartz crystal. As the sample container is acrylic only aqueous solutions can be used with this cell.

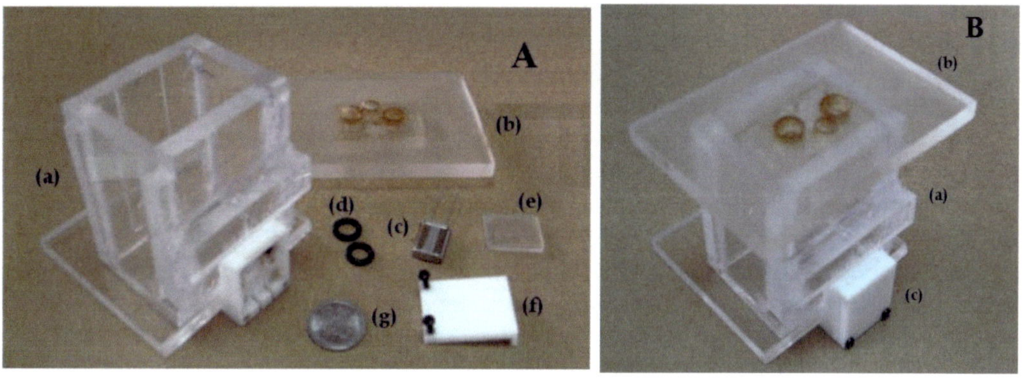

Fig. 2. Photograph of A) diassembled and B) assembled cell: (a) acrylic sample container, (b) acrylic cell cover, (c) teflon quartz crystal and his container, (d) O-rings, (e) acrylic cover of quartz cristal container, (f) teflon cover of quartz crista container and (g) coin (ϕ = 0.8 cm).

2.3 Electrochemical measurements

Electrochemical experiments were carried out in a three-compartment cell in a high purity nitrogen atmosphere at room temperature. The electrode reference consisting of Ag|AgCl|3 moldm^{-3} NaCl (BAS) and the counter electrode was a platinum wire. The working electrodes were: carbon paste electrode (0.1452 cm^2), glassy carbon, Au and Pt (0.0707 cm^2) and a 9 MHz AT-cut quartz crystal coated with gold (0.1963 cm^2). Two types of carbon paste electrode were prepared the first by intimately mixing 1.0 g of finely ground graphite powder with 1.0 g of Nujol. The second type of electrode is characterized by included the monomer into the carbon paste bulk, the electrode is prepared mixing graphite powder, nujol and the monomer. In both cases the resulting paste was then packed into a plastic 1 mL syringe in which a piece of coop wire was wound to produce the electrical contact. The surface was smoothed by a weight paper before each experience. The glassy carbon, Au and Pt electrodes were systematically polished successively with 3 and 1 μm polish diamond (BAS) on a Microcloth felt disk (BAS). Following this, the electrodes were thoroughly rinsed with deionised water and sonicated in an ultrasonic bath for 10 minutes. The precoated PANI film for electrodeposition of polyANS was carried out by realizing 10 cycles of potential between -200 and 1200 mV, except for the three first cycles for which the upper limit was 1100 mV. The scan rate was 50 mVs^{-1}, and the electrolyte solution was 0.5 M H_2SO_4 + 0.5 M aniline.

3. Results and discussion

3.1 Electrosynthesis and characterization of polyANS

The electrosynthesis of a film of polyANS was tested by chronopotenciometry and cyclic voltametry in five different electrodes: Au, Pt, glassy carbon, carbon paste and Au modified with PANI. No deposition of a polymer onto the Au, Pt and glassy carbon electrodes was observed, in agreement to reported for aminonaphtalen disulfonics acids (Mažeikienė & Malinauskas, 2004) probably because the products formed during the oxidation are very solubles. On the contrary film growth was presented onto carbon paste and Au/PANI electrodes, the syntheses and properties obtained in each electrode are presented to continuation.

3.1.1 Electrosynthesis and electrogravimetric study of polyANS synthesized onto Au/PANI electrodes

The syntheses electrochemical of film of PolyANS onto the Au/PANI electrode was carried out by cyclic voltametry, this modified electrode was used as working electrode for the reason that in earlier work it was demonstrated that contrary to the metallic electrodes a film of sulfonated PANI is obtained onto Au/PANI electrodes since, during electroxidation of PANI positive charges are generated at the polymer chain, which are compensated by incorporation of anions from the solution. Consequently, the $-SO_3$ group of sulfonated PANI is incorporated to the PANI film as a dopant anion, anchoring sulfonated monomers on the PANI surface. Then, when the oxidation potential of monomers is attained the monomer is polymerized on PANI. During the reduction the anion cannot be ejected of PANI because the polymer synthetised is trapped within the PANI chains. (Cano-Márquez et al., 2007). Figure 3 shows curves obtained in diverse stages of growth of polyANS. At the very beginning of the electrodeposition the curve obtained is very similar to those of PANI (Shin-Jung & Su-Moon, 2002), in fact, in this stage the polyANS produced is small in comparison to those of PANI as a result the response obtained is those of PANI, which is characterized by three well redox centered pair centered at around 250, 780 and 500 mV corresponding to leucoemeraldine to emeraldine, emeraldine to pernigraniline redox transitions and sobreoxidation products respectively. When the cycle increase the response change gradually until become in those of polyANS, in fact the peaks attributed to oxido-reduction of film shift until overlapping (Figure 3). The reduced separation of the two peaks for sulfonated polyaniline has been associated with steric effects caused by the bulky sulfonic acid substituent (Wei et al., 1996; Yue et al., 1991). Additionally to the two oxido-reduction process of polyANS situated between 200 and 500 mV, the cyclic voltammogram of polyANS present a redox pair centered at around 540 mV, at present time no clear assignation can be proposed for this peak, however it can be speculated that have as origin the oxido-reduction of degradation products because is located of same potential of degradation products of PANI.

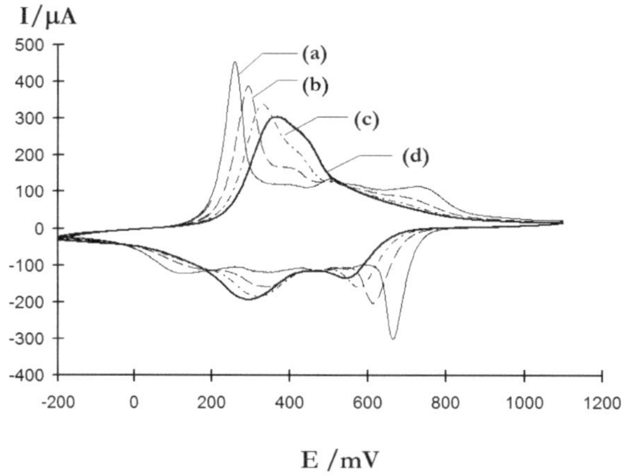

Fig. 3. Cyclic voltammograms of the 14th (a), 28th (b) 42th (c) and 56th (d) cycle of the electropolymerization of 1 mM ANS in 0.5 M H_2SO_4, obtained onto Au/PANI electrode. The potential was scanned from -200 to 1090 mV at 100 mVs^{-1}.

From the perspectives of electrochemical applications, arguably the most important underlying process of conducting polymers is the exchange between the films and ions molecules that accompany film redox switching. In case of sulfonated polyanilines is accepted that the charge compensation is accomplished mainly by the ejection of cations (Mello et al., 2000; Varela et al., 2000a, 2000b, 2001; Cano-Márquez et al.; 2007) in contrast with the behavior of PANI, which is carried out by the incorporation of anions (Hillman & Mohamud, 2006). In order to define as is realized the charge compensation of polyANS electrochemical quartz crystal microbalance (EQCM) measurements were carried out in monomer free solution, the curves obtained were analyzed considering that the change of frequency is correlated to change of mass for the Sauerbrey equation (Donjuan-Medrano & Montes-Rojas, 2008):

$$\Delta f = -C_f \Delta m = -\left(\frac{2f_0^2}{\rho_q \vartheta_q}\right)\Delta m \quad (1)$$

Where Δf and Δm are the change of frequency and the change of mass respectively, C_f is the proportionality constant which depend of the base frequency (f_0); density of quartz (ρ_q) and the wave velocity within the quartz crystal (υ_q). The negative sign in the equation indicate that when the frequency increase the mass diminishes and when the frequency decreases the mass augment. In order to use the Sauerbrey equation it is necessary probe that the film deposited in the quartz is "rigid", it is the mass loading of the quartz crystal microbalance show ideal acoustic coupling to the crystal surface (Buttry & Ward, 1992) and determinate experimentally the value of C_f (Donjuan-Medrano & Montes-Rojas, 2008). In raison that in this work only a qualitative analysis was carried out, it was not required to validate the Souerbrey equation, neither it was indispensable determinate experimentally the value of C_f, only was assumed of Sauerbrey equation that:

$$\Delta m = -C_f \Delta \quad (2)$$

It is that the change of mass is proportional to -Δf. The obtained results for H_2SO_4 are shown in Figure 4, in a first time the potentiodinamic profile of I and -Δf was analyzed over the range -200-900 mV (Figure 4a), it was observed that when the scan begin there are not redox process, so the frequency remain constant, to 300 mV appear the oxidation peak this indicate that the polyANS is oxidate and positives charges in the polymer chain are generated, simultaneously a decrease of mass is presented, this mean that for maintain the electrical neutrality of the doped polyANS the charge compensation is carried out predominantly for ejection of cations from the polymer phase to the solution as is attaint for a self-doped polymer, when it was reach the potentials for the full oxidation of polymer an increase of mass was observed, contrary to decrease of mass observed for PANI in similar conditions (Orata & Buttry, 1987). During the reverse sweep the mass decrease until finalized the reduction, it is the anion incorporated during the oxidation are ejected when the film back to netral condition, to end of scan the mass augment to until reach nearly the initial value. For the reason that the two peak of oxido-reduction of film are before 500 mV, the same experiment was repeated fixing the upper limit potential to 500 mV, as can be been in Figure 4b, the behavior is similar to those obtained for the extended range, in fact the mass of polyANS remains constant until the film commence the oxidation process in this moment the mass descend, this behavior indicate that the electroneutralization of film is achieved principally by expulsion of protons containing in the $-SO_3H$ group of polyANS, to higher potentials the mass increase, a similar behavior has been presented for sulfonated PANI (Varela et al., 2000). After the reverse of the scan direction the mass continue increase until the process of reduction begin, in this point the mass decrease, finally the mass increase attaint the value original. The experimental evidence presented above demonstrates that the charge process during the charge compensation is carried out principally by expulsion of cations of polyANS, however is not possible establish whether or not anion participation. In addition the evolution of frequency in the direct scan is completely different to those obtained in the inverted scan, it is the change of mass obtained is not reversible resulting in a broad frequency curve. This behavior is different to those obtained for sulfonated PANI (Mello et al, 2000; Varela et al. 2001) where the curves of frequency were more reversible. This behavior suggest that the electroneutralization is more complexes for the polyANS that for the films synthesized to anilines ring substituted by sulfonic groups, probably due the distance of the sulfonic group and nitrogen is much longer in polyANS that the anilines substituted directly in the ring for sulfonic groups, so in this case the electroneutralization is more easy.

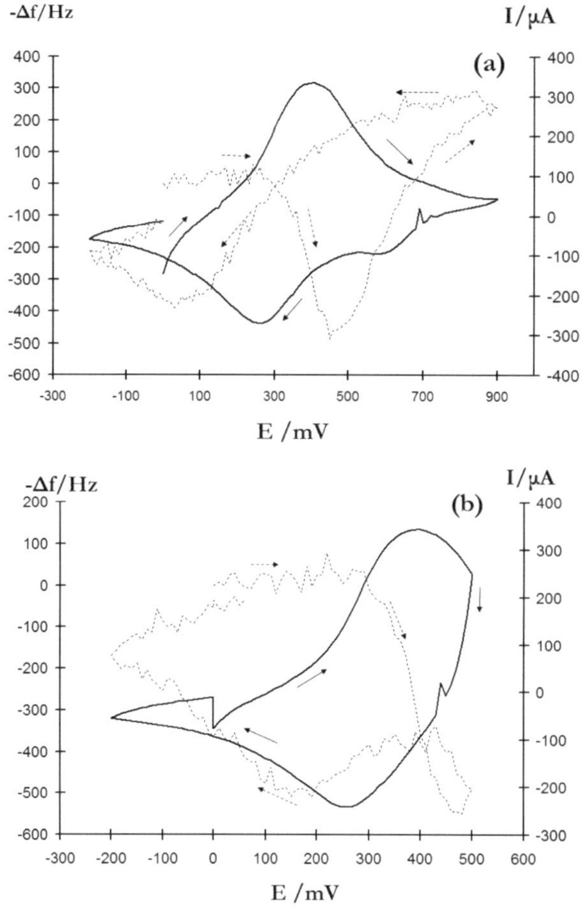

Fig. 4. Plot I/E (full line) and -Δf/E (doted line) potentiodinamic profil for polyANS obtained for two upper limit (a) 900 and (b) 500 mV in H_2SO_4 0.5 M electrolyte solution at 100 mVs^{-1}.

In order to determinate whether there are or not anion participation in the electroneutralization of PolyANS, EQCM measurements were carried out in solutions for acids with anions of different molar mass: H_2SO_4 (96 gmol^{-1}) (Figure 4), HCl (35.45 gmol^{-1}), $HClO_4$ (99.4 gmol^{-1}), HNO_3 (62 gmol^{-1}) and camphorsulfonic acid (HCS) which molar mass of anion is 231.1 gmol^{-1} (Figure 5). A decrease of mass is presented during the oxidation in H_2SO_4, HCl, $HClO_4$ and HNO_3 showed that during the electroneutralization, protons are ejected to polyANS. However, a participation of anion can be proposed in reason of the marked dependence of frequency profile and cyclic voltammogram of anion of acids as is showed in Figure 4 and 5. The decrease of mass (-Δf) change with the acid used and not relation between the anion molar mass and the variation of frequency was present. These results show that the charge compensation of polyANS is carried out by both expulsion of protons and additions of anions, but the predominant process is the cation expulsion. In the HCS case also the electroneutralization is realized principally by proton expulsion despite the fact that the mass remain constant during the oxidation, in fact the change of mass registered are the sum of two contribution incorporation of anions and ejection of protons, as the frequency remains constant during the oxidation this mean that the augmentation of mass by incorporation of anions has the same value that the diminution of mass by expulsion of cations, this resultant is due to the molar mass of CS is more grand that those of the anions of other acids. Finally, it is important to note that the process of electroneutralization of polyANS is different of those of homopolymer of ortanilic acid, in

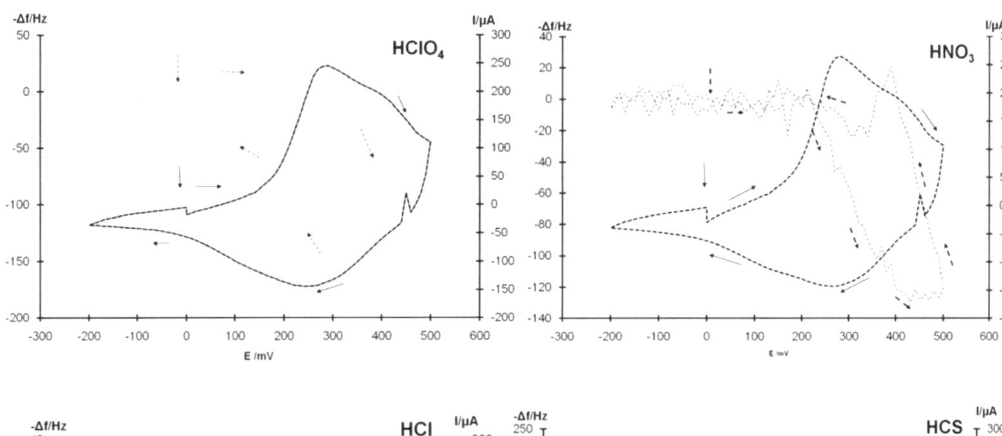

Fig. 5. Cyclic voltammograms and frequency responses recorded simultaneously of a polyANS film in dfferents 0. 5 M acids solutions. The scan rate was 100 mVs^{-1}.

fact for this polymer the compensation of charge is carried out exclusively by ejection of protons and the profiles of change of frequency are very reversible (Cano-Márquez et al, 2007), which suggest that the process is more simple. The differences between these sulfonated homopolymers proved that the distance between the sulfonic group and the nitrogen impact in the electroneutralization process, when these groups are closer the cation participation is more important.

3.1.2 Electrosyntheses of polyANS using carbon paste electrodes

The electrodeposition of polyANS was also evaluated using carbon paste electrodes, the electrosynthesis was carried out by two methods with the monomer in solution and incorporated onto the carbon paste electrode. The results found employed the first methodology are presented in a subsequent section, in the case of the second mode the electrodeposition was realized potenciodinamic in a solution containing only H_2SO_4 1 M and the working electrode was carbon paste modified with ANS, the curve registered during the successive scans are presented in the Figure 6, this cyclic voltammogram is similar to those obtained with Au/PANI electrodes, in fact the curve can be show a shoulder and a broad peak situated around 309 and 424 mV correspondingly, in the counterpart cathodic are two overlapping peak centered in 269 and 319 mV. Following an analogy to the parent polyaniline these peaks can be assigned to the leucoemeraldine to emeraldine transition and the emeraldine to perigraniline transition, respectively. Additionally a redox process appear around 100 mV, it not was possible the assignation of these peaks to an specific reaction, however it is possible that the peak correspond to the oxido-reduction of oligomers accumulate in the electrode, in fact these peaks not are observed when the response is analyzed in solutions basic. Subsequent cycles show anodic and cathodic current maxima with increasing currents, indicating progressive film formation. The electrochemical behavior of films synthetised was examined by cyclic voltametry, for this experience was necessary shift the upper limit potential to 600 mV, with the finality of avoid the oxidation of monomer and more film synthesis, it is examine only

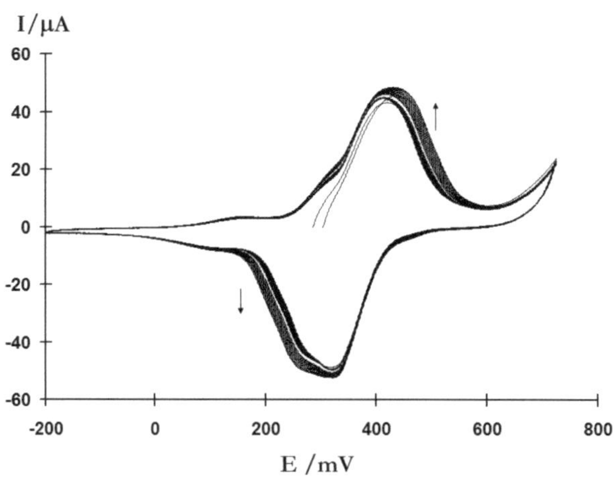

Fig. 6. Succesive cyclic voltammograms of a paste carbone electrode modified with ANS (0.1452 cm^2) in H_2SO_4 (1 M). The composition of electrode was graphite powder, nujol and ANS in percentages of mass of 80, 18.5 and 1.5 respectively. Scan rate: 100 mVs^{-1}.

the response of polyANS. The polymers which dissolve very slowly, it is when are cycled repeatedly diminish slightly the current intensity. The peak heights scale linearly with the sweep rate as is expected for a surface bound species.

The electropolymerization of ANS in carbon paste electrodes not modified with this monomer in solution were carried out by chronoamperometry and cyclic voltammeter. The results obtained with potentiodinamic methods are presented in the copolymer section, in the case of chronoamperometry the electroxidation was achieved with a solution formed only by ANS, since the monomer was used itself as an electrolyte, the potential was stepped in 1250 mV during 60 s. After the growth was terminated, the film was studied by recording cyclic voltamograms; the curves obtained are show in figure 7. We note that the locations of the redox peak and the general appearance of the voltammogram are similar to sulfonated polyanilines o (Royappa et al., 2001; Nguyen et al., 1994) and the polyANS synthetised with Au/PANI and carbon paste modified with ANS. The cyclic voltamogram obtained exhibit two broad and overlapping oxidation peak between 200 and 600 mV. The oxidation peak correspond to the oxidation of the neutral nitrogen atoms in the polymer backbone to form the radical carbon and dication ant they compare well with the peaks for PANI films (Buttry & Ward, 1992). A process centered in 100, similar to those obtained with the monomer contain in the bulk paste electrode also appear when the electrodeposition is realized with the monomer in solution. A linear relationship was found between the peak current and scan rate, indicating that the elecroactive polymer film is well adhered to electrode. To investigate the influence of the pH on the redox behavior of the film, cyclic voltammograms of electrochemically synthesized films prepared under the same conditions as above, were recorded in solutions of different pH values. The voltammograms obtained are shown in Figure 7. One of the most obvious changes in the voltammetric behaviour of polymer in neutral and alkaline media in comparison to pH acid is the diminution in current, which is produced by the haut solubility of sulfonated polymer in pH neuter. In addition the peak attributed to polyANS shift not much to negative potentials. Thus the electroactivity of polyANS is practically independent of pH as is attaint for a self-doped polymer.

The same experiments were made with glassy carbon, Au and Pt as working electrode, the chronoamperograms obtained showed no rising transients, and no evidence of a film was obtained by cyclic voltammetry, in agreement with Mažeikienė et al, 2004. These results show that the polymerization of sulfonated monomers only can be carried out in carbon paste electrodes. The reasons for this positive deposition are unknown. However physic adsorption of monomers and oligomers in the paste carbone, it may be speculated

In this part of work was showed that the ANS can be polymerized in carbon paste and Au/PANI electrodes, this result is contrary to those obtained for other amino naphtalensulfonic, in fact in this study is proved that is not possible the electrochemical homopolymerization of these type of compounds (Mažeikienė & Malinauskas, 2004). Additionally is well established that Sulphonated anilines are difficult to polymerise under conventional conditions, in fact the haut solubility of film in aqueous milieu made difficult the synthesis of a film, the strategies employed for the electrodepositon of sulfonated homopolymer has been oriented to the diminution of solubility of oligomer, for this , the electropolymerization has carried out to bass temperatures and the combination of organic and aqueous solvents (Krishnamoorthy et al., 2002). In other works is reported the formation of sulfonated polymers using metallic electrodes however the polymerization rate is very bass (Zhang, L; 2006). The advantages of the use of Au/PANI and carbon paste electrodes for the electrosynthesis of sulfonated polymers in relation to other strategies is the facility to obtain thick films rapidly in working to room temperatures and with aqueous solutions.

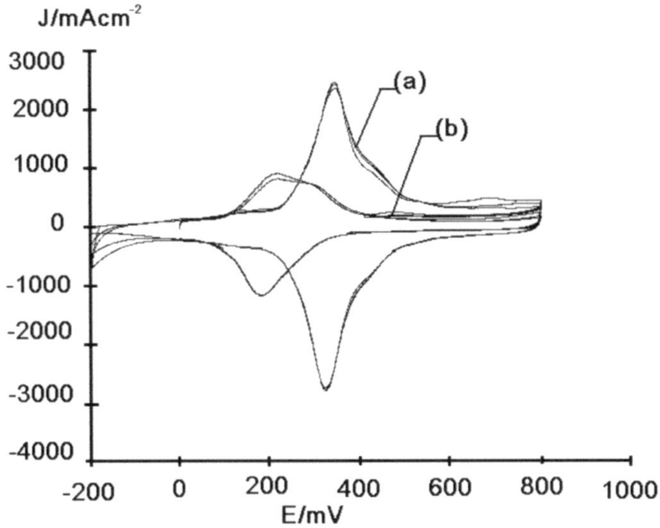

Fig. 7. Cyclic voltammograms of polyANS in: (a) H_2SO_4 0.1 M and (b) Na_2SO_4 0.1 M. Synthesis of the films was performed by potential step at 1240 mV using as working solution ANS 8 mM. Scan rate: 100 mV/s.

3.2 Electrodeposition and characterization of polyluminol

3.2.1 Electrosyntheses of polyluminol

The films obtained of electrooxidation of luminol have been denominated polyluminol. However, different works have showed that the product obtained are dimmers and no polymers (Ferreira et al., 2008; Robertis et al., 2008), the chemical composition of film has been attributed to the bass solubility of monomer in solution. In order to obtain a polymer of luminol we propose increase the monomer concentration in the interface electrode/electrolyte using paste carbon electrodes bulk modified with luminol. The curves obtained during successive scan for two different upper limit potential (E_λ) are presented in figure 8, in both cases the peaks due to the oxidation and reduction increase in intensity for each cycles as it is characteristic of growth of a film. The voltammogram acquired are different of those obtained with metallic electrodes in similar conditions, for the reason that additionally to process B_1/B_1' presented in metallic electrodes (Chang et al., 2005; Kumar et al., 2009), a second process B_2/B_2' is presented. A similar process to B_2/B_2' is obtained for the copolymers of aniline and luminol synthesised using solutions with more concentration of luminol that aniline (Roberti et al., 2008; Ferreira et al., 2008) and during the oxidation of luminiol in higher potentials 1.2 V (Zhang & Chen, 2000). Additionally the aspect of voltammogram presented in Figure 8 is very similar to those of copolymer therefore probably this peak are associated to polymer formation. The current of process B_2/B_2' by rapport to B_1/B_1' is strongly dependent of upper limit potential, for E_λ= 0.9 V the intensity of current B_2/B_2' is very small in comparaison with those B_1/B_1', while that when E_λ= 0.8 the

current of B_1/B_1' is only environ three fold more grand that B_2/B_2'. This dependence show that product correspondent to peaks B_1/B_1' are produced in more quantity when E_λ is more higher. Two difference important were observed in voltammogramas obtain using two differents E_λ, first, the evolution of the intensity of current, For E_λ = 0.9 V the current of peaks B_1 and B_1' increase significantly the first cycles, after remain almost constant, while that for E_λ = 0.8V, the current augment slowly but continually, this last evolution of current is more congruent with a polymerization. The second is relationated with the peak B_1/B_1', for E_λ = 0.8 V, the peak is more broader that for E_λ = 0.9 V in addition in some cycles is possible to observe that the peak is the resultant of the overlap of two peaks, probably those of the dimmer or degradation oxidation and those of the bipolaron state of polymer. It is we propose that the peak B_1/B_1' and B_2/B_2' can be assigned to polaron and bipolaron states of polymer respectively; because the voltammograms are very similar to those of copolymer formed by aniline and luminol and those of sulfonated polyanilines.
Finally the film were analysed in H_2SO_4 solution by cyclic voltammetry, for this E_λ was shifted to more negatives potentials in order to avoid the oxidation of monomer in the bulk electrode paste. Voltamograms were carried out for different scan rates, the peak heights scale linearly with the sweep rate and the cathodic and anodic peak separation remains constant as it is attaint for an electroactive specie fix to surface electrode.

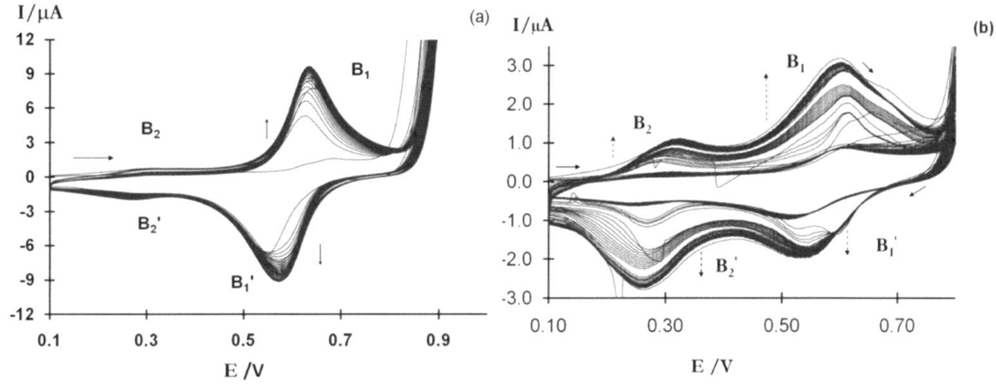

Fig. 8. Subsequent multisweep cyclic voltammograms obtained in a H_2SO_4 (0.5M), the upper limit were: (a) 0.9 and (b) 0.8 V. The working electrode was paste carbon electrodes formed by: graphite (60 %), nujol (26%) and luminol (14%). The scan rate was 100 mVs^{-1}.

In order to determinate weather the monomer concentration affect the electrosyntheses of film of polyluminol, electrodeposition were carried out employing carbon paste electrodes modified with different percentages of luminol, the results are presented in Figure 9, the voltamogramms have the same aspect, only they are differenced by the intensity in current which increase with the quantity of monomer in the bulk carbon paste electrode. This means that only the quantity of film is affected by the quantity of momomer. A similar experiment was achieved using a carbon paste electrode not modified and a solution with luminol, the curve obtained has an aspect equal to those obtained with modified electrode. It can be speculated that the curve not change with luminol concentration in the interface electrode/electrolyte, since when the monomer is in solution, the luminol is adsorbed in the electrode surface and the concentration in the interface is increase by preconcentration, so the concentration of luminol is more elevated than the metallic electrodes.

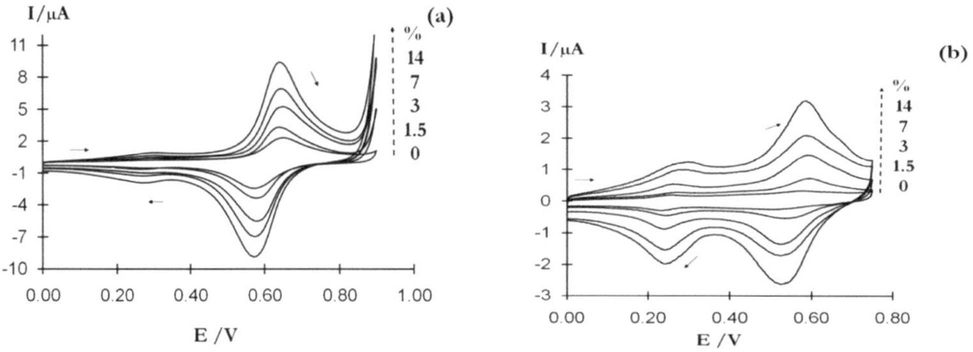

Fig. 9. Cyclic voltammograms (50 mVs^{-1}) of electrochemically synthesised polyluminol in 0.5 M aqueous H_2SO_4. The films were synthetised with the conditions of Figure 8 with (a) $E_\lambda = 0.9$ and (b) $E_\lambda = 0.8$ V. The composition of carbon paste was graphite (60 %), luminol (indicated in the Figure) and nujol (the necessary to reach the 100%). El 0 % indicate the film synthesised with the luminol in solution (1 mM).

3.2.2 Electrochemical activity of polyluminol to pH neutral

It is proved that the films obtained to oxidation of luminol are self-doped (Ferreira et al., 2008), this implicate that they are electroactive to 4 < pH. Figure 10 shows the voltammograms for polyluminol synthesised with $E_\lambda = 0.8$ V at acid and neutral pH. It can be observed that the peaks shift slightly to lower potentials in a Na_2SO_4 aqueous solution (1 M), when the same experiment is carried out with posphate buffers solution, pH = 7, (PBS) the peaks of polyluminol overlap into only one, while the current of this process remain in similar values. The difference between the voltamperometric responses at two electrolytes of a similar pH, probably is due to the charge/discharge process because the anion and cations are different. This behaviour in neutral milieu is similar to those observed in sulfonated polyanilines (Sanchís et at., 2008; Kariakin et al., 1994), and as a consequence congruent with the behaviour of a polymer. It is important note that an irreversible oxidation is observed in buffer solution (inset Figure 10), correspondent to the oxidation of monomer in the bulk electrode.

To compare the properties of polyluminol synthesized to monomer in solution and in the bulk carbon paste, cyclic voltammogramms in supporting electrolyte solutions acid and neutral pH were investigated. Figure 11 shows voltammograms for polyluminol films obtained in acid and neutral electrolytes. The films are quite electrochemicaly active in both neutral solutions Na_2SO_4 and PBS, in Na_2SO_4 a shift to more positives potentials was observed and in PBS an overlap of the two peak was presented. However we note that the more defined and reversible peak in the case of films synthesised with the monomer in the bulk paste carbon electrode, in addition the relation between the anodic current peak in acid solution (I_0) respect to those obtained in milieu neuter PBS (I_{PBS}) or Na_2SO_4 (I_s), were $I_{PBS}/I_0 = 0.579$ and $I_{PBS}/I_0 = 1.249$, while that when the films were made to monomer in solution $I_{PBS}/I_0 = 0.487$ and $I_{PBS}/I_0 = 0.412$. A comparable tendance was showed for films obtained with other percentages of luminol in carbon paste 1.5, 3 and 7 %. These result demonstrate that in terms of extension of its electrochemical properties to high pH, are better the films elaborated to modified carbon paste electrodes than those carried out the monomer in solution.

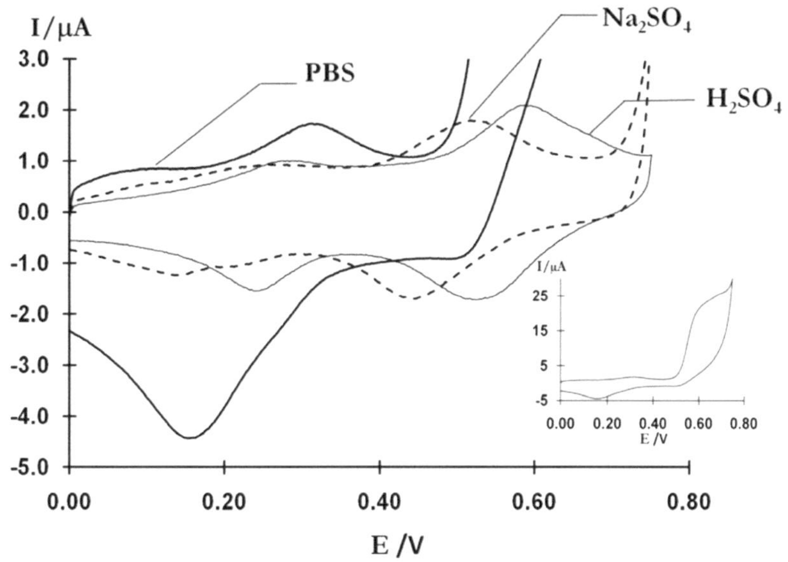

Fig. 10. Cyclic voltammograms for polyluminol synthesised using E_λ = 0.8 V and carbon paste electrode graphite (60 %), luminol (7 %) and nujol (33 %) at differents electrolytes. Scan rate = 100mVs-1. Inset complete cyclic voltammogram for polyluminol at PBS.

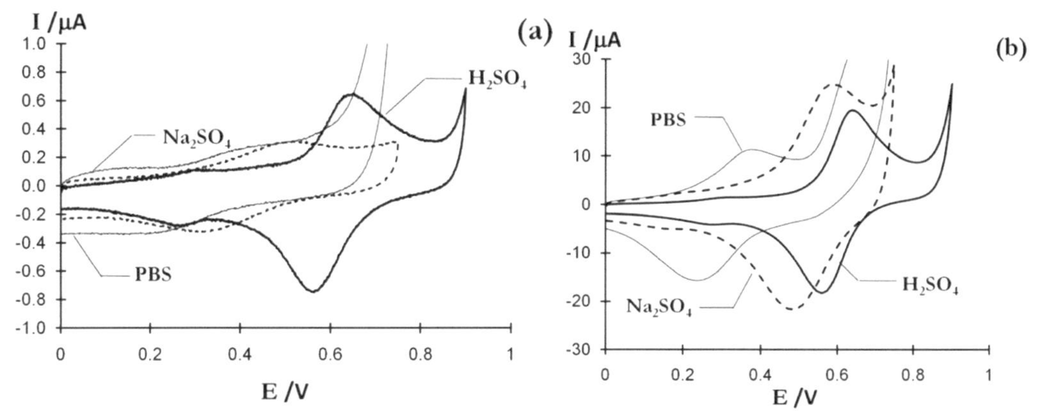

Fig. 11. Cyclic voltammetry of polyluminol, obtained in different supporting electrolytes containing: H_2SO_4 (0.5 M), Na_2SO_4 (1 M) and phosphate buffer solution (pH = 7). Synthesis of the films was performed as indicate in Figure 8a, using: (a) luminol (1 Mm) in H_2SO_4 (0.5 M) and (b) a modified carbon paste electrode (14%).

3.2.3 Electrochemical oxidation of ascorbic acid

A more extended applications of polyluminol films and its derivatives is the electrocatalysis of ascorbic acid (Chen & Lin, 2002; Ashok et al., 2009; Ma et al., 2010; Kumar et al., 2009a; Kumar et al., 2009b). As the potentiodinamic response of films obtained using carbon paste electrodes are different to those synthesised with metallic electrodes; in fact the cyclic voltammogram is more similar to those of copolymers of aniline and luminol, we think that the chemical composition of films in carbon paste correspond to those of a polymer, this supposition is supported by the behaviour of films in neutral solutions. In order to determinate weather the chemical composition of deposits of luminol affect or not the catalytic properties of films, the electoxidation of ascorbic acid (aa) was studied with the films obtained and compared with those of metallic electrodes, the curves obtained are presented in Figure 12, as can be seen, the oxidation peak current augment with increasing aa concentration. The inset of Figure 12 shows that the anode peak current is linearly dependent on the aa concentration in the studied range, to slope of line was obtaining the sensibility that was 43.2 μAmM^{-1}, this value is the double of those obtained using a polyluminol synthetised in a metallic electrode. This result show that the electrocatalytic activite of polyluminol film is enhanced with is obtained in carbon paste electrodes.

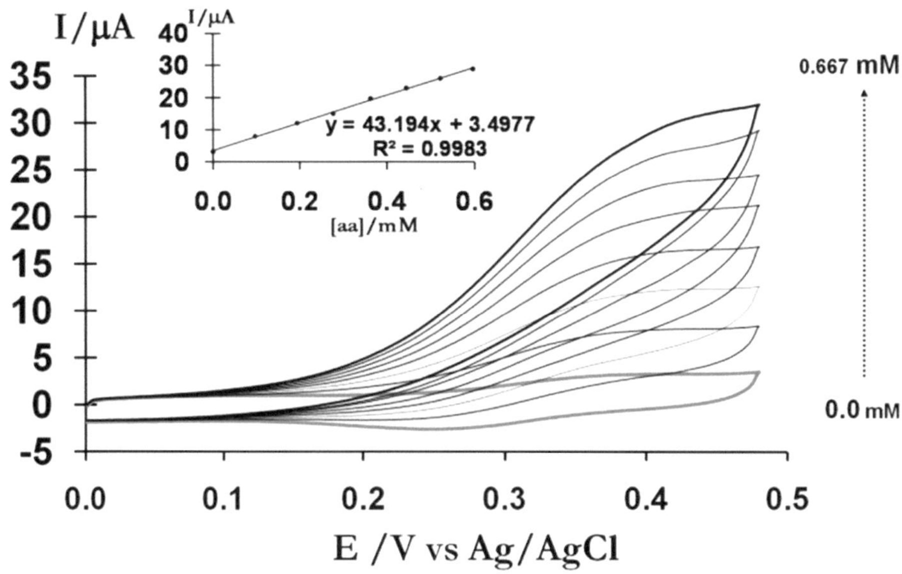

Fig. 12. Cyclic voltammetry of polyluminol in PBS containing different concentrations of aa: 0, 0.098, 0.195, 0.279, 0.364, 0.444, 0.522, 0.596 and 0.667 mM. The film was synthesised like those of Figure 8b. The scan rate was 100 mVs^{-1}.

3.3 Electrodeposition of copolymer

Finally we have tried the electrosynthesis of a novel self-doped polymer formed by ANS and luminol, in fact the electroxidation of both compound can to produce self-doping polymers, However the polyANS is soluble in milieu neuter, this characteristic limit its use in biosensors, while the polyluminol is insoluble but the electrochemical activity is more dependent of pH in comparison with the polyANS in reason of the $-SO_3H$ is a strong acid. The combination of both monomers can be produced an insoluble polymer which present electrochemical activity in a wide pH range.

3.3.1 Electrochemical properties of monomers

In order to determinate the potential which both monomers are oxided, potenciodynamic experiments were carried out (Figure 13a), the voltammograms exhibit a broad irreversible anodic peak situated in the 0.7 -1.2 Vand 0.9 and 1.2 V for ANS and luminol respectively. The anodic response correspond to the irreversible oxidation of monomers entities, the potential where the oxidation of both monomers is carried out is environ 0.9 V. The effect of the scan rate in the intensity of current was studied, in the Figure 13b can be observed that in the case of ANS the peak heights are linear with the square root of sweep rate in all range studied, this mean that the oxidation of ANS is controlled by diffusion; while in the luminol case the peak heights scale linearly with the square root of sweep rate only in the range between 5 and 100 mVs^{-1}, in consequence for higher sweep rate, the process of oxidation can be controlled by adsorption or a combination of adsorption and diffusion. It is important note that the slopes of both curves are essentially identical, in considering the Randles-Sevcik can be established that the Diffusion coefficient of both monomers should be very similar. The oxidation of both monomers is 0.9 V and in order to both monomes are in the same regime of mass transfer it is convenient working to sweep rate ≤ 100 mVs^{-1}.

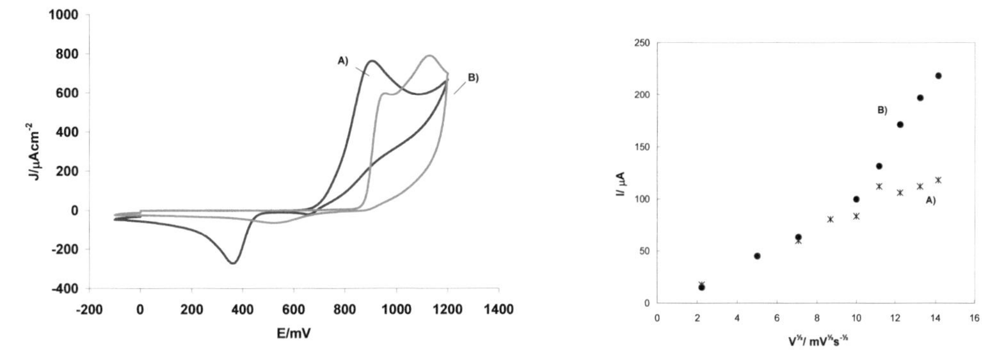

(a) (b)

Fig. 13. (a) Cyclic voltammetric response of: (A) ANS (0.8 mM) and (B) luminol (0.8 mM) on a carbon paste electrode (0.1452 cm^2) in H_2SO_4 1 M. The carbon paste was formed by a mixture 1:1 of luminol and graphite. Potential scan rate 100 mVs^{-1}. (b) Dependence of anodic peak current with the square root of sweep rate

3.3.2 Electrochemical synthesis and characterization of copolymers

The copolymers were synthesised using a H_2SO_4 (1 M) solution contain luminol (0.8 mM) and ANS (0.8 mM) as working electrode carbon paste electrode was used, 50 potential cycles were carried out between 0 y 0.9 V. Concurrently the same experiment was repeated with a solution contain only a monomer. After their electrosyntesis, copolymers and homopolymers were characterised by cyclic voltametry in 1 M H_2SO_4 (Figure 14), in orden to avoid the sobreoxidation of films the upper potentials were shifted to lower values. The cyclic voltammogram of copolymer was characterised by present two redox process, attributed in analogy with polyluminol and polyANS to partial and total oxidation of polymer. Comparison of the peak of copolymer with those of luminol and ANS taken under similar conditions indicate that the potential and the current of copolymers peak lies between them. This suggest that the films prepared by this procedure are indeed copolymers and not just mixtures of polyluminol and polyANS. We have tried to modulate the amount of lumiol in the copolymer film. To do this, the electrosynthesis of copolymers was performed using different ratios of luminol and ANS, in addition the potential scan rate was changed. Concerning the ratio of monomer when the amount of luminol increase in solution the cyclic voltammogram of copolymer present current and potentials mores similar to those of luminol, while that when it is augmented the amount of ANS in comparison to those of luminol in solution, the potentiodynamic response of copolymer is more near to those of polyANS. Thus, it is possible modulate the monomer proportion in the film changing the ratio of monomer in the electrosynthesis solution.

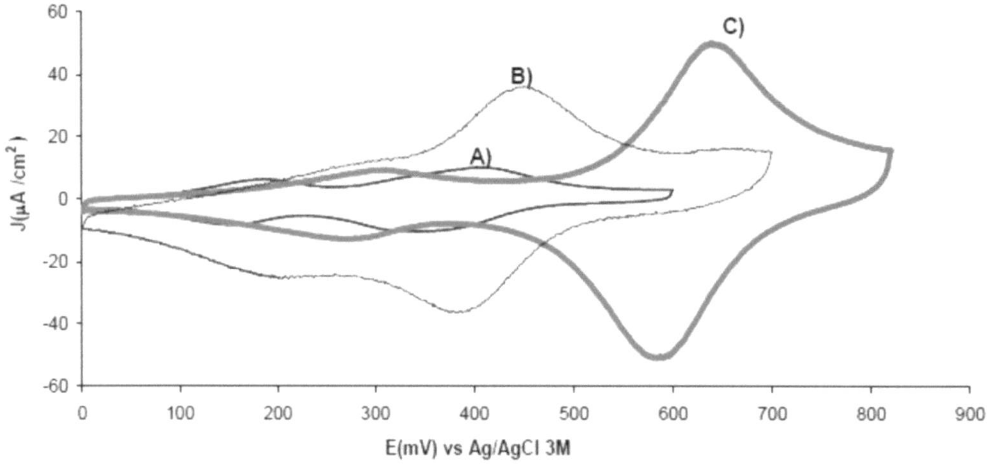

Fig. 14. Voltammograms for (A) polyANS, (B) copolymer and (C) polyluminol films obtained in H_2SO_4 (1 M). Potential scan rate 100 mVs^{-1}.

On the other hand, as can be been in Figure 14 and Figure 15 the potentiodynamic response is dependent of the scan rate used during the electropolymerization, in the case of homopolymers the current change with this parameter in contrast the potential is the same for all scan rate studied. This indicate that only change the amount of film obtained and the chemical composition is the same, in the copolymer case both current and potential are dependent of potential scan rate Therefore the amount and composition of films synthesized are modified with the scan rate. These facts could be related with the solubility of films, because the products of oxidation of ANS are more soluble that those of luminol, since to lower scan rate the oligomers formed have time for a probable adsorption in the surface of paste carbon, since the polymerization of ANS is enhanced to low scan rate (Figure 15a). The products of luminol are insoluble and are less dependent to the scan rate, since the proportion of luminol and ANS onto the film are dependent of this variable.

The electroactivity of copolymers was analysed in neutral solutions by cyclic voltammetry, as can be showed in Figure 16, the curves obtained show process redox, this indicate that are electrochemical activity to pH neutral. This result is congruent with the behaviour of a self-doped polyaniline. The response was affected by the ratio of monomers in the films, for the films obtained to 5 mVs^{-1} the behaviour was similar to those of polyANS, while the film synthetised to 100 mVs^{-1} the cyclic voltammogram was similar to those of polyluminol.

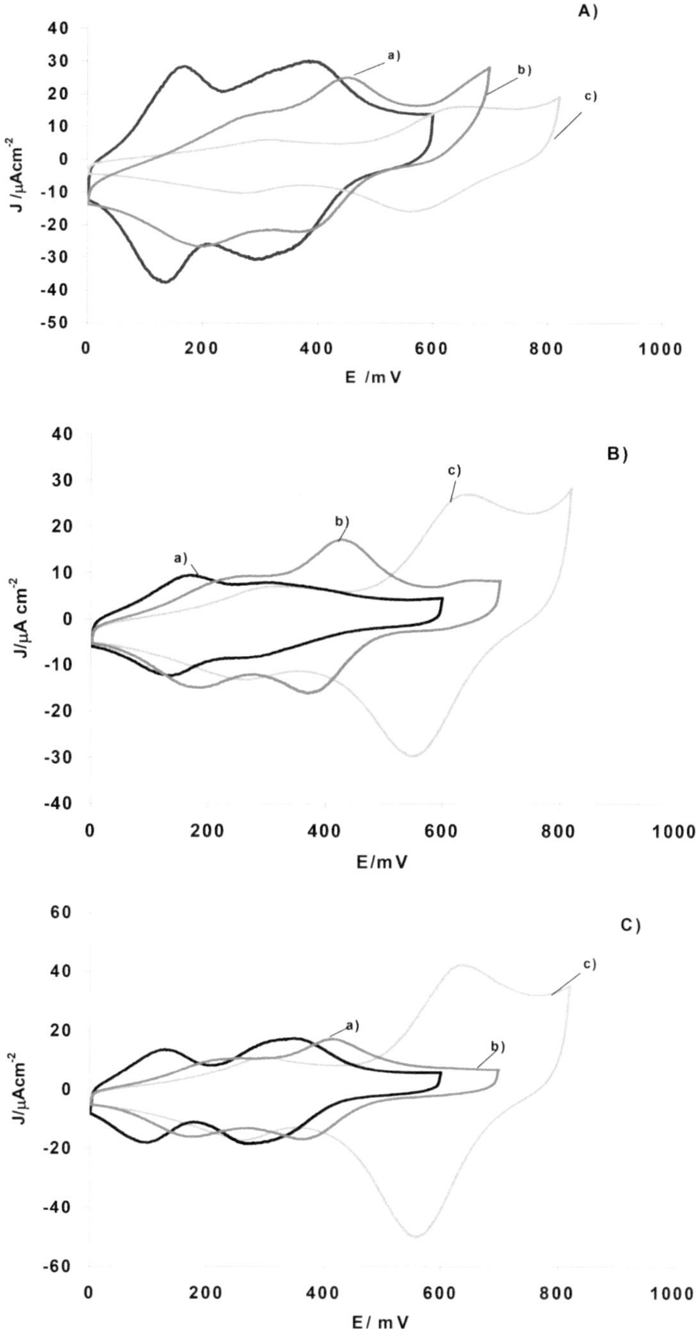

Fig. 15. Voltammograms for (a) polyANS, (b) copolymer and (c) polyluminol films obtained in H_2SO_4 (1 M). Potential scan rate 100 mVs^{-1}. The scan rates used during the electrodeposition were: (A) 5, (B) 25 and (c) 50 mVs^{-1}.

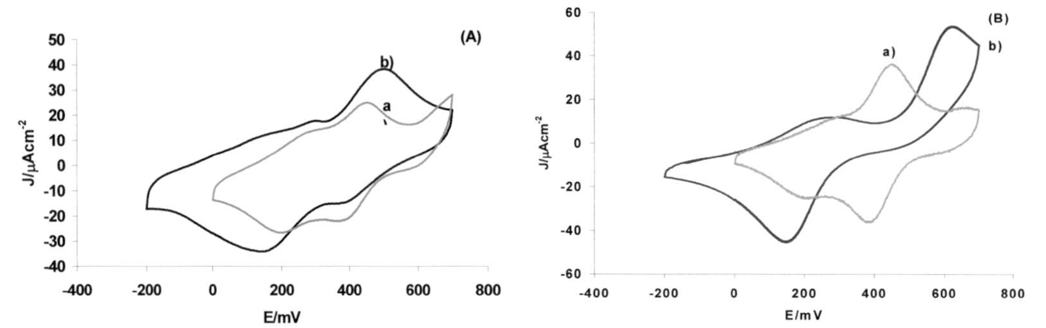

Fig. 16. Cyclic voltammograms of copolymer synthesised to (A) 5 and (B) 100 mVs^{-1} in (a) H_2SO_4 (1 M) and Na_2SO_4 (1 M). The scan rate was 100 mVs^{-1}.

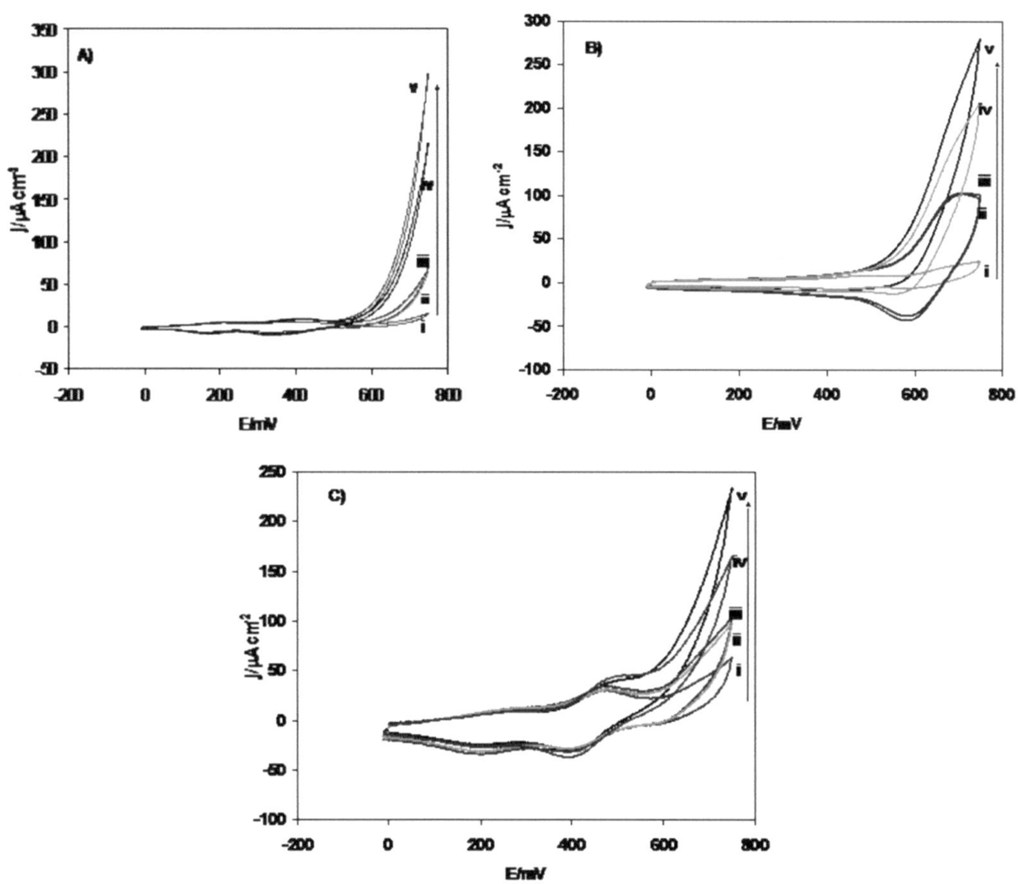

Fig. 17. Cyclic voltammograms curves for (A) polyANS, (B) polyluminol and (C) copolymer electrode in the prescence of (i) 0.1, (ii) 0.2, (iii) 0.4, (iv) 0.6 and (v) 0.8 mM. Supporting electrolyte H_2SO_4 1M. The scan rate was 100 mVs^{-1}.

3.3.3 Electrocatalytic oxidation of ascorbic acid

Finally the oxidation of aa was analysed in the copolymer and comparated with those of homopolymers, Figure 17 show the ascorbic acid oxidation on the different films, immersed in 1 M H_2SO_4 solution in the absence and in the presence of different concentrations of ascorbic acid. An increase in intensity of the anodic current peak, as the acid ascorbic concentration was increased is an indication of catalytic oxidation of ascorbic acid mediated for each film. The effect catalytic of polyluminol in the oxidation of aa is well established (Chen & Lin, 2002), however the electrocatalysis in polyANS and copolymer has been not reported. The current obtained for a same concentration of aa is higher in the polyANS this suggest that the catalysis is more favourable for this film.

4. Conclusion

In conclusion we have found that it is posibble the formation of polymer of ANS, when the electroxidation of monomer is carried out in PANI/Au or carbon paste electrodes. The films obtained are electrochemically active in neutral pH. The charge compensation of this film is carried out principally by ejection of cation, but anion insertion is simultaneuslly presented. On other hand, the electrochemical polymerization of luminol in carbon paste electrodes give a film with characteristic of a seld-doped polymer contraty to the film obtained using metallic electrodes which are dimmers. A polymer was synthetised to luminol and ANS, the cyclic voltammogram obtain show peak intermediate beetween luminol and ANS, the film can catalize the oxidation of aa and is electroachemically active to neutral pH, the proportion of monomers in the film can be modulate by the solution composition and the scan rate.

5. Acknowledgment

This work was supported by the SEP (C06-PIFI-03.18.18) and CONACYT (C01-AINAT-01-39.39).

6. References

Barbero, C. & Kötz, R. (1994). Electrochemical formation of a selfdoped conductive polymer in the absence of an supporting electrolyte. The copolymerization of *o*-aminobenzensulfonic acid and aniline, *Advanced Materials,* Vol. 6, No. 7-8; (July-August 1994), pp. 577-580, ISSN 1521-4095

Buttry, D.A. & Ward, M.D. (1992). Measurement of interfacial processes at electrode surfaces with the electrochemical quartz crystal microbalance. *Chemical Review* Vol. 92, No. 6, (September 1992), pp. 1355-1379, ISSN 0365-6675

Cano-Márquez, A.G.; Torres-Rodríguez, L.M. & Montes-Rojas, A. (2007). Synthesis of fully and partially sulfonated polyanilines derived of ortanilic acid: an electrochemical and electromicrogravimetric study, *Electrochimica Acta,* Vol. 52, No. 16, (April 2007), pp. 5294-5303, ISSN 0013-4686

Chang, Y.T. ; Lin K.C. & Chen, S.M. (2005). Preparation characterization and electrocatalytic properties of poly(luminol) and polyoxometalate hybride film modified electrodes, *Electrochimica Acta*, Vol. 51, No. 3; (October, 2005) pp. 450-461. ISSN 0013-4686.

Chen, S. M. & Lin, K.C. (2002). The electrocatalytic properties of biological molecules using polymerized luminol film-modified electrodes, *Journal of Electroanalytical Chemistry*, Vol. 523, No. 1-2, (April 2002), pp. 93-105, ISSN 0022-0728

De Robertis, E. ; Neves, R.S. & Motheo, A.J. (2008). Electropolymerization Studies of Pani/polyluminol over platinum electrodes, *Molecular Crystals and Liquid Cristals*, Vol. 484, pp. 322-334, ISSN 1542-1406

Donjuan-Medrano, A.L. & Montes-Rojas, A. (2008). Effect of the thickness of thallium deposits on the values of EQCM sensitivity constant, *New Journal of Chemistry*, Vol. 32, No. 11, pp. 1935-1944, ISSN 1144-0546

Ferreira, D. ; Cascalheira, A.C. & Abrantes, L.M. (2008). Electrochemical copolymerisation of luminol with aniline: A new route for the preparation of self-doped polyanilines, *Electrochimica Acta*, Vol. 53, No. 11, (April 2008), pp.3803-3811 ; ISSN 0013-4686

Hillman, A.R & Mohamoud, M.A. (2006). Ion, solvent and polymer dynamics in polyaniline conducting polymers films, *Electrochimica Acta*, Vol. 51, No. 27; (August 2007) pp. 6018-6024. ISSN 0013-4686

Karyakin, A.A. ; Strakhova, A.K. & Yatsimirsky, A.K. (1994). Self-doped polyanilines electrochemically active in neutral and basic aqueous solutions. Electropolymerization of substituted anilines. *Journal of Electroanalytical Chemistry*, Vol. 371, No. 1-2, (June, 1994), pp. 259-265. ISSN 0022-0728

Krishnamoorthy, K ; Contractor, A.Q & Kumar, A. (2002). Electrochemical Synthesis of fully sulfonated polyanilines n-dopable polyaniline : poly(metanillic acid), *Chemical Communications*, No. 3, (December 2001), pp. 240-241, ISSN 1359-7345

Kumar, A.S ; Chen, H.W & Chen, S.M. (2009a). Electroanalysis of ascorbic acid (vitamine C using nano-ZnO/poly(luminol) hybrid film modified electrode, *Reactive & Functional Polymers*, Vol. 69, No. 6, (June 2009), pp. 364-370, ISSN 1381-5148.

Kumar, A.S. ; Chen, H.W ; Chen, S.M. (2009b). Selective detection of uric acid in the presence of ascorbic acid using polymerized luminol film modified glassy carbon electrode, *Electroanalysis*, Vol. 21, No. 2, (Octobre, 2009), pp. 2281-2286, ISSN 1040-0397

Ma, J. ; Wang, S.R. ; Wang, M.Q. ; Zhang, J.L. ; Wang, L.A. & Du, X.Y. (2010). Polyluminol/Single walled carbon nanotubes composites film-modified electrodes for simultaneous determination of propyl gallate and ascorbic acid, *Sensor letters*, Vol. 8, No. 4; (August 2010) pp. 672-676. ISSN 1546-198X

Malinauskas, A. (2004). Self-doped polyanilines, *Journal of Power Sources*, Vol. 126, No. 1-2, (February 2004), pp. 214-220, ISSN 0378-7753

Mažeikienė, R. & Malinauskas, A. (2004). Electrochemical preparation and study of novel self-doped polyanilines, *Materials Chemistry and Physics*, Vol. 83 ; No. 1 ; (January, 2004), pp. 184-192 ; ISSN 0254-0584.

Mello, R.M.Q ; Torresi, R.M. ; Córdoba de Torresi, S.I. ; Ticianelli ; E.A. (2000). Ellipsometric, electrogravimetric and spectroelectrochemical studies of the redox process of sulfonated polyaniline, *Langmuir*, Vol. 16, No. 20, (June, 2000), pp. 7835-7841, ISSN 0743-7463.

Nguyen, M.T. ; Kasai, P. ; Miller, J.L. ; Díaz, A. F. (1994). Synthesis and Properties of Novel Water-Soluble Conducting Polyanilines copolymers, *Macromolecules*, Vol. 27, No. 3, (June 1994), pp. 3625-3631, ISSN 0024-9297

Orata, D. & Buttry, A.D. (1987). Determination of ion populations and solvent content as functions of redox state and pH in polyaniline, *Journal of American Chemical Society.* Vol. 109, No. 12, (June 1987), pp. 3574-3581, ISSN 0002-7863

Robertis, E.D. ; Neves, R.S. & Motheo, A.J. (2008). Electropolymerization studies of PAni/(poly)luminol over platinum electrodes, Mol. Cryst. Liq. Crist. Vol. 484, pp. 322-334, ISSN 1542-1406

Royapa, A. T. ; Steadman, D. D. ; Tran, T.L. ; Nguyen, T. ; Prayaga, C.S. ; Cage, B. & Dalal, N. (2001). Synthesis of sulfonated polyaniline by polymerization of the aniline heterodimer 4-aminodiphenilamine-2-sulfonic acid ; *Synthetic Metals*, Vol. 123, No. 2, (September 2001), pp. 273-277, ISSN 0379-6779

Sanchís, C. ; Salavagione, H.J. ; Arias-Padilla, J. & Morallón, E. (2007). Tuning the electroactivity of conductive polymer at physiological pH. *Electrochimica Acta,* Vol. 52, No. 9, (February, 2007), pp. 2978-2986., ISSN 0013-4686

Shin-Jung, C. & Su-Moon, P.; (2002). Electrochemistry of conductive polymers XXVI. Effects of electrolytes and growth methods on polyaniline morphology; *Journal of Electrochemical Society*; Vol. 149, No. 2, (February 2002), pp. E26-E34, ISSN 0013-4651

Varela, H. ; Torresi, R.M. & Buttry, D.A. (2000). Mixed cation and anion transport during redox cycling of a self-doped polyaniline derivative in nonaqueous media. *Journal of Electrochemical Society,* Vol. 147, No. 11, (Novembre 2000), pp. 4217-4223, ISSN 0013-4651

Varela, H. ; Torresi, R.M & Buttry, D.A. (2000). Study of charge compensation during the redox process of self-doped polyaniline in aqueous media. *Journal of Brazilien Chemical Society,* Vol. 11, No. 1, (January-february), pp. 32-38, ISSN : 0103-5053

Varela, H. ; de Albuquerque Maranhão, S.L. ; Mello, R.Q.M ; Ticianelli, A.E. & Torresi, R.M. (2001). Comparisons of charge compensation process in aqueous media of polyaniline and self-doped polyanilines ; *Synthetic Metals* ,Vol. 122 ; No. 2 , (June, 2001), pp. 321-327, ISSN 0379-6779

Wei, X.L.; Wang, Y.Z.; Long, S.M.; Bobeczko, C. & Epstein, A.J. (1996). Synthesis and Physical Properties of Highly Sulfonated Polyaniline, *Journal of American Chemical Society*, Vol. 118, No. 11, (December 1996), pp.2545-2555, ISSN 2156-8251

Yue, J.; Wang; Z.H.; Cromack, K.R.; Epstein, A.J.; MacDiarmid, A.G. (1991). Effect of sulfonic group on polyaniline backbone, *Journal of American Chemical Society*, Vol. 113, No. 7, (March 1996), pp.2665-X, ISSN 2156-8251

Zhang, G.F.; Chen, H.Y. (2000) Studies of polyluminol modified electrode and its application in electrochemiluminiescence analysis with flow system, Analytica Chimica Acta, Vol. 419, No. 1, (August 2000), pp. 25-31, ISSN 0003-2670

Zhang, L.; Jiang, X.; Niu, L & Dong, S., (2006). Syntheses of fully sulfonated polyaniline and its application to the direct electrochemistry of cytochrome C, *Biosensors & Bioelectronics*, Vol. 21, No. 7, pp. 1107-1115, ISSN 0956-5663

6

Electrosynthesis and Characterization of Polypyrrole in the Presence of 2,5-di-(2-thienyl)-Pyrrole (SNS)

Nasser Arsalani[1], Amir Mohammad Goganian[1], Gholam Reza Kiani[3], Mir Ghasem Hosseini[2] and Ali Akbar Entezami[1]

[1]*Polymer Research Laboratory, Department of Organic Chemistry, Faculty of Chemistry, University of Tabriz, Tabriz,*
[2]*Electrochemistry Research Laboratory, Department of Physical Chemistry, Faculty of Chemistry, University of Tabriz, Tabriz,*
[3]*School of Engineering-Emerging Technologies, University of Tabriz, Tabriz*
Iran

1. Introduction

Electronically conducting polymers are a very popular research field among the chemists due to their use in a wide variety of marketable applications such as electrochromic devices [Mortimer et al., 2006; Sahin et al., 2005], polymer light-emitting diodes (LEDs) [Kraft et al., 1998], artificial muscles [Cortes & Moreno, 2003], gas sensors [Nicolas-Debarnot & Poncin-Epaillard, 2003], bio sensors [Geetha et al., 2006; Malinauskas et al., 2004] and corrosion protection of metals [Hosseini et al., 2007, 2008; Oco´n et al., 2005]. The preparation, characterization and application of electrochemically active, electronically conjugated polymeric systems are in the foreground of research activities in electrochemistry [Heinze et al., 2010]. Among the conducting polymers, polypyrrole has attracted a lot more interests [Jang & Oh, 2004; Zhang et al., 2006; Chang et al., 2009]. This polymer is easy to synthesize both chemically and electrochemically, exhibiting good electrical conductivity and relatively good stability under ambient conditions, but lacking good electroactivity and redoxability.

In order to improve the electroactivity and redoxability of the electro-synthesized polypyrrole, another molecule containing conjugated system can be used during the electropolymerization of pyrrole. 2,5-di-(2-thienyl)-pyrrole (SNS) is one of the molecules containing conjugated system and have been studied by various electrochemical methods such as cyclic voltammetry, chronopotentiometry, and chronoamperometry under different conditions (changing the electrolyte, electrode, electrochemical potential range and etc) [Otero et al., 1998; Brillas et al., 2000; McLeod et al., 1986].

Entezami et al. have studied the electropolymerization of pyrrole and *N*-methyl pyrrole in the presence of 1-(2-pyrrolyl)-2(2-thienyl) ethylene (PTE) and 2-(2-thienyl) pyrrole (TP) by cyclic voltammetry in different conditions [Kiani et al., 2001]. Recently, we have studied the electropolymerization of thiophene and 3-Methyl thiophene in the presence of small amount of 1-(2-pyrrolyl)-2-(2-thienyl) ethylene (PTE). It was found that the conductivity, electroactivity and redoxability of polythiophene and poly(3-methyl thiophene) are improved in the presence of PTE [Kiani et al., 2008a, 2008b].

In this work, the effects of conjugated molecule (SNS) on the electropolymerization and electrochemical behaviour of pyrrole was investigated. Firstly, the electropolymerization of pyrrole and SNS were carried out separately by CV method. Secondly, the electropolymerization of pyrrole in the presence of small amount of SNS was carried out and then the influence of SNS on the electropolymerization and electrochemical behaviour of pyrrole was investigated. In addition, the effect of SNS addition on the electron transfer reaction of ferro/ferricyanide redox system on the polypyrrole film was studied and finally the conductivity of poly(Py-SNS) was determined by electrochemical impedance spectroscopy (EIS) method.

2. Experimental

2.1 Materials

Solvents were purified and dried according to the common procedures in the literature [Perin & Armarego, 1998]. Acetic anhydride, ammonium acetate, ferro/ferricyanide were purchased from Merck and lithium perchlorate were bought from Fluka and all of them were used directly.

2.2 Preparation of 2,5-di-(2-thienyl)-pyrrole (SNS)

2,5-di-(2-thienyl)-pyrrole (SNS) was prepared by the method described by Wynberg and Metselaar [Wynberg & Metselaar, 1984]. The yield was improved by refluxing the solution of the intermediate 1,4-di-(-2-thienyl)-1,4-butanedione (3 *g*) with ammonium acetate (40.3 *g*), glacial acetic acid (120 *ml*) and acetic anhydride (24 *ml*) overnight under a nitrogen atmosphere. The reaction mixture was then poured into 250 *ml* of distilled water and the resulting dark-green solid was chromatographed over a silica gel column with dichloromethane: hexane (3:2) elution to give SNS with 75% yield, as pale yellow crystals of melting point 82-83 °C.

IR(KBr): 3490 cm^{-1} (N-H); ^1H NMR($CDCl_3$): δ: 6.2(2H,d), 6.8(6H,m), 8.0(1H,s).

2.3 Electropolymerization method and conductivity measurements

The electropolymerization was carried out using digital potentiostate/galvanostate (Autolab PGSTAT 30). A glassy carbon (GC) disk (2 *mm* diameter) as working electrode, a platinum wire as a counter electrode and Ag/AgCl as a reference electrode were used. Acetonitrile was distilled over P_2O_5 and lithium perchlorate was used as an electrolyte in 0.1 *M* concentration. The electropolymerization of pyrrole (0.5 *ml*, 7.4 *mmole*) and SNS (16 *mg*, 0.074 *mmole*) in 0.1 *M* $LiClO_4$/acetonitrile electrolyte were performed separately in two different potential ranges vs. Ag/AgCl at the scan rate of 50 $mV.s^{-1}$. Similar conditions were adopted for the electropolymerization of pyrrole in the presence of SNS (7.4 *mmole*: 0.074 *mmole*). In the study of the cyclic voltammetry experiment of electron transfer, 1 mM ferro/ferricyanide in the 1 *M* H_2SO_4 media at 50 $mV.s^{-1}$ scan rate was used.

For the EIS measurements, the AC frequency range extended from 100 *kHz* to 10 *mHz*, a 10 *mV* peak-to-peak sine wave was as the excitation signal. Both real and imaginary components of the EIS in the complex plane were analyzed using the Zview(II) software to estimate the parameters of the equivalent electrical circuit. A computer-controlled potentiostate (PARSTATE 2263 EG&G) was used for EIS measurements.

3. Results and discussion

3.1 Electrochemical synthesize of polymers

The electropolymerization of pyrrole was performed by cyclic voltammetry in the potential range of -100 to 900 *mV* through 15 scans. As shown in figure 1, at the first scan, there is an anodic peak at ca. 800 *mV*. By continuing electropolymerization through second scan, another anodic peak was observed at 550 *mV* indicating formation of polypyrrole. After the formation of black colored polymer film on the GC electrode surface, the electrode was taken out from electrochemical cell and was washed with acetonitrile. For the resulted polymer, the cyclic voltammograms at various scan rates were shown in figure 2 indicating a quasi-reversible behaviour.

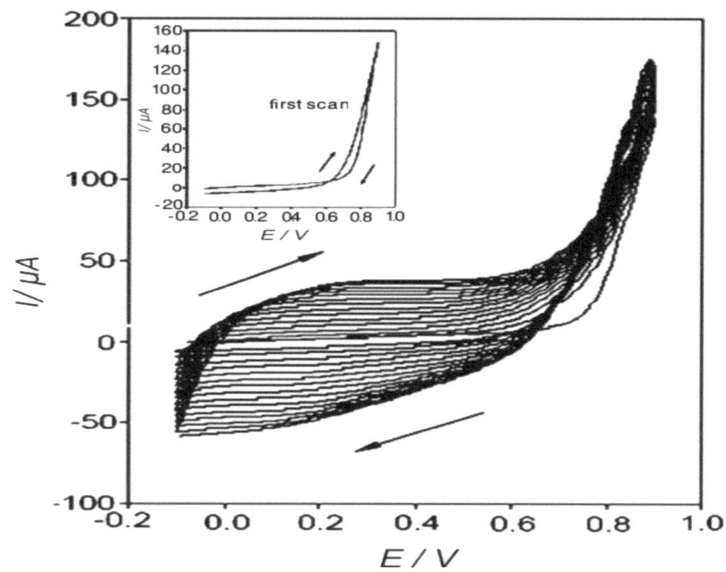

Fig. 1. Cyclic voltammograms of 7.4×10^{-3} M pyrrole in 0.1 M $LiClO_4$ / CH_3CN electrolyte at scan rate 50 mV/s vs. Ag/AgCl

The cyclic voltammetry investigations of SNS were carried out in the potential range of -400 to 1500 *mV* (Fig. 3). At the first scan two anodic peaks at ca. 570 and 1300 *mV* were observed resulting from the oxidation of SNS. In the backward scan from 1500 to -400 *mV*, there is one peak indicating a quasi-reversible reaction. At the second scan, a new anodic peak current was observed indicating formation of the electroactive poly(SNS) with an ionic structure. As shown in figure 3, after 7 scans, the second anodic peak at 1300 mV was eliminated. The cyclic voltammogram of poly(SNS) in the potential ranges between -400 to 1000 *mV* at various scan rates was shown in figure 4.

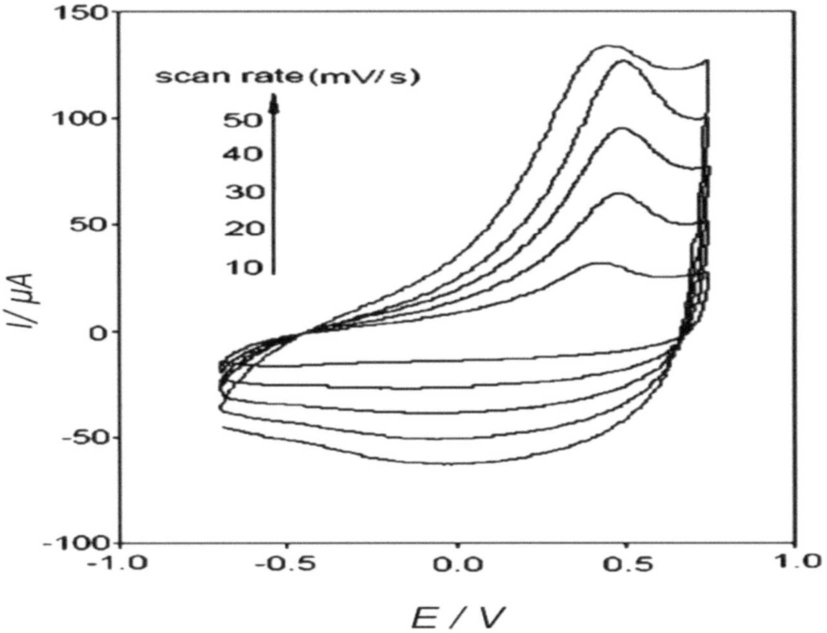

Fig. 2. Cyclic voltammograms of poly(Py)in 0.1 M $LiClO_4$ / CH_3CN electrolyteat various scan rates

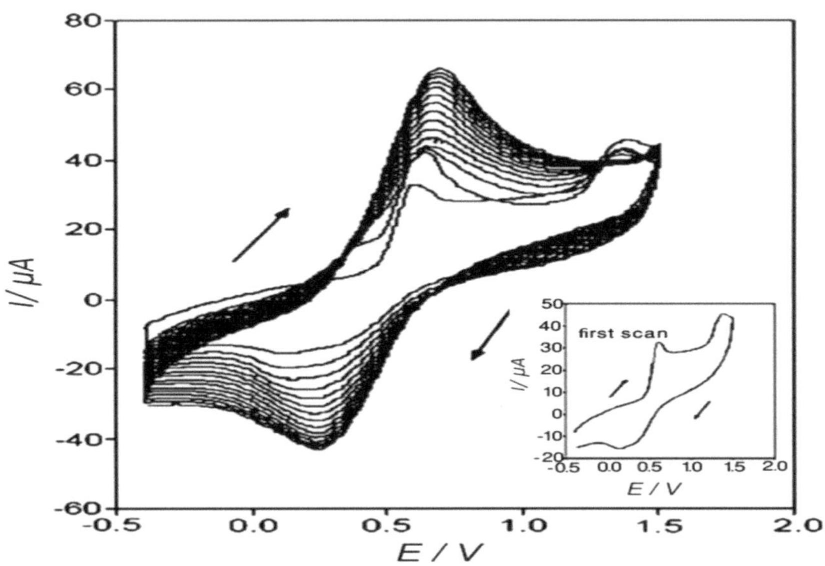

Fig. 3. Cyclic voltammograms of 7.4×10^{-5} M SNS in 0.1 M $LiClO_4$ / CH_3CN electrolyte at scan rate 50 mV/s vs. Ag/AgCl

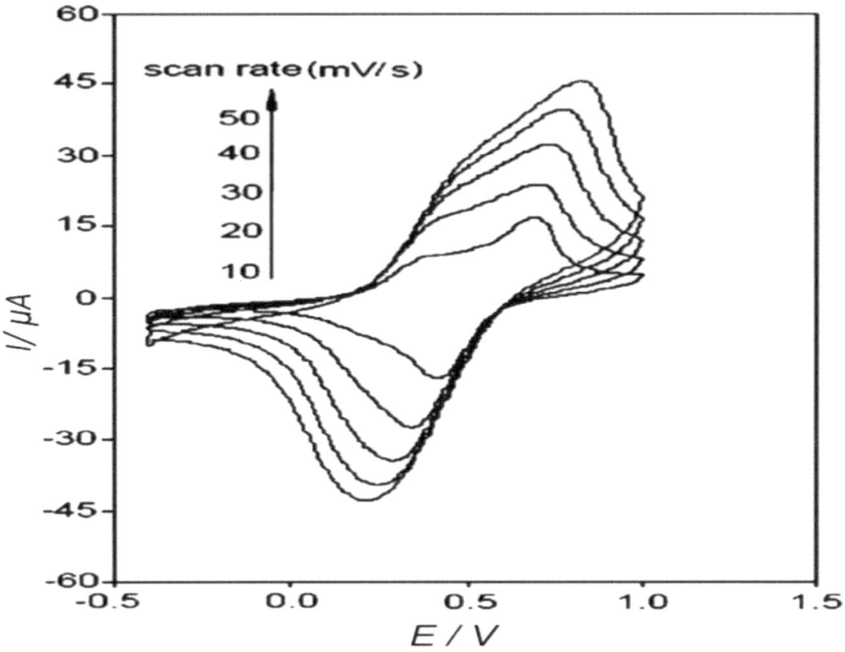

Fig. 4. Cyclic voltammograms of poly(SNS) in 0.1 M $LiClO_4$ / CH_3CN electrolyte at various scan rates

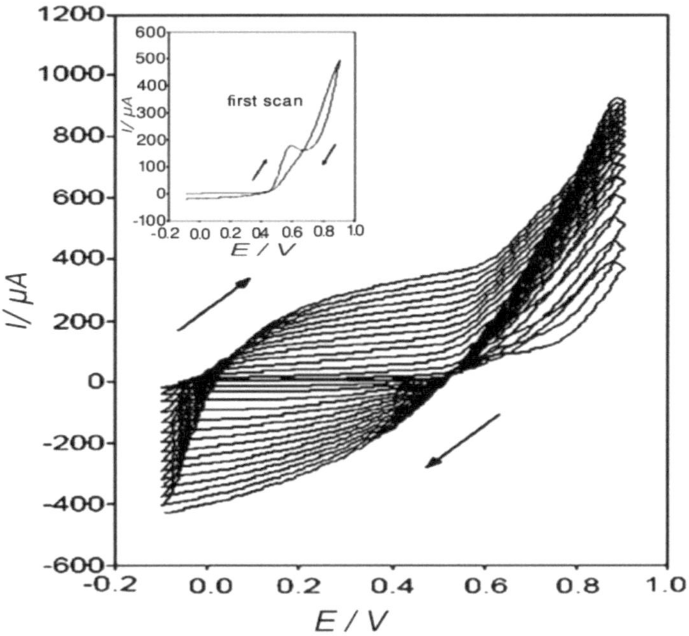

Fig. 5. Cyclic voltammograms of Py-SNS (100:1 mole ratio) in 0.1 M $LiClO_4$ / CH_3CN electrolyte at scan rate 50 mV/s vs. Ag/AgCl

During the electropolymerization of pyrrole in the presence of the SNS (7.4 *mmole*: 0.074 *mmole*) two anodic peaks appeared at 570 and 800 *mV*. These peaks are due to the oxidation of SNS and pyrrole, respectively. In addition, the anodic peak at 570 *mV* was absent during electropolymerization of pyrrole without SNS (Fig. 5). Because of conjugated backbone of SNS, the oxidation potential of this monomer is less than pyrrole. The cyclic voltammogram for the resulted polymer in various scan rates showed a relatively reversible behaviour (Fig. 6).

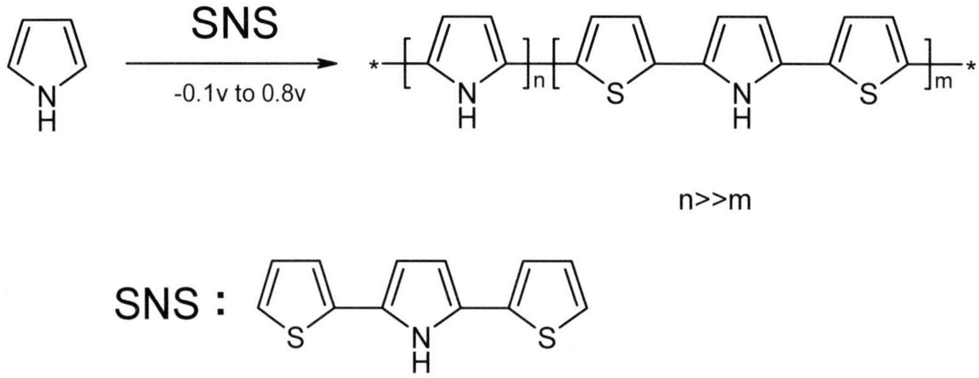

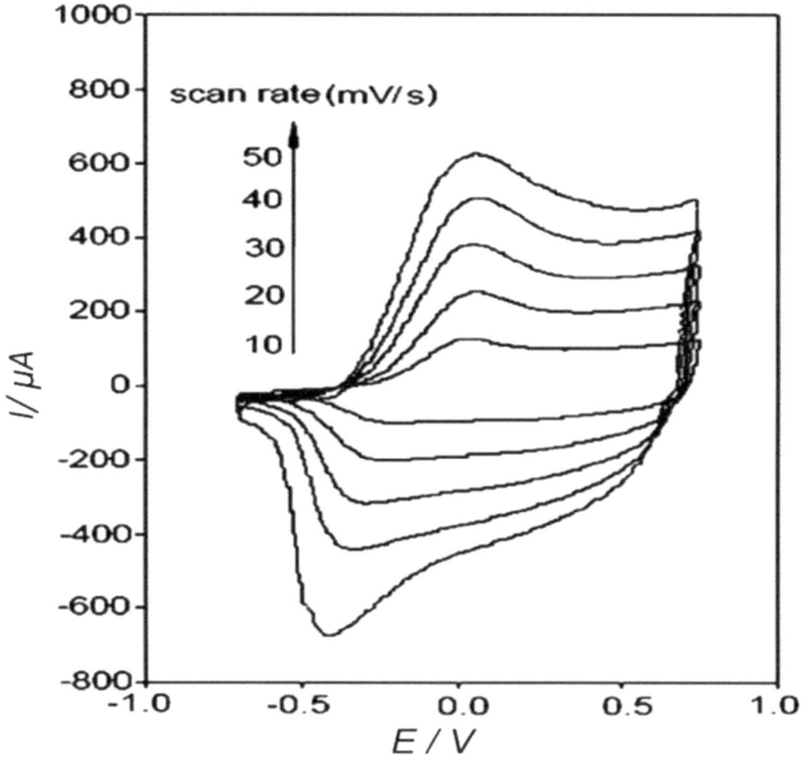

Fig. 6. Cyclic voltammograms of poly(Py-SNS) in 0.1 M $LiClO_4$ / CH_3CN electrolyte at various scan rates

Figure 7 presents the plot of anodic peak currents vs. different scan rates for obtained polymers. These curves show that the slop for poly(Py) and poly(Py-SNS) increases from 2.83 to 12.37 $mAs.mV^{-1}$. These results indicated considerable increase in the electroactivity and rate of electropolymerization of polypyrrole in the presence of a small amount of SNS compared to the those of polypyrrole and poly(SNS). According to extracted data from cyclic voltammetries of polymers (see Table1), it can be seen that at various scan rates, E_{Pa} for poly(Py-SNS) is less than those two for other polymers, but i_{pa} for former polymer is more than those for two others. In other words, the conductivity of SNS included polypyrrole is better than polypyrrole alone and poly(SNS). Also, it is evident that at scan rates of less than 50 $mV.s^{-1}$, ΔE_p for poly(Py-SNS) is lower those for than two others, indicating improvement of redoxability for poly(Py-SNS) in comparison with poly(Py) and poly(SNS). At scan rate of 50 $mV.s^{-1}$ the redoxability of poly(Py-SNS) is relatively similar to that of poly(Py).
The overall scheme of electrosynthesis of polypyrrole in the presence of SNS as shown Fig. 6.

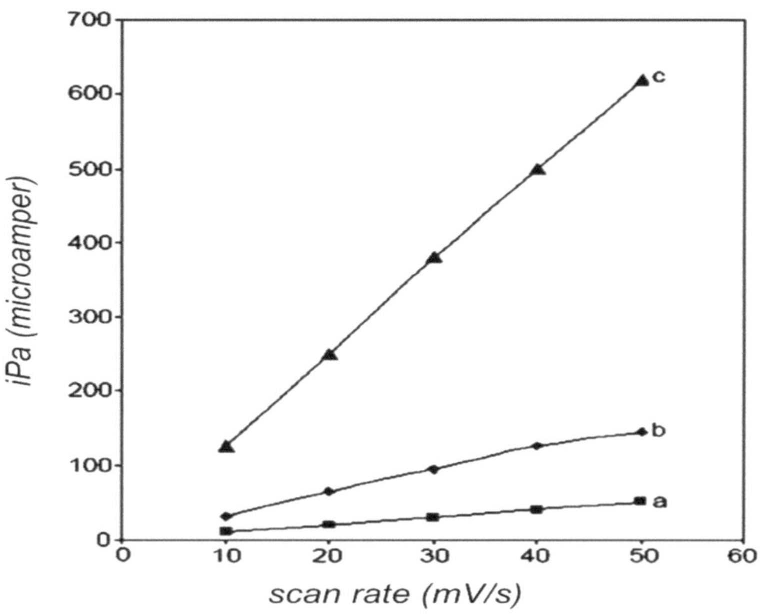

Fig. 7. Plots of anodic peak currents vs. scan rates for (a) poly(SNS), (b) poly(Py), (c) poly(Py-SNS)

Scan	rates	Poly(Py)	Poly(SNS)	Poly(Py-SNS)
E_{pa}(V)	10	0.450	0.715	0.052
	30	0.500	0.750	0.075
	50	0.460	0.841	0132
E_{pc}(V)	10	-0.123	0.405	-0.240
	30	-0.070	0.277	-0.329
	50	-0.050	0.201	-0.431
ΔE_p(V)	10	0.573	0.310	0.292
	30	0.570	0.473	0.404
	50	0.510	0.640	0.563
i_{pa}(μA)	10	32	19	120
	30	94	33	380
	50	103	47	620

E_{Pa}: potential of anodic peak
E_{Pc}: potential of catidic peak
i_{Pa}: anodic peak currents

Table 1. Obtained date for poly(Py), poly(SNS) and poly(Py-SNS) at scan rates 10 and 50 mV/s

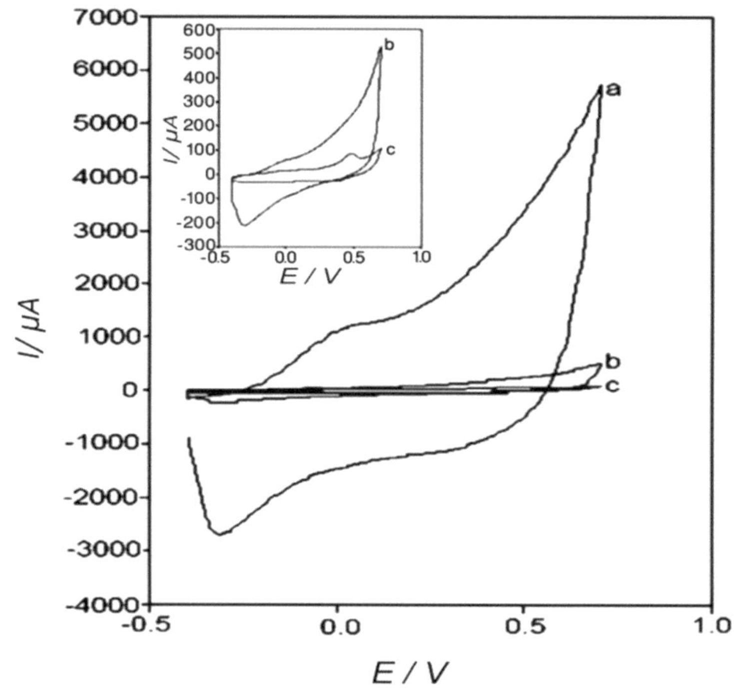

Fig. 8. Cyclic voltammograms of (a) poly(Py-SNS), (b) poly(SNS), and (c) poly(Py) on the GC electrode in 1 M H_2SO_4 and 1 mM $Fe(CN)_6{}^{4-/3-}$ redox system at 50 mV/s scan rate

The CV experiments were performed to study the effect of SNS in the polypyrrole film in the electron transfer of ferro/ferricyanide redox system. Figure 8 shows the CV of electron transfer ferro/ferricyanide redox on different modified GC electrodes with poly(Py), poly(SNS), and poly(Py-SNS). This figure indicates that the electron transfer of ferro/ferricyanide on polypyrrole in the presence of the SNS is more feasible than that of polypyrrole alone, because the conductivity of polypyrrole increases in the presence of SNS.

3.2 Evaluation of conductivity and electrochemical behaviour by electrochemical impedance spectroscopy (EIS)

Electrochemical impedance spectroscopy (EIS) is a measurement technique which allows for a wide variety of coating evaluations. EIS is an effective method to probe the interfacial properties of surface-modified electrodes. EIS has been used to characterize the electrical properties of the electropolymerized films [Kiani et al., 2008a, 2008b]. The electrochemical behaviour of polypyrrole changes in the presence of SNS .In order to choose a suitable electrical equivalent circuit for EIS experimental data fitting, one must take in consideration the physicochemical picture of the system under study. In other words, each element of the equivalent circuit should have a physicochemical aspect attributable to it. In the model circuit chosen, R_s presents the uncompensated resistance of the solution between working and reference electrode. CPE_1 and R_1 stand for the dielectric and resistive characteristics of the conductive polymer on the GC electrode, respectively. In this case R_1 is a reverse measure of polymer conductivity. CPE_2 and R_2 show the capacitance and resistance of the polymer/GC interface. As is evident from the high values of R_2 for all as well as the Nyquis plot of studied polymers, due to the fact that polymer layer is impermeable to the ionic charge carrier species, low frequency behaviour of the polymer/GC interface tends to be of capacitive nature. Again more evidence for this fact is reflected in the values of n_{CPE2} which for all studied samples is not very far than unity. Our main aim by EIS studies was to determine the polymer layer bulk resistivity (or its reciprocal i.e conductivity). In the selected equivalent circuit R_1 corresponds to this parameter. The studied electrical parameters were calculated using Zview(II) software. All fitting results are presented in Table 2.

Sample	$R_s(\Omega)$	$R_1(\Omega.cm^2)$	$R_2(\Omega.cm^2)$ Y_0	CPE_1 n	CPE_1 Y_0	CPE_2 n	CPE_2
Poly(Py)	43.86	5.81	2.5E4	8.0E-4	0.64	8.3E-4	0.98
Poly(SNS)	38.94	2282	7.2E4	1.3E-4	0.72	1E-4	1.00
Poly(Py-SNS)	30.50	1.46	7.2E4	0.67	0.58	0.01	0.98

R_s : uncompensated resistance of the solution
R_1 and CPE_1 : dielectric and resistive characteristics of the conductive polymer
R_2 and CPE_2 : capacitance and resistance of the polymer/ GC interface
Y_0 : CPE Admittance
n : CPE exponent

Table 2. Impedance parameters obtained by fitting the EIS data of poly(Py), poly(SNS) and poly(Py-SNS) on the GC electrode in 3.5% NaCl

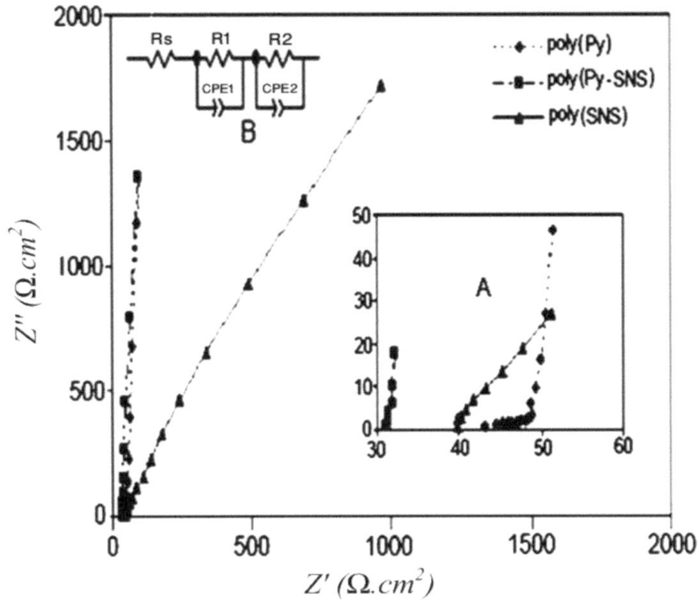

Fig. 9. Nyquist plots for poly(Py), poly(Py-SNS)(100:1 mole ratio) and poly(SNS) in 3.5% (W/V) NaCl solution: A) Exploded view in the high frequency range, B) Proposed equivalent circuit

According to these results (Table 2), we can notice a decrease in the charge transfer resistance value in the case of the polypyrrole in the presence of SNS systems as compared to polypyrrole alone. The R_{ct} (R_{ct}: charge transfer resistance) values obtained for polypyrrole and poly(SNS) are 5.81 and 2282 $\Omega.cm^2$ respectively. This value decreases in the presence of SNS to 1.46 $\Omega.cm^2$. The polypyrrole film formed in the presence of SNS is more conductive. On the other hand, in the presence of SNS, value of the capacitance of the double layer, CPE_1, rises from 8.0E-4 to 0.67 $\mu F.cm^{-2}$ which can be attributed to an increase in the electrode surface area. This change in the capacitance strongly supports the hypothesis of the incorporation of SNS in the polypyrrole film. Also, these results support the results of CV in the Figure 9. In the presence of SNS, the conductivity of polypyrrole is improved. Increased value of CPE_1 for polypyrrole in the presence of SNS compared to pure polypyrrole confirmed the easy electron transfer of ferro/ferricyanide redox system for poly(Py-SNS) (Fig. 8). Improvement of the conductivity, electroactivity and redoxability of polypyrrole containing SNS leads it to extensive applications in many fields.

4. Conclusions

The resulted poly(Py-SNS)(100:1 mole ratio) showed a considerable increase in the electroactivity, redoxability, and the rate of polymerization in comparison to polypyrrole alone. The cyclic voltammograms of electron transfer ferro/ferricyanide redox system on different modified GC electrode showed that the rate of charge transfer for polypyrrole in the presence of SNS increased in comparison to pure polypyrrole. In addition, the conductivity of polypyrrole was studied by electrochemical impedance spectroscopy. The obtained R_{ct} value for polypyrrole is 5.81 $\Omega.cm^2$, whereas the value decreases to 1.46 $\Omega.cm^2$ in the presence of SNS. By considering the fact that decreasing the R_{ct} leads to an increase in conductivity, it is predictable that the film of polypyrrole formed in the presence of SNS will be more conductive. In the presence of SNS, value of electrical double layer capacitance (CPE_1) rises, indicating a probable increase in the electrode surface area. There is a good complementary agreement between the results of CV and EIS measurements. From these results it can be concluded that the produced polypyrrole containing small amount of SNS has better performance compared to polypyrrole alone for production of batteries, capacitors, diodes, electrochromic devices, sensors and etc.

5. Acknowledgment

The authors acknowledge Mr. I. Ahadzadeh, for his kind help in EIS measurements.

6. References

Brillas, E. Carrasco, J. Oliver, R. Estrany, F. Vilar, J. & Morlans, J.M. (2000). Electropolymerization of 2,5-di-(-2-thienyl)-pyrrole in ethanolic medium. Effect of solution stirring on doping with perchlorate and chloride ions. *Electrochimica Acta,* Vol. 45, pp. 4049-4057.

Chang, C.H. Son, P.S. Yang, G.H. & Choi, S.H. (2009). Electrochemical Synthesis of the Functionalized Poly(pyrrole) Conducting Polymers. *Journal of the Korean Chemical Society,* Vol. 53, No.2, pp. 111-117.

Cortés, M.T. & Moreno, J.C. (2003). Artificial muscles based on conducting polymers. *e-Polymers,* No. 041, pp. 1-42.

Geetha, S. Chepuri, R.K.R. Vijayan, M. & Trivedi, D.C. (2006). Biosensing and drug delivery by polypyrrole. *Analytica Chimica Acta, Vol.* 568, pp. 119-125.

Heinze, J. Frontana-Uribe, B.A. & Ludwigs, S. (2010). Electrochemistry of Conducting Polymers—Persistent Models and New Concepts. *Chemical Reviews,* Vol. 110, No.8, pp. 4724-4771.

Hosseini, M.G. Sabouri, M. & Shahrabi, T. (2007). Corrosion protection of mild steel by polypyrrole phosphate composite coating. *Progress in Organic Coatings,* Vol. 60, pp. 178-185.

Hosseini, M.G. Sabouri, M. & Shahrabi, T. (2008). Comparison of the corrosion protection of mild steel by polypyrrole-phosphate and polypyrrole-tungstenate coatings. *Journal of Applied Polymer Science.* Vol. 110, No. 5, pp. 2733-2741.

Jang, J. & Oh, J.H. (2004). Morphogenesis of Evaporation-Induced Self-Assemblies of Polypyrrole Nanoparticles Dispersed in a Liquid Medium. *Langmuir,* Vol. 20, pp. 8419-8422.

Kiani, G.R. Arsalani, N. & Entezami, A.A. (2001). The Influence of the Catalytic Amount of 1-(2-Pyrrolyl)-2-(2-Thienyl) Ethylene and 2-(2-Thienyl) Pyrrole on Electropolymerization of Pyrrole and N-Methylpyrrole. *Iranian Polymer Journal,* Vol. 10, pp. 135-142.

Kiani, G.R. Arsalani, N. & Entezami, A.A. (2008). Synthesis of Poly(3-methylthiophene) in the Presence of 1-(2-Pyrrolyl)-2-(2-thienyl) Ethylene by Electropolymerization. *Journal of Iranian Chemical Society,* Vol. 5, pp. 559-565.

Kiani, G.R. Arsalani, N. Hosseini, M.G. & Entezami, A.A. (2008). Improvement of the conductivity, electroactivity, and redoxability of polythiophene by electropolymerization of thiophene in the presence of catalytic amount of 1-(2-pyrrolyl)-2-(2-thienyl) ethylene (PTE). *Journal of Applied Polymer Science,* Vol. 108, No. 4, pp. 2700-2706.

Kraft, A. Grimsdale, A.C. & Holmes, A.B. (1998). Electroluminescent Conjugated Polymers-Seeing Polymers in a New Light. *Angewandte Chemie International Edition,* Vol. 37, No.4, pp. 402-428, ISSN 14337851

Malinauskas, A. Garjonyte, R. Mazeikiene, R. & Jureviciute, I. (2004). Electrochemical response of ascorbic acid at conducting and electrogenerated polymer modified electrodes for electroanalytical applications: a review. *Talanta, Vol.* 64, pp. 121-129.

McLeod, G.G. Mahboubian-Jones, M.G.B. Pethrick, R.A. Watson, S.D. Truong, N.D. Galin, G.C. & Francois, J. (1986). Synthesis, electrochemical polymerization and properties of poly(2,5-di-(-2-thienyl)-pyrrole). *Polymer,* Vol. 27, pp. 455-458.

Mortimer, R.J. Dyer, A.L. & Reynolds, J.R. (2006). Electrochromic organic and polymeric materials for display applications. *Displays,* Vol. 27, No.1, pp. 2-18.

Nicolas-Debarnot, D. & Poncin-Epaillard, F. (2003). Polyaniline as a new sensitive layer for gas sensors. *Analytica Chimica Acta,* Vol. 475, pp. 1-15.

Oco´n, P. Cristobal, A.B. Herrasti, P. & Fatas, E. (2005). Corrosion performance of conducting polymer coatings applied on mild steel. *Corrosion Science,* Vol. 47, pp. 649–662.

Otero, T.F. Villanueva, S. Brillas, E. & Carrasco, J. (1998). Electrochemical processing of the conducting polymer poly(SNS). *Acta Polymerica,* Vol. 49, pp. 433-438. IISSN 0323-7648

Perin, D.D. & Armarego, W.L.F. (1998). Purification of Laboratory Chemicals, Edition, Pergamon publisher, Oxford.

Sahin, E. Camurlu, P. Toppare, L. Mercore, V.M. Cianga, I. & Yagci, Y.J. (2005). Conducting copolymers of thiophene functionalized polystyrenes with thiophene. *Journal of Electroanalytical Chemistry,* Vol. 579, pp. 189-197.

Wynberg, H.& Metselaar, J. (1984). A Convenient Route To Polythiophenes. *Synthetic Communications,* Vol. 14, pp. 1-9.

Zhang, X. Zhang, J. Song, W. & Liu, Z. (2006). Controllable Synthesis of Conducting Polypyrrole Nanostructures. *Journal of Physical Chemistry B,* Vol. 110, pp. 1158-1165, ISNN 16471658.

7

Polypyrrole Composites: Electrochemical Synthesis, Characterizations and Applications

R. N. Singh, Madhu and R. Awasthi
Banaras Hindu University
India

1. Introduction

The electronically conducting polymers (ECPs), such as polypyrrole (PPy), polythiophene (PT) and polyaniline (PANI) are known to possess unusually high electrical conductivity in the doped state. Due to this, these materials have been of great interests for chemists as well as physicists since their electrical properties were reported (Diaz et al., 1979). The ECP films behave like a redox polymer and have potential applications in electrocatalysis, solar energy conversion, corrosion, electronics, etc. The redox polymer reaction is accompanied by a change in the electrical properties of the film from an insulator to an electrical conductor involving both electron and ion transport within the film (Kaplin & Qutubuddin, 1995).

Conducting polymers can be synthesized either chemically or electrochemically. Electrochemical synthesis is the most common method as it is simpler, quick and perfectly controllable. PPy is one of the most interesting conducting polymers since it is easily deposited from aqueous and non-aqueous media, very adherent to many types of substrates, and is well-conducting and stable. Electrochemical polymerization produces thin films with a thickness of few micrometers on an electrode surface (Diaz et al., 1979), while a chemical oxidation yields a fine-grained material. However,the yield and quality of the resulting polymer films are influenced by several factors, such as nature and concentration of monomer and the counter ion, solvent, cell conditions (e.g. electrode and applied potential), temperature and pH (Sadki et al. 2000; Ansari, 2006 & Pina et al. 2011).

ECPs can be modified in several ways (Juttner et al., 2004) to obtain tailored materials with special functions: (i) derivatization of the monomer by introducing aliphatic chains with functional groups; (ii) variation of the counterion, incorporated for charge compensation during the polymerization process; (iii) inclusion of neutral molecules with special chemical functions and (iv) formation of compounds with noble metal nanoparticles as catalyst for electrochemical oxidation and reduction processes.

The electropolymerization reaction is a complex process and its mechanism is still not fully understood. A number of mechanisms have been proposed (Genies et al., 1983; Kim et al., 1988; Asavapiriyanont et al., 1984; Qui & Reynolds, 1992) and are comprehensively reviewed (Sadki et al. 2000; Ansari, 2006 & Pina et al. 2011). Among these, Diaz's mechanism is the most accepted one (Genies et al., 1983) and supported by Waltman and Bargon (1984 & 1985) also. In this mechanism, the pyrrole (Py) activation occurs through electron transfer from the monomer forming a radical cation-rich solution near the electrode in several steps. Details of steps involved are given in Scheme 1.

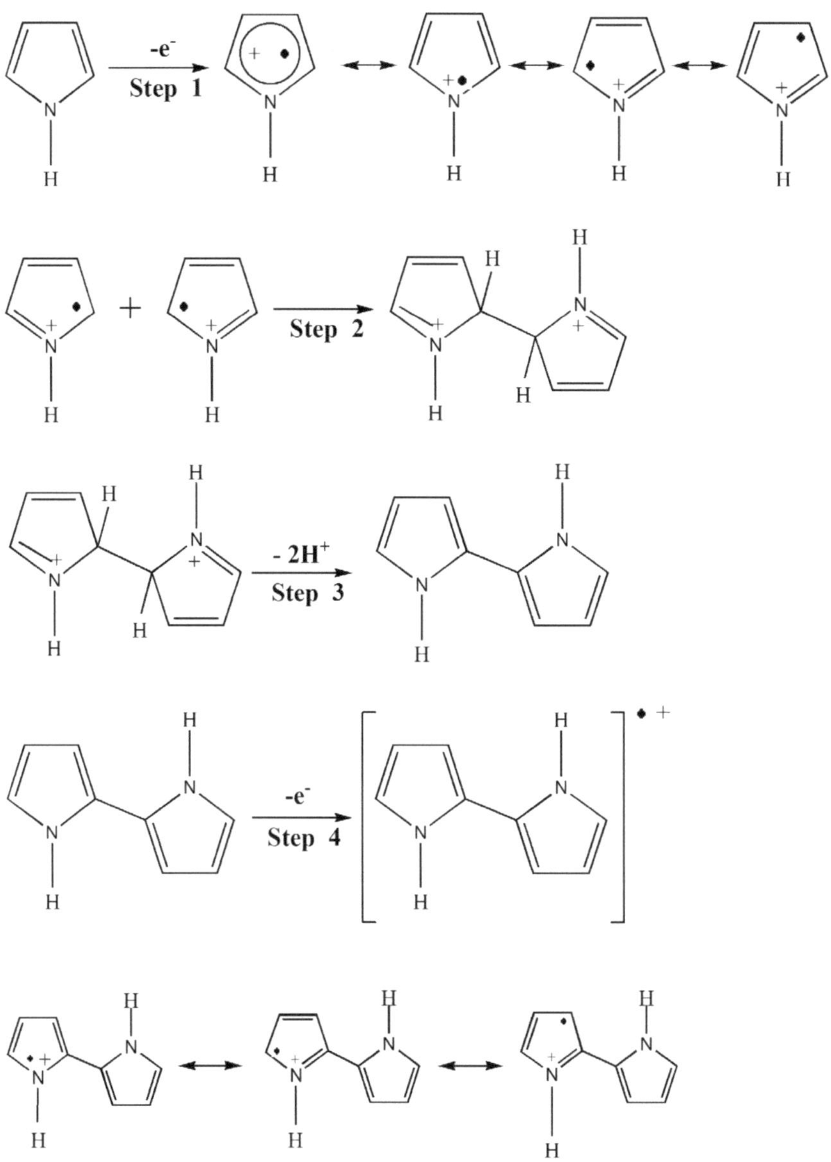

Scheme 1.

The propagation continues via the same sequences: oxidation, coupling, deprotonation until the final polymer product is formed (Fig.1a). The electropolymerization does not give the neutral nonconducting polymer but its oxidized conducting form (doped). The final polymer chain, in fact, carries a positive charge every 3-4 Py units, which is counter balanced by an anion. The structure of the doped polymer can be given as shown in Fig.1b.

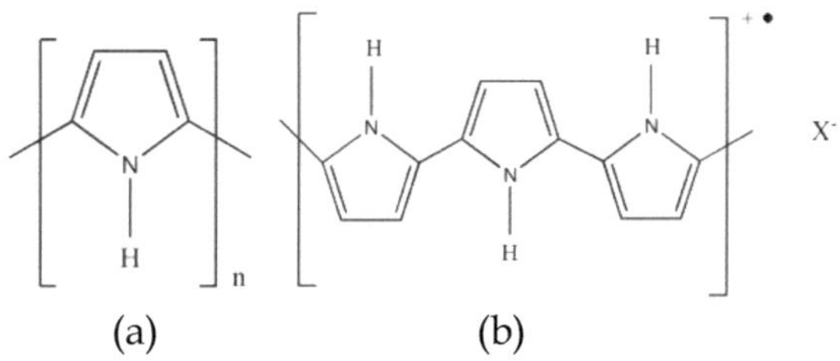

Fig. 1.

Electrochemical film formation is often followed by stoichiometric determination of the number of electrons donated by each molecule. This value is generally found to be in between 2 and 2.7, where 2 electrons serve in the film formation and the excess charge is consumed by the polymer oxidation.

The final step in the polymerization is not clear and different hypotheses have been proposed. Funt and Diaz (1991) believe that the reaction with water could be responsible for the polymerization quench. While, Street (1986) believes that the growth of the chain stops because the radical cation becomes relatively unreactive towards the chain propagation or because the reactive chain ends become sterically blocked.

This article presents an overview of electrochemical synthesis of PPy nanocomposites and their structural and electrochemical properties and applications in electrochemical devices. Applications of nanocomposites described include solar cells, fuel cells, batteries, corrosion protection coatings and super capacitors. Nanocomposites consists of PPy and one or more components, which can be carbon nanotubes, metals, oxide nanomaterials, etc.

2. Synthesis

2.1 Polypyrrole (PPy) films

PPy films are obtained through electropolymerization of Py on suitable substrates/working electrodes by using different electrochemical techniques such as cyclic voltammetry (CV), potentiodynamic, galvanostatic, potentiostatic, reversal potential pulsing technique, etc. To carry out the electropolymerization of Py, a three-electrode one-compartment cell (50 or 100 ml capacity) with provision of passing an inert gas (N_2/Ar) into the electrolyte before the start of the oxidation and above the surface of the electrolyte during the polymerization process is employed. Saturated calomel electrode (SCE) or Ag/AgCl (saturated KCl) is used as the reference electrode. PPy composites prepared were either in the form of sandwich-type or simple films.

Generally, the electroplymerization is carried out on noble metals or inert materials such as Pt, Au, glassy carbon (GC), or indium tin oxide (ITO). It is because of the fact that the standard oxidation potential of pyrrole is fairly high (E_{ox} = 0.70 V vs. SCE) and so, when scanning the potential the dissolution of most of metal supports takes place prior to the oxidation of monomer is reached and, thus, the electropolymerization reaction is inhibited. However, when we are using supports which are oxidizable, for example, Fe, Zn, Al, etc, as the working electrode, it is necessary to find new electrochemical conditions for slowing the dissolution of the working electrode without preventing electropolymerization. Recently, a new electrosynthesis process of PPy films on oxidizable metals Al and Fe has been reported (Bazzaoui et al. 2005). In this process, the formation of homogeneous and strongly adherent

PPy films is achieved in only one step using an aqueous medium containing saccharin and Py. Similarly, Bazzaoui et al. (2006b) obtained PPy coating on Fe by the use of an aqueous solution of 0.1M sodium saccharinate and 0.5M Py.
Various techniques have been employed to prepare composite films of PPy with pure metals, oxides (pure as well as mixed valence ones), anions (simple as well as complex ones), etc. It is difficult to to describe all of them here. However, efforts would be made to include as much as possible those techniques which have repeatedly been used in preparation of different composites.

2.2 Functionalized PPy film

The functionalized PPy films can be obtained in mainly two ways. The first involves the polymerization of a functionalized monomeric molecule. The second technique makes use of the anion exchange properties of the oxidized form of the polymer.
The functionalization of a monomer Py molecule can be carried out by introducing a substituent on the nitrogen atom or on its 3, 4 positions. N-substituted pyrroles are excellent electropolymerizable monomers for the preparation of polymer films containing active centres. A series of Py-substituted pyridine, 2,2′-bipyridine, l,l0-phenanthroline and 2,2′:6-2″-terpyridine ligands and their corresponding Ru(II), Fe(II), Re(I), Cu(I), Cu(II), Co(II), Ni(II), Rh(III), Mn(III), Pd(II), Zn(II), Ag(I) and Ir(III) complexes have been synthesized. In most cases, films of PPy bearing these complexes can be obtained by anodic electropolymerization of the monomer. There are also two types of Py-substituted metallic tetraphenylporphyrins (M=Ni^{+}, Zn^{+}, Co^{+}, Cu^{2+}, Mn^{+2}); one involving the connection of the Py group directly on the phenyl group, the other through a flexible chain. The efficiency of electropolymerization seems to be lower for complexes in which the Py is connected directly to the macrocycle. With this configuration, steric hindrance and mesomeric effects limit the efficiency of the electrochemical polymerization process (Deronzier & Moutet, 1996). Recently, Diab and Schuhmann (2001) synthesized Mn^{+2}-meso-tetracarboxyphenylporphyrin linked via a spacer chain to a Py unit using electropolymerization method. Abrantes et al. (2000) prepared substituted Mn and Fe metalloporphyrins with two Py groups bonded in lateral chains through the electropolymerization under potentiodynamic condition.
The literature suggests that 3-substituted Py molecules would require a less anodic potential for their oxidation and would be more conductive than the N-substituted PPy. However, Py monomers substituted at the 3-position with metallic complexes are limited in literature; only two examples are cited here, the first one involves ferrocene derivatives, and the second one an entwining phenanthroline Cu(I) complex (Deronzier & Moutet, 1996).
The functionalized PPy films can also be obtained by carrying out electropolymerization of Py in a medium containing an anionic species as the supporting electrolyte. Following this procedure PPy has been doped by several metallic tetraphenylporphyrin, tetrasulphonates or carboxylates (Co^{II}, Fe^{II}, Mn^{II}) and metallic phthalocyanine, mono and tetrasulphonates (Co^{II}, Fe^{II}, Cu^{II}). Some other complexes have also been incorporated by this technique. For instance, PPy films containing cobalt salts, thiomolybdate anions, nickel and palladium complexes of maleonitriledithiol, heteropolyanions, a chromium oxalate complex, an iron trisphenanthroline sulphonic acid complex or ferri/ferrocyanide and Prussian blue have been prepared by electropolymerization in a medium containing these anions as supporting electrolyte. All these works has been cited in (Deronzier & Moutet, 1996). It is expected that the polymer frame would be formed around the trapped anion and retains it strongly.

However, such electrode materials are less stable than those in which the substituents are covalently attached to the polymeric skeleton, owing to some exchange of dopant anions with counter-anions of the electrolyte.

2.3 PPy-metal composite films

Several methods have been reported on modification of conducting PPy by metal catalysts (Juttner et al., 2004). They can be put into the following four main categories:

1. Composite layers of metal particles embedded in PPy are formed in a two-step process: electropolymerization of Py followed by an electrochemical deposition of the metal from a solution containing its own ions. Electrodeposition of the metal by this method leads to the formation of metal crystallites on the surface of PPy film, while the amount of the metal tends to zero inside the film. Using this technique PPy/metal composites prepared are: PPy/Pd (Mangold et al., 2004; Chen et al., 2006) PPy/Pt (Vork & Barenrecht, 1989; Bouzek et al., 2000; 2001), PPy/Pt or PPy/Pt+Pb (Del Valle et al., 1998), Pt-based alloy/PPy (Becerik & Kadirgan, 2001).
2. Insertion of colloidal metal particles in the course of electropolymerization of the PPy film. For the purpose, a solution of nanodispersed metal particles is separately prepared by suitable chemical reduction method. To this solution, the polymer monomer is added for the electropolymerization. Both PPy and metal particles are deposited simultaneously. Using this technique PPy/Pt electrocatalysts were obtained by Bose & Rajeshwar (1992) and Del Valle et al. (1998).
3. Insertion of a suitable metal comlex anion as a counterion in the PPy film during the electropolymerization process and subsequent electrochemical reduction of the complex metal anion into the metal. Using this technique PPy/metal composites prepared are: PPy/Pt (Vork &Barenrecht, 1989; Bouzek et al., 2000; 2001) Cu/functionalized PPy (Pournaghi-Azar et al., 1999), PPy/$PtClO_4^{-2}$ (Zhang et al., 2005a), PPy/$[Fe(CN)_6]^{3-}$ (Pournaghi-Azar et al., 2000), PPy/$[Fe(CN)_6]^{4-}$ (Raoof et al., 2004).
   Methods, 2 and 3, results a homogeneous distribution of the metal within the whole PPy film.
4. PPy-metal nanoparticle composites were electro-co-deposited from a solution containing Py, metal salt as source of metal ions, a suitable neutral salt as supporting electrolyte and small amount of a suitable additive to achieve stability of the metal complex salt. For examples, PPy-Au (Rapecki et al., 2010), PPy/Ni (Haseko et al., 2006), PPy/Fe (Chipara et al., 2007), PPy/Co (Ikeda et al., 1983) and Pd/PPy/foam-Ni (Sun et al., 2010) are some examples of nano composites obtained by this technique.

2.4 PPy-Oxide composite films

Sandwich-type composite films of PPy and an oxide (Ox) having composition, PPy/PPy(Ox)/PPy, onto a GC electrode were obtained by a sequential electrodeposition method. For the purpose, two electrolyte solutions (A and B) were used. Solution A contained 0.10M Py and 0.05M K_2SO_4 and that solution B contained 0.10M Py, 0.05M K_2SO_4 and 8.33 g L^{-1} oxide powders. The first layer of PPy (~2.1 μm thick) was obtained onto graphite (G)/GC electrode in solution A by carrying out electrolysis at j (current density) = 5.0 mA cm^{-2} for 100 s under unstirred condition. After electrolysis, G/PPy electrode was removed from the cell, washed with distilled water, dried in air and then introduced into the cell containing solution B and electrolysis was again carried out at j = 20 mA cm^{-2} for 200 s under stirred condition so as to obtain a second layer of PPy(Ox) (~16 µm thick). The polymer coated G electrode [G/PPy/PPy(Ox)], so obtained, was washed with distilled water, dried and reintroduced in the former cell containing solution A to electrodeposit the third and final layer of PPy (~4.2 µm thick); electrolysis condition being j = 5.0 mA cm^{-2}, t=200 s, unstirred. Prior to electrodeposition of PPy, the electrolytes (solution A & B) were degassed for 45 min by bubbling Ar and maintained under Ar atmosphere during the electrodeposition process also.

Following similar method several sandwich-type binary composite electrodes of PPy and a mixed oxide belonging to both spinel and perovskite families, namely $CoFe_2O_4$ (Singh et al., 2004; Malviya et al., 2005a & b), $Cu_xMn_{3-x}O_4$ (x = 1.0 – 1.4) (Nguyen Cong et al., 2005; 2002a; 2000), $Ni_xCo_{3-x}O_4$ (x = 0.3 & 1.0) (Nguyen Cong et al., 2002b; 2003, Gautier et al., 2002), $La_{1-x}Sr_xCoO_3$ ($0 \le x \le 0.4$) (Singh et al., 2007a), $LaNiO_3$ (Singh et al., 2007b) and $La_{1-x}Sr_xMnO_3$ ($0 \le x \le 0.4$) (Singh et al., 2007c) were prepared.

2.5 Other composites

Preparation of PPy/Chitosan: PPy/Chitosan composite films have been electrochemically synthesized on the Pt electrode from an aqueous solution containing 4.0 mg ml^{-1} chitosan, 0.3M oxalic acid and 5 mmol Py (as monomer) by using cyclic voltammetry method (Yalcinkaya, 2010). A standard 3-electrode cell containing two platinum sheets for use as the counter and working electrodes and an Ag/AgCl (saturated with KCl) electrode as the reference electrode was employed to carry out the electrolysis. The potential was scanned from – 0.60 to + 0.90 V at scan rate of 50 mV s^{-1}. Chitosan is a natural polymer and exhibits characteristic properties, such as chemical inertness, high mechanical strength, biodegradability, biocompatibility, high quality film forming properties and low cost (Yalcinkaya et al., 2010). The proposed structure of the composite is shown in Fig.2.

Preparation of PPy/PANI: 0.02M Py was polymerized in 0.25M H_2SO_4 solution at 1.0 V under inert atmosphere for 30 min. The electrode was washed with distilled water and dried at 60°C. 0.1M pure ANI (aniline) was then electrochemically polymerized in 0.25M H_2SO_4 solution at 0.8 V using PPy coated electrode as working electrode under inert atmosphere for 30 min (Hacaloglu et al., 2009). Similarly, the PANI/PPy composite was prepared. In this case first of all, 0.1M pure ANI solution was polymerized. The polymer films obtained were washed with distilled water several times to remove unreacted monomer as well as the electrolyte and subsequently dried in vacuum.

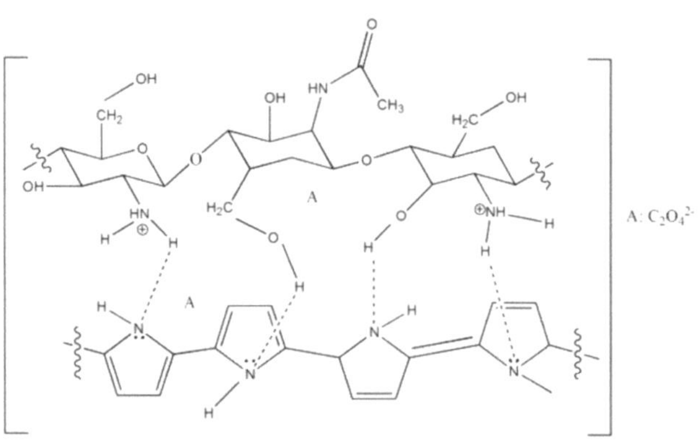

Fig. 2. Structural representation of the PPy/Chitosan composite

3. Characterization

3.1 Thermal Gravimetric Analysis (TGA)

TGA is frequently used to quantify the amount and thermal stability of PPy in the composite and also to know whether there occurs some interaction between PPy and the other constituent of the composite. The TGA curve of the electrochemically synthesized PPy/chitosan composite on Pt electrode in air showed two stages of the weight loss (Yalcinkaya et al., 2010). The first stage of the thermal degradation was observed at 150-200°C and was attributed to removal of the dopant molecule (oxalate ion) from the polymer structure. The decomposition of chitosan chain was indicated by minimum weight loss in temperature range 300-370°C. On the other hand, the maximum weight loss observed at 380-400°C was attributed to the degradation and intrerchain crosslink of the composite. The comparison of results of the PPy/chitosan composite with those obtained from chitosan suggests that an interaction occurs between chitosan and PPy.

Similarly, The TGA analysis of SnO_2-PPy nanocomposite from 25 to 700°C exhibited two weight losses (Cui et al., 2011). The first weight loss, in the temperature range of 25-250°C, attributed to desorption of physisorbed water, while the second in the range of 250-700°C, attributed to the oxidation of PPy. Bare SnO_2 does not show any weight change in the whole temperature range of the investigation, while the pure PPy is burnt off. So, on the basis of weight change before and after the oxidation of PPy, the SnO_2 content in the composite was estimated. TGA analyses of the S/PPy composite, bare S and PPy powder indicated (Wang et al, 2006) that S is burnt completely on temperature up to 340°C, followed by the oxidation of PPy in the second stage on temperature above 340°C. The weight loss in the second stage was about 40wt% which represents the amount of PPy in the S-PPy composite. TGA measurements were performed in air. The nanocomposites, SnO_2-PPy and S/PPy were prepared by chemical polymerization.

3.2 Scanning Electron Microscopy (SEM)

The morphology of the composite strongly depends upon the nature as well as the method of preparation. From the SEM analyses of pure PPy and its composites it was observed that morphology of the PPy changed significantly when the composite was formed. As, in the case of the PPy/chitosen composite, the SEM image of composite was significantly different compared to that of PPy (cauliflower-like spherical shape) or chitosan (smooth surface) (Yalcinkaya et al, 2010). On the other hand, HRTEM (High Resolution Transmission Electron Microscope) images of pure PPy and graphene nano sheet (GNS)/PPy composite obtained by chemical method (Zhang et.al., 2011) showed that pure PPy has the amorphous structure, while the PPy is homogenously surrounded by GNS in the composite. The particle size of PPy/GNS was found to be smaller than pure PPy.

3.3 X-Ray Diffraction (XRD)

The XRD pattern of GNS/PPy exhibited diffraction peaks at $2\theta \approx 24.5°$, 26° and 42.8°. The diffraction peaks at $2\theta \approx 24.5°$ and 42.8° correspond to (002) and (100) planes of graphite like structure while that the peak at $2\theta \approx 26°$ corresponds to amorphous PPy (Zhang et al., 2011). In the GNS/PPy composite, as GNS percentage increased, the broad peak shifted from $2\theta \approx 26°$ to 24.8°, implying that interaction occurs between GNS and PPy. The Au/PPy core-shell nanocomposites (Liu and Chuang, 2003) displayed a broad maximum at $2\theta \approx 25.1°$ (d = 3.5 Å) which was ascribed to the closest distance of approach of the planar aromatic rings of Py like face to face Py rings. However, the XRD of pure PPy film electrodeposited on Au substrate showed the broad maximum at the lower angle, i.e. at $2\theta \approx 19.0^o$ (d = 4.7Å). This may be caused due to scattering from side-by-side Py rings. The estimated values of the coherence length from the Scherrer equation were 11.03 and 9.73Å for Au/PPy nanocomposite and electrodeposited PPy on Au, respectively. The higher coherence value obtained for Au/PPy nanocomposite clearly indicates increased crystallinity and crystalline coherence, which would contribute to the higher conductivity of PPy.

The pure PPy exhibited a broad diffraction peak at $2\theta \approx 24.6°$, which is characteristic peak of amorphous like PPy, while the diffraction pattern of PPy/Y_2O_3 (30wt% oxide) composite was the same as Y_2O_3 and PPy. It indicated that PPy deposited on the surface of Y_2O_3 particles has a little effect on crystallization performance of Y_2O_3 (Vishnuvardhan et al., 2006).

3.4 Infra Red (IR)/Raman/UV-visible absorption spectroscopy

Zhang et al., (2011) recorded the FTIR of pure PPy and PPy/GNS composite and observed that most of peaks, which correspond to PPy, get shifted towards the left when GNS was introduced. This was considered to an association of graphene to the nitrogenous functional group of PPy backbone. The FTIR spectra of PPy-Ag composite showed (Ayad et al., 2009) peaks at 1560 and 1475 cm^{-1}, which correspond to the C-C and C-N stretching vibrations of PPy ring, respectively. The peak position of the composite showed a little shift towards the higher wave number compared to pure PPy. It was probably due to interaction between PPy and Ag particle and improvement of doping level of the polymer.

Raman spectra of PPy/GNS composite exhibited two prominent peaks, at ~1590 (G band) and 1350 (D band) cm^{-1} which get broadened with increasing amount of GNS in the composite suggesting that the nanocrystalite size decreases due to the phonon confinement (Zhang et al., 2011). A broad peak at 1051 cm^{-1} and two small peaks at 933 and 981 cm^{-1} observed for pure PPy and PPy/GNS composite are the characteristic peaks of PPy. In the case of S-PPy composite, peak at ~500 cm^{-1} corresponds to the S particle and the peak between 800 and 1700 cm^{-1} corresponds to the PPy nanoparticle suggesting that S-PPy contain both the sulphur (S) and conductive PPy elements (Wang et al., 2006). The Raman spectra of MWCNT exhibited the bands at ~1350 cm^{-1} (D band) and ~1580 cm^{-1} (G band), (Fang et. al., 2010). The ratio of D to G band is associated with extent of defect present in the MWCNT and is sensitive to molecular interaction. After PPy deposition D/G ratio decreased which was prominent in short-pulse deposition than that of continuous deposition method. This decrease in D/G ratio is associated with reduction of disorder at the MWCNT surface. Further, Raman spectra of PPy/CNT composite did not produce any additional peaks except the characteristic peaks of CNT and PPy (Kim et al., 2008a). This suggestes that no new chemical bond is formed between CNT and PPy in composite and no chemical change in PPy composite occurs.

UV-visible absorption spectra of the composite PPy-CdS with different CdS contents clearly indicated two absorption bands at 300-400 nm and 700-1000 nm. The former absorption band was assigned to the π-π^* transition and the latter, to the oxidation state of the film. There occurs an increase in absorption of the composite material with CdS concentration, which implies that electronic structure of PPy is affected by CdS (Madani et al., 2011). Sharifirad et al. (2010) prepared PPy on Cu in three aqueous media (citric acid: CA, sodium acetate: SA and sodium benzoate: SB) and recorded their UV-visible specta. The UV spectra of PPy/Cu in CA showed two absorption peaks at 295 and 460 nm, assigned to the π-π^*

transition and to the formation of perninfreaniline form (PB) and the spectra of PPy/Cu synthesized in SA and SB indicated a shoulder respectively at 360 and 350 nm, which was assigned to the formation of polybenzoic salt. $AuCl_4^-$ nano complex showed a band at ~308 nm which disappeared after addition of Py monomer and a new band of π-π* transition of PPy appeared in the region 400-500 nm with absorption maximum at 463 nm, showing the formation of Au/PPy nanocomposite with core-shell structure (Liu & Chuang, 2003). The UV-visible absorption spectrum of pure PPy indicated (Konwer et al, 2011) a weaker absorption at 330 nm (π-π* transition) and stronger absorption at 570 nm (bipolar state of PPy). The PPy/EG (expanded graphite) composite showed a red shift in UV visible spectrum suggesting that EG assists the polymerization in such a way that it maintains a higher conjugation length in the chain of the PPy and there may be possible some coupling between conjugation length of PPy and EG. Estimate of the optical band gap of PPy found by using equation, E_g^{opt} (ev) = 1240/λ_{edge}(nm), was 2.17 eV, while it decreased to 1.85-1.93 eV in PPy/EG with addition of EG.

3.5 Electrochemical impedance spectroscopy (EIS)

A conductive polymer electrode is a porous material. Various equivalent circuits have been applied to present impedance behavior of the conducting polymer electrodes (Passiniemi & Vakiparta, 1995). But till now, no general equivalent circuit that can satisfy all electrolyte-conducting polymer configurations has been reported. In literature, (Levi & Aurabch, 1997, 2004; Lang & Inzelt, 1999; Mohamedi et al., 2001) different equivalent circuit models have been used to treat the experimental data and some of them are given in Fig.3; wherein subscripts 'sl' and 'ct' represent the surface layer and the charge transfer and symbols: Z_w, C_{int} and Z_{FLW} and Z_{FSW} represent Warburg impedance, capacitor due to intercalation and finite length and finite space Warburg type elements, respectively; other symbols have their usual meaning.

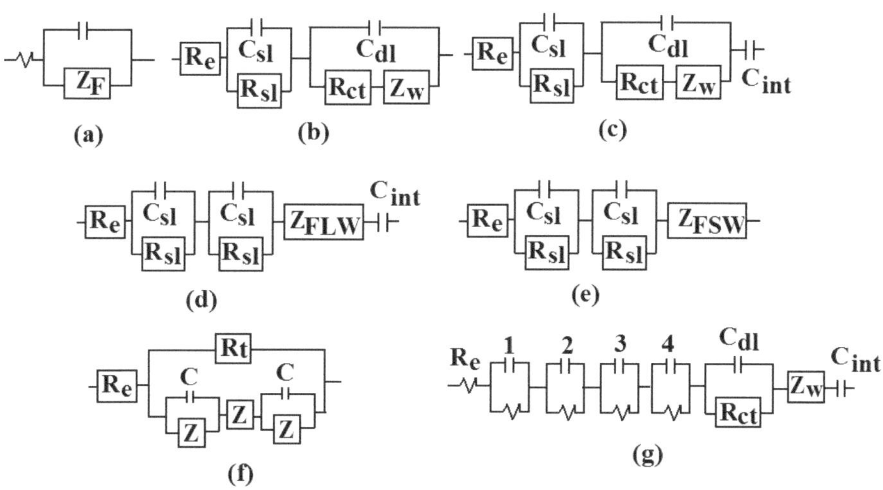

Fig. 3. Equivalents circuit (EC) models: (a)Randle`s EC model; (b-d) Modified Randle`s EC (analogs for ion-intercalated electrodes); (e) Meyers EC Model (analog for mixed particle porous electrode); (f) EC model of polymer film electrodes; (g) Voigt and Frumkin and Melik-Gaykazyan (FMG) model (for interfacial boundaries of lithiated graphite electrode).

The EIS spectra (i.e. Nyquist plots) of several composites such as S/T-PPy (Tubular PPy) and S/G-PPy (Granular PPy) (Liang et al., 2010), PEDOT/h-PPy (Poly 3,4-ethylenedioxythiophene/horn like-PPy) (5:1) (Wang et al., 2007), PPy films on DuPont 7012 carbon composite (Li et al., 2005), S-PPy(Wang et al., 2006) and Si/PPy (9:1) (Guo et al. 2005) reported in literature displayed a small semi-circle at high frequencies and a straight line at lower frequencies. The observed linearity at low frequencies has been ascribed to the anion diffusion in the composite matrix and the semicircle in the high frequency region, to the contact resistance and charge transfer resistance. Only in the case of PEDOT/h-PPy, a line indicating a capacitive behavior related to the film charging mechanism at low frequencies was observed.
The EIS study of sandwich-type composite film electrodes of PPy and a mixed valence oxide have generally produced more or less similar Nyquist plots. Each Nyquist curve indicated a capacitive behavior at low frequencies, a small semicircle (or an arc) at high frequencies and more or less a linear curve at intermediate frequencies. The observed semicircle at high frequencies is considered to be produced due to contribution of the charge transfer reaction (Singh et al., 2004). In some cases, no semicircle corresponding to the charge transfer reaction is observed (Malviya et al., 2005; Singh et al., 2007a); in these cases the charge transfer resistance is thought to be much smaller than the sum of the film and solution resistances. In fact, at higher frequencies, the diffusion of anions (ClO_4^-, Cl^- or SO_4^{2-}) into the polymer matrix dominates the impedance results and that a conventional semi-infinite Warburg response is observed. At relatively lower frequencies, the $\Delta Z_{im}/\Delta Z_{real}$ value increases with decreasing frequency and that the impedance response results a capacitive-type behavior. Thus, a change over seems to take place from the semi-infinite Warburg regime to the finite Warburg regime with the variation of frequency from high (100 KHz) to to a low value (20 mHz).
The spectra obtained were analyzed by fitting the equivalent circuits, $R_s(RQ)Q'$, $R_s(R_1Q_1)(R_2C_2)(R_3C_3)C_4W_4$, R_sQWC, LR(RQ)(RQ)Q and LR(RQ)(RQ), where R_s is solution resistance, Q is constant phase elemnt, W is Warburg resistance and C is capacitance.The apparent diffusion coefficient (D_a) for transport of anions particularly SO_4^{2-} in the polymer matrix were also estimated at varying potentials. Estimates of D_a values for SO_4^{2-} anions at a constant potential, E = -0.1 V vs SCE in deoxygenated 0.5M K_2SO_4 + 5mM KOH at 25°C were 1.1, 12.1, 2.5, 0.7, 0.60 × 10^{-7} cm^2 s^{-1} in binary composites of PPy and $LaNiO_3$, $LaMnO_3$, $La_{0.8}Sr_{0.2}MnO_3$, $La_{0.7}Sr_{0.3}MnO_3$ and $La_{0.6}Sr_{0.4}MnO_3$, respectively.

3.6 Cyclic Voltammetry (CV)

During the CV, the PPy film electrode-electrolyte system involves the redox couple, PPy^+/ PPy (Kaplin & Qutubuddin, 1995), where PPy is the neutral species (associated with three to four monomer units) and PPy^+ is the radical cationic species or polaron (one positive charge localized over three to four monomer units). In the oxidized state, the PPy is positively charged and so, some anions get migrated from solution into the polymer matrix so as to maintain electroneutrality. Similarly, on reduction, as electrons are injected and the positive charge on the polymer chain disappears, some anions leave the PPy matrix and enter into the solution. The formation of a second redox couple, PPy^{++}/ PPy^+, has also been proposed (Genies and Pernaut, 1985) to explain two reduction peaks observed in some CVs. PPy^{++} is the dicationic species or bipolaron. The counterion concentration will be greater for a PPy film containing both polaron and bipolaron species than for a film containing only polaron species (Kaplin and Qutubuddin, 1995).

The preparation method strongly influences the electrochemical reaction activity of PPy. CVs of PPy films deposited on GC by using three different methods, constant current, constant potential electrolysis and CV, at the scan rate of 100 mV/s in the potential region from -0.20 to 1.02 V vs. Ag/AgCl in 0.1M PBS (Phosphate buffer saline) have shown that the galvanostatically deposited polymer film had the greatest electrochemical activity than that of potentiostatically deposited film. A triple lyer of PPy films on the GC electrode, obtained by sequential electrodeposition method, has the worst redox capacity (Lee et al., 2005).

The charge transport properties in the PPy matrix depends upon the nature and concentration of the doping anions. The charge transport properties of a triple lyer of PPy films on the GC electrode, obtained by sequential electrodeposition method, were examined by CV under Ar atmosphere in Py-free 0.15M KA (A = Cl^-, PF_6^-, ClO_4^-, NO_3^-, SO_4^{2-}) solution at a scan rate of 10 mVs^{-1} from -0.8 to +0.8 V/SCE (Nguyen Cong et al., 2005). The changes in shape and characteristics of CVs were observed with doping anions. For instance, the increase of the anodic peak potential (E_{pa}) in sequence Cl^- (150 mV/SCE) < ClO_4^- (380 mV/SCE) < PF_6^- (440 mV/SCE < NO_3^- (490 mV/SCE) < SO_4^- (520 mV/SCE) is accompanied by a drop in the peak current intensities. In fact, in its oxidized state the PPy incorporates the doping anions in order to neutralize the positive charges (dications, i.e. bipolaron) created on its backbone by the electropolymerization process. The switching between the oxidized and reduced states implies the deintercalation/intercalation of these anions, whose ability to move depends on their nature. This explains the changes in shape of the CVs.

Recently, CVs of sandwich-type electrodeposited composite films of PPy and $CoFe_2O_4$ on graphite (G) having structure G/PPy/(PPy($CoFe_2O_4$)/PPy in Ar-deoxygenated 0.5 M KOH containing K_2SO_4 as a dopant anion indicated a broad anodic and a broad cathodic maxima corresponding to establishment of the redox couple,[PPy^+. SO_4^{2-}]/[PPy + SO_4^{2-}] (Singh et al., 2004), transport of doping anion being a diffusion controlled one. Features of CVs of pure PPy and composite electrodes were similar. Similar results were also found from the study of CVs of similar sandwich type composites of PPy with $LaNiO_3$ (Singh et al., 2007b), $La_{1-x}Sr_xMnO_3$ ($0 \leq x \leq 0.4$) (Singh et al., 2007a), $Cu_{1.4}Mn_{1.6}O_4$ (Nguyen Cong et al., 2002a), $NixCo_{3-x}O_4$ (x = 0.3 and 1.0) (Nguyen Cong et al., 2002b).

Very recently, CVs of nanocomposite systems, MnO_2/PPy, SO_4^{2-} (Sharma et al, 2008), PPy_{Cl} (PPy films doped with Cl^-), Cl^- and PPy_{ClO4} (PPy films doped with ClO_4^-), ClO_4^- (Sun et al., 2009) and PEDOT/c-PPy, ClO_4^- and PEDOT/h-PPy, ClO_4^- (Wang et al., 2007) have shown that the specific capacitance of composites are greatly enhanced in comparison with those obtained for their respective constituents elements/materials. Further, the CV curves of PEDOT/h-PPy and PPy doped with ClO_4^- have rectangle like (i.e. ideal) shapes while CV curves for MnO_2/PPy, PPy doped with Cl^- and PEDOT/c-PPy are non ideal.

4. Applications

4.1 Corrosion protection coatings

Conductive polymers are presently being considered as potential materials for corrosion protection of metals. They can substitute conventional protection materials such as chromates and phosphates used in the electroplating and paint industries with stronger more resistant and environmentally friendly coatings. The traditional anticorrosion materials provide excellent corrosion protection coatings but their toxicity has been severely questioned.

A conductive organic coating forms a physical barrier against corrosive agents and a passive layer on the metal surface. The corrosion protection efficiency of this organic coating can be greatly improved by introducing organic or inorganic materials embedded within the polymer structure by electropolymerization. It is reported that the composites with micron- and submicron ceramic particles such as TiO_2 (Ferreira et al., 2001), WO_3 (Yoneyama et al., 1990), Fe_3O_4 (Garcia et al., 2002), MnO_2 (Yoneyama et al., 1991) and $Zn_3(PO_4)_2$ (Lenz et al., 2007) with PPy improved the mechanical and corrosion resistance of the coatings.

Electrochemically polymer (PANI+PPy) coated stainless steel plates showed improved corrosion resistance with acceptable contact resistance under proton-exchange membrane fuel cell (PEMFC) condition (Joseph et al., 2005). Recently, it has been demonstrated (Ren & Zeng, 2008) that a bilayer conducting polymer coating, composed of an inner layer of PPy with large dodecylsulfate ionic groups and an external PANI layer with small SO_4^{2-} groups, reduced the corrosion of the type 304 stainless steel, used for bipolar plates of a PEMFC, much more effectively than the single PPy coatings in 0.3M HCl. Besides these, some PPy-based composites such as PPy-CMC (carboximethylcellulose), PPy-SDS (sodium dodecylsulfate), PPy-TiNT (Titanate nanotube) and PPy-zinc phosphate were observed to protect steel (Herrasti & Ocoan, 2001), 1Cr18Ni9Ti stainless steel (Zhang &Zeng, 2005), stainless steel type 904L (Herrasti et al., 2011), and AISI 1010 steel (Lenz et al., 2007) surfaces effectively from corrosion, respectively. Electrochemically prepared PPy-WO_4^{2-} (Sabouri et al., 2009), PPy/PANI (Panah & Danaee, 2010) and Pt/PPy (Rahman, 2011) composite films provided noticeable corrosion inhibition for carbon steel, whereas PPy-TiO_2 (Lenz et al., 2003), PPhe (polyphenol)/PPy (Tuken et al., 2004), PPy-P (Hosseini et al., 2007), PPy-DGEBA (Riaz et al., 2007), PPy-$PMo_{12}O_4^{3-}$ (Kowalski et al., 2008), and PPy-PANAP (poly(5-amino-1-naphthol)) (Bereket & Hur, 2009) provided better protection for corrosion of mild steel.

PPy-based coatings were also found to reduce the corrosion of oxidizable metals such as Fe (Bazzaoui et al., 2006; Bazzaoui et al., 2005; Lee et al., 2005), Al (Bazzaoui et al., 2005; Lehr & Saidman, 2006a; 2006b)and Al alloys(Lehr & Saidman, 2006a; 2006b). The efficiency of corrosion protection of PPy depends on the nature of the doping agents (Balaskas et al., 2011).

Electrochemically deposited PPy films on Cu displayed good protection against Cu corrosion in a 3.5% NaCl solution (Herrasti et al., 2007). The effectiveness of the protection is enhanced when the PPy film is electrosynthesized from a solution of dihydrogen phosphate (Redondo & Breslin, 2007) or sodium saccharinate (Bazzaoui et al., 2007).

PPy films electrodeposited onto Ni-Ti alloy employing sodium bis(2-ethylhexyl) sulfosuccinate (Aerosol OT or AOT) solutions improved the corrosion performance of the alloy at the open circuit potential and at potentials where the bare substrate suffers pitting attack (Flamini & Saidman 2010). PPy coatings on Mg alloy AZ91D, obtained respectively from aqueous solutions of a dicarboxylic organic acid salt (Turhan et al., 2011a) and sodiumsalicylate (Turhan et al., 2011b) by cyclic voltammetry (CV) method, demonstrated good corrosion protections of the alloy.

4.2 Fuel cell

In a fuel cell, the fuel (hydrogen, natural gas, methanol, ethanol, etc.) is electrochemically oxidized at the anode, whereas the oxidant (oxygen from the air) is reduced at the cathode. Because of difficulties involved in the production, storage, and distribution of hydrogen, the use of alcohols (methanol, ethanol, etc.) as hydrogen carrier is preferred and the resulting cell is called as the direct alcohol fuel cell (DAFC). Thus, the overall efficiency of a DAFC depends upon the efficiencies of both the electrode reactions (e.g. alcohol oxidation and oxygen reduction). The mechanism of oxygen reduction reaction (ORR) is very complex. The most accepted mechanisms in acid media are, the direct four electron pathway ($O_2 + 4H^+ + 4e^- \leftrightarrow 2H_2O$, $E^o = 1.229\ V_{SHE}$) and a series two-electron pathway ($O_2 + 2H^+ + 2e^- \leftrightarrow H_2O_2 + 2H^+ + 2e^- \leftrightarrow 2H_2O$). The ORR and Methanol/Ethanol oxidation reaction (MOR/EOR) are traditionally catalyzed by Pt or Pt-based alloys dispersed on high surface area carbon support materials.

PPy has been investigated as carbon-substitute supports for fuel cell catalysts. Polymer-supported metal particles present the higher specific surface area and the higher tolerance to poisoning due to the adsorption of CO species, in comparison to the serious problem of poisoning of bulk and carbon supported metals. Moreover conducting polymers are not only electron conducting, but also proton conducting materials, so they can replace Nafion in the catalyst layer of the fuel cell electrode and provide enhanced performance also.

4.2.1 PPy-based anodes

Several PPy-based anodes for fuel cells have recently been prepared and investigated for their applications in DAFCs. Most of them are decribed briefly in following lines.

In the MOR, Pt nano-particles decorated PPy-MWCNT$_S$ composite electrodes showed higher catalytic stability than Pt/MWCNTs binary catalyst, due to the synergic interaction between PPy and the carrier (Qu et al, 2010). Zhao et al. (2008a) demonstrated that the Pt nano-particles deposited on PPy-C with naphthalene sulfonic acid as dopant exhibit better catalytic activity than those on plane carbon in fuel cells. They also reported (Zhao et al., 2009) that bimetallic Pt-Co nano- particles co-deposited on PPy-MWCNT composite via over-oxidation treatment had higher catalytic activity towards methanol oxidation. Further, the MOR was observed to improve on the Pt-Fe/PPy-C catalyst compared to commercial Pt/C catalyst (Zhao et al., 2008b).

Mohana Reddy et al. (2008) examined the suitability of the cobalt-polymer-MWCNT composite electrode for the ORR in DMFCs and DEFCs by using Pt-Ru/MWCNT and Pt-Sn/MWCNT, respectively, as anode electrocatalysts. The study indicated an improved power densities for hydrogen, methanol and ethanol based fuel cells compared to the previously non Pt based electro-catlysts. Hammache et al. (2001) observed a higher catalytic activity for MOR on the dispersed gold micro particle on PPy coatings on Fe than a bare gold electrode in acidic media.

The application of PPy film containing nanometer-sized Pt and Pt/Pd bimetallic particles on ITO glass plates has also been investigated as anode for MOR. The modified electrode was found to exhibit significant electrocatalytic activity. The enhanced electrocatalytic activities may be due to the uniform dispersion of nanoparticles in the PPy film and a synergistic effect of the highlydispersed metal particles so that the PPy film reduces electrode poisoning by adsorbed CO species (Selvaraj et al. 2006). A Pd/PPy/Pd/G generated by sequential electrodeposition method (Ding et al., 2011) showed a satisfactory electrocatalysis toward the formic acid oxidation.

4.2.2 Oxygen Reduction Reaction (ORR)

As early as in 1983, Bull et al. observed that Fe tetrasulfonated phthalocyanines-doped PPy film on GC electrode catalyzes the reduction of O_2 at potentials 250-800 mV less negative than at bare GC or at PPy-coated GC electrode. Co-doped PPy films on metal electrodes also showed the electrocatalytic activity for O_2 reduction (Ikeda et al., 1983). Osaka et. al. (1984)

studied the ORR at the PPy film electrode containing Co(III) tetrakis(sulfophenyl)porphine and observed a 4 electron reduction of O_2 to H_2O in H_2SO_4, contrary to the 2 electron reduction which usually occurs in this system in homogeneous state. The ORR at the GC/PPy/CoTSP ((tetrasulfonatopthalocyaninato)cobalt) electrode in 0.05 M H_2SO_4 showed apprx. 0.5 V more anodic onset potential than the value obtained with a GC electrode in 0.05 M H_2SO_4 containing 10^{-3} M CoTSP (Osaka et al., 1986). The electrodeposited Pt particles on PPy films were also active for the ORR (Vork and Barendrecht, 1989). PPy films containing nanometer-sized Pt particles (PPy/Pt), electrosynthesized from a solution containing Py and colloidal Pt particles, exhibited high catalytic activity towards the ORR (Bose & Rajeshwar, 1992).The ORR study has also been carried out on PPy films doped with anionic Co-species (Seeliger & Hamnett, 1992), tungstophosphate anions (Dong & Liu, 1994), Fe/Co phthalocyanines (Coutanceau et al., 1995), ferriporphyrin (FeTPPS) (Wu et al., 1999), meso-tetra(4-sulfonatophenyl)porphine (TPPS4) (Johanson et al., 2005), and anthraquinone-2,7-disulfonate (AQDS) (Zhang et al., 2007). These modified electrodes exhibited good activity towards the ORR. The PPy-cobalt complex-modified C particles displayed electrocatalytic activity for four-electron reduction of O_2 and the catalyst showed high stability against degradation after use for several hours (Yuasa et al., 2005). The catalytic reduction of molecolar oxygen on the PPy-Mn phthalocyanine film (GC/ITO-coated glass support) indicated the participation of the Mn center of the PPy in the reduction of molecular oxygen (Rodrigues et al., 2005). The ORR displayed a pathway of irreversible 2-electron reduction to form H_2O_2 on the PPy/AQDS (anthraquinonedisulfonate) composite film on GC at all electrolyte pH employed, the pH 6.0 buffer solution being a more suitable medium for the reduction of dioxygen (Zhang & Yang, 2007).

Recently, it has shown (Ding & Cheng, 2009) that electrochemically produced MnO_2-PPy composite material is ORR active in 0.5M H_2SO_4 when compared to the pure PPy. Template-synthesized Co porphyrin/PPy nanocomposite in a neutral medium catalyzes the ORR mainly through a 4-electron pathway, exhibiting excellent electrocatalytic activity (Zhou et al., 2007). Also, Pd-PPy/C nanocomposite efficiently catalyzes reduction of oxygen, with resistance to methanol oxidation, directly to water through a four-electron pathway (Jeyabharathi et al., 2010). C-supported CoPPy composite material exhibits the Tafel slopes, -110 -120 mV (Millan et al., 2009). Very recently, Zhao et al. (2011) observed an enhanced electrocatalytic performance for the ORR on the AQS (anthraquinone-2-sulfonate)/PPy composite modified graphite electrode.

Investigations on the ORR at sandwich composite electrodes of PPy and $Cu_{1.4}Mn_{1.6}O_4$ (Nguyen Cong et al., 2000; 2002a), PPy and $Ni_xCo_{3-x}O_4$ (Gautier et al., 2002; Nguyen Cong et al. 2002b), PPy and $CoFe_2O_4$ (Singh et al., 2004), PPy and $LaNiO_3$ (Singh et al., 2007a), PPy and $La_{1-x}Sr_xMnO_3$ ($0 \le x \le 0.4$) (Singh et al., 2007b) and $La_{1-x}Sr_xCoO_3$ (Singh et al., 2007c) were also carried out. The results have shown that the composite electrode had excellent catalytic activities as well as remarkable stability even in acid solutions wherin mixed oxide cathodes normally undergo deactivation.

4.3 Rechargeable batteries

Among the various types of rechargeable batteries, the Li/S battery system is a very attractive candidate for rechargeable Li-batteries due to its high theoretical specific capacity of 1672 mAhg^{-1} and theoretical power density of 2600 Wh kg^{-1} based on S active materials. The use of S as a cathode material is advantageous because of its abundance, low cost, and environmental friendliness. As S is an insulating material, the cathode material is combined with 30-55% carbon black (Jin et al., 2003; Song et al., 2004) of the total weight of the electrode materials. Recently, PPy has been used as an additive to improve the performance of anode and cathode materials in Li-ion batteries (Guo et al., 2005; Pasquier et al., 1999; Veeraraghavan et al., 2002; Wang et al., 2006).

Very recently, Liang et al., (2010) observed that the morphology of the PPy combined with S shows a significant effect on the dispersion status and electrochemical behaviour of S. When S was highly combined with two types of PPy, G-PPy and T-PPy, by in-situ oxidation and co-heating methods, the cycle durability of the composite was more favourable with the T-PPy matrix in comparison with G-PPy matrix. Guo et al. (2005), for the first time, prepared a series of novel Si/PPy composites by high-energy mechanical milling techniques. These anode materials had high capacities characteristic of the Li-Si alloy sytem but substantially improved cyclability compared to bare Si anodes.

Metal oxide powders such as $LiCoO_2$ (Ohzuku & Ueda, 1994), $LiMn_2O_4$ (Tarascon et al., 1994), $LiNiO_2$ (Kanno et al., 1994), and V_2O_5 (Leroux et al, 1996) have been considered as promising candidates for positive electrode materials in rechargeable Li batteries having high energy densities. In order to prepare these oxide electrodes, a conducting matrix and a binder are usually mixed with the oxide powder. PPy can serve as both as the conductor and binder. In view of this, composite electrodes of $LiMn_2O_4$ and PPy were prepared and studied for the charge-discharge properties of electrodes for 3V class (Gemeay et al., 1995) and also 4V class (Kuwabata et al., 1999) of Li-batteries. Results show that the PPy works well as a conducting matrix for the redox reaction of $LiMn_2O_4 \leftrightarrow Li_{1-x}Mn_2O_4 + x\,Li^+ + x\,e^-$ as well as like a capacitor and contributed to the capacity of the $LiMn_2O_4$/PPy composite.

Recently, $LiFePO_4$ has emerged as an important cathode material for lithium-ion batteries due to its high theoretical capacity (170 mAh/g), high potential (3.4 V versus Li/Li+), low cost, natural abundance and environmental friendliness of Fe. Bare $LiFePO_4$ is an insulator with an electrical conductivity of about ~10^{-11} S/cm. To improve the electrical conductivity efforts have been made. Since PPy is a conductive polymer and also has lithium storage capacity in lithium-ion cells, a coating of PPy on the $LiFePO_4$ particles would increase the electrical conductivity of the $LiFePO_4$. With this idea, Wang et al (2005) prepared a series of PPy-$LiFePO_4$ composite materials. The PPy-$LiFePO_4$ composite electrodes demonstrated an increased reversible capacity and better cyclability, compared to the bare $LiFePO_4$ electrode. The thin PPy film has also been used to make Li-batteries lighter and more flexible than the existing ones for portable electronic equipments. Wang et al. (2008) prepared highly flexible, paper-like, free-standing PPy and PPy-$LiFePO_4$ composite film electrodes and observed that the cell with PPy-$LiFePO_4$ composite film had a higher discharge capacity beyond 50 cycles (80 mAh/g) than that of the cell with pure PPy (60 mAh/g).

To ensure long cycle life and safety, the use of graphite, among carbon materials, as anode for the Li-ion battery is favoured, however carbonaceous anodes exhibit capacity loss during the first intercalanation step. To minimize the capacity loss the graphite electrode was coated by a thin PPy film, the later is found to decrease the initial E_{rev} capacity loss of the graphite anode (Veeraraghavan et al., 2002). The decrease in the E_{rev} capacity loss has been ascribed to the reduction in the thickness of the solid electrolyte interface (SEI) layer. PPy/C (7.8%) gives the optimum performance based on the E_{rev} capacity loss and the discharge capacity of the composite.

The research work on the use of PPy as electrode material of the aqueous based power sources has recently been also commenced. Grgur et al. (2008) obtained PPy thin film on graphite electrode in 0.1M HCl galvanostatically and characterized as cathode material for the aqueous based rechargeable zinc batteries. Results have shown that Zn/PPy cell have potentially promising characteristic.

4.4 Supercapacitor

PPy offers a greater degree of flexibility in electrochemical processing than most conducting polymers, and consequently the material has been the subject of much research as a super capacitor or battery electrode (Snook et al., 2011). Due to greater density, polymer had a high capacitance per unit volume (400-500 Fcm^{-3}). Combining PPy with polyimide (a dopant of high molecular weight) is claimed to improve the charge storage property (Iroh & Levine, 2003), due to the polyimide matrix protecting the PPy from oxidative degradation. Polyimide is cathodically electroactive, whereas PPy is anodically electroactive.

MWCNT coated with PPy has also been used in supercapacitor. The maximum value of 163 F g^{-1} has been obtained for $MWCNT_S$ prepared at 600°C and modified by a PPy layer of 5 nm, whereas it is only 50 F g^{-1} for the pristine nanotube (Jurewicz et al., 2001). PPy/GNS composite is a promising candidate for supercapacitor to have higher specific capacitance, better rate capability and cycling stability than those of pure PPy. The specific capacitance of PPy/GNS composite based on the three electrode cell configuration is as high as 482 F g^{-1} at a current density of 0.5 A g^{-1} (Zhang et al., 2011).

PPy doped with nafion ions or ClO_4^- exhibits a specific capacitance of 344 or 355 F g^{-1}. Cycle life experiments revealed that the nafion material retains 98% of the original capacitance after 3000 cycles whereas that doped with ClO_4^- retains only 70% over this period (Kim et al., 2008b).The PPy active films doped with ClO_4^- ion in NaCl solution demonstrates an ideal supercapacitor behavior, i.e. rectangle-like CV shape at scan rates from 5-200 mV s^{-1}, linear galvanoststic charge/discharge curves at currents loads from 0.5 to 2 mA and stable cyclic property. However, dopings with Cl^- ions give rise to nonideal property of supercapacitor (Sun et al., 2009).

PEDOT/PPy composite can be used as an electrochemical supercapacitor electrode material. This composite electrode produced the specific capacitance as 230 Fg^{-1} in 1M lithium perchlorate aqueous solution and 290 F g^{-1} in 1M KCl aq solution (Wang et al., 2007). A pulsed polymerized Py film obtained on the polished graphite plate has been reported to exhibit very high capacitance (400 Fg^{-1}) and high energy density (250 Wh h kg^{-1}) and can be used as supercapacitor. Stability tests performed on this pulse polymerized PPy electrode yield long cycle life upto 10,000 cycles, the charge/discharge current density being 5mA cm^{-2} (Sharma et al., 2008).

4.5 Solar cells

Literature reveals that the increased application of the PPy material has been made, during recent years, in solar cells, particularly in dye-sensitized solar cells (DSSCs) to improve the overall energy conversion efficiency and also to reduce the cost of the cell. Recently, PPy was synthesized by vapour phase polymerization and traditional electrochemical polymerization and employed as counter electrode in a DSSC. Both PPy and Pt electrodes showed good catalytic behaviour in DSSCs (Xia et al., 2011). The composite, PPy/graphite on ITO glass also showed favourable catalytic activity for I_2/I^- redox reaction (Xiaoming et al., 2011).The overall energy conversion efficiency of the DSSCs based on the ruthenium dye (N719) with PPy/graphite composite counter electrode reached 6.01% under simulated AM

1.5 irradiation (100 mW/cm^2), which was 92% of the energy conversion efficiency of the DSSCs with Pt electrode.

Photoelectrochemical and electrochemical behavior of gold electrode modified with bilayers of PPy and PANI have been investigated in acid solutions (Upadhyay et al., 1995). Both PPy and PANI films on gold exhibit photo electrochemical activity, with the former showing a considerably high activity than the latter. PEC solar cells based on nanostructured ZnO/dye/PPy/ film electrode display excellent properties as anode in the conversion process of light to electricity (Hao et al., 2000). PPy was prepared on Ru-dye sensitized TiO_2 nanoporous film and solar cell was constructed using gold as the counter electrode with PPy acting as the hole conductor (Cervini et al., 2004). Photodevices comprising covalently grafted PPy films on surface modified mesoporous TiO_2 substrates via 3-(trimethoxysilyl) propyl methacrylate were fabricated and tested their performances with a counter electrode having a thin layer of gold (Senadeera et al., 2006). Significant enhancements in photoresponses were observed with the above additives in PPy than the reported devices comprising TiO_2/PPy. Hybrid Cu-In disulphide/PPy photovoltaic structures prepared by electrodeposition exhibited (Bereznev et al., 2005) significant photovoltage and photocurrent under standard white light illumination. Electrodeposited Cu-In-Se/PPy PV structures exhibited the formation of a n-p barrier between the n-CuInSe and p-PPy layers (Bereznev et al., 2006). The PPy encapsulated TiO_2 nanotube array (PPy/TiO_2 NTs) electrode was also synthesized (Zhang et al., 2008) by electropolymerization to encapsulate PPy inside the TiO_2 nanotube channels and walls in order to enhance the photocurrent density.

5. Conclusion

Different electrochemical methods for obtaining pure PPy, functionalized PPy and PPy composites are briefly described. Structural and electrochemical characterizations of new conductive materials and their application in solar cells, fuel cells, batteries, super capacitors and in corrosion protection have been highlighted. Studies available in literature have shown that the PPy has been used as carbon-substitute, particularly in fuel cells and batteries and found to greatly improve their performances.The use of active films or composites of PPy is very effective in enhancement of the capacitance of the supercapacitors. The PPy and its composites such as PPy/graphite have also shown favourable catalytic activity for I_2/I^- redox reaction. The PPy film has also improved the efficiency of the photo anodes in dye-sensitized solar cells significantly. PPy and PPy-based coatings are proved to be very effective inhibitors in the corrosion of oxidizable metals and alloys, stainless steels, mild steel, etc. Besides these, PPy films are being used in other areas such as sensors, bio-fuel cells, etc.

6. Acknowledgement

Authors gratefully acknowledge the Council of Scientific and Industrial Research (CSIR), New Delhi, India for the financial support through the research project (01/2320/09-EMR-II).

7. References

Abrantes, L. M.; Cordas, C. M.; Correia, J. P.; Montforts, F-P. & Wedel, M. (2000). Electropolymerization of pyrrole substituted metalloporphyrins-synthesis and characterization. *Portu. Electrochim. Acta,* Vol.18, No.1, pp. 3-12, ISSN 1647-1571

Ansari, R. (2006). Polypyrrole conducting electroactive polymers: Synthesis and stability studies. *E- Journal of Chemistry*, Vol.3, No.13, pp. 186-201, ISSN 0973-4945

Asavapiriyanont, S.; Chandler, G.K.; Gunawardena, G.A. & Pletcher, D. (1984). The electrodeposition of polypyrrole films from aqueous solutions. *J. Electroanal. Chem.*, Vol.177, No.1-2, pp.229-244. ISSN 1572-6657

Ayad, M.M. & Zaki, E. (2009). Synthesis and characterization of silver-polypyrrole film composite. *Appl. Surf. Sci.* Vol.256, No.3, pp. 787-791, ISSN 0169-4332

Balaskas, A.C.; Kartsonakis, I.A.; Kordas, G.; Cabral, A.M.; Morais, P.J. (2011). Influence of the doping agent on the corrosion protection properties of polypyrrole grown on aluminium alloy 2024-T3 *Prog. Org. Coat.*, Vol.71, No.2, pp. 181-187, ISSN 0300-9440

Bazzaoui, M.; Martins, J.I.; Bazzaoui, E.A.; Martins, L. & Machnikova, E. (2007). Sweet aqueous solution for electrochemical synthesis of polypyrrole part 1B: On copper and its alloys. *Electrochim. Acta*, Vol.52, No.11, pp. 3568–3581, ISSN 0013-4686

Bazzaoui, M.; Martins, J.I.; Costa, S.C.; Bazzaoui, E.A.; Reis, T.C. & Martins, L. (2006). Sweet aqueous solution for electrochemical synthesis of polyrrole: Part 2. On ferrous metals. *Electrochim. Acta*, Vol.51, No.21, pp. 4516-4527, ISSN 0013-4686

Bazzaoui, M.; Martinsa, J.I.; Reis, T.C.; Bazzaoui, E.A.; Nunes, M.C. & Martins, L. (2005). Electrochemical synthesis of polypyrrole on ferrous and non-ferrous metals from sweet aqueous electrolytic medium. *Thin Solid Films*, Vol.485, No.1-2, pp. 155 – 159, ISSN 0040-6090

Becerik, I. & Kadirgan, F. (2001). Glucose sensitivity of platinum-based alloys incorporated in polypyrrole films at neutral media. *Synth. Met.*, vol.124, No.2-3, pp. 379-384, ISSN 1379-6779

Bereket, G. & Hur, E. (2009). The corrosion protection of mild steel by single layered polypyrrole and multilayered polypyrrole/poly(5-amino-1-naphthol) coatings. *Prog. Org. Coat.*, Vol.65, No.1, pp. 116-124, ISSN 0300-9440

Bereznev, S.; Kois, J.; Golovtsov, I.; Opik, A. & Mellikov, E. (2006). Electrodeposited (Cu–In–Se)/polypyrrole PV structures. *Thin Solid Films*, Vol.511–512, pp. 425–429, ISSN 0040-6090

Bereznev, S.; Konovalov, I.; Opik, A.; Kois, J. & Mellikov, E. (2005). Hybrid copper indium disulfide/polypyrrole photovoltaic structures prepared by electrodeposition. *Sol. Energy Mater. Sol. Cells*, Vol.87, No.1-4, pp. 197–206, ISSN 0927-0248

Bose, C.S.C.; Rajeshwar, K. (1992). Efficient electrocatalyst assemblies for proton and oxygen reduction: the electrosynthesis and characterization of polypyrrole films containing nanodispersed platinum particles. *J. Electroanal. Chem.*, Vol. 333, No.1-2, pp. 235-256, ISSN 1572-6657

Bouzek, K.; Mangold, K.M. & Juttner K. (2000). Platinum distribution and electrocatalytic properties of modified polypyrrole films. *Electrochim. Acta*, Vol.46, No.5, pp.661-670, ISSN 0013-4686

Bouzek, K.; Mangold, K.M. & Juttner K. (2001). Electrocatalytic activity of platinum modified polypyrrole films fr the methanol oxidation reaction. *J. Appl. Electrochem.*, Vol.31, No.5, pp.501-507, ISSN 1572-8838

Bull, R.A.; Fan, F.R. & Bard, A.J. Polymer films on electrodes. 13. Incorporation of catalysts into electronically conductive polymers: iron phthalocyanine in polypyrrole. NTIS.

Report (1983), (TR-31; Order No. AD-A127700), 22 pp. From: Gov. Rep. Announce. Index (U. S.) 1983, 83(17), 4213,

Cervini, R.; Cheng Y.; & Simon, G. (2004). Solid-state Ru-dye solar cells using polypyrrole as a hole conductor. *J. Phys. D: Appl. Phys.* Vol.37, No.1, pp. 13–20, ISSN 1361-6463

Chen, Ge; Wang, Z.; Yang, T.; Huang, D. & Xia, D. (2006). Electrocatalytic Hydrogenation of 4-Chlorophenol on the Glassy Carbon Electrode Modified by Composite Polypyrrole/Palladium Film. *J Phys. Chem. B*, Vol.110, No.10, pp. 4863-4868, ISSN 1520-5207

Chipara, M.; Skomski, R. & Sellmyer. D. J. (2007). Electrodeposition and magnetic properties of polypyrrole–Fe nanocomposites. *Mater. Lett.*, Vol.61, No.11-12, pp. 2412-2415, ISSN 0167-577X

Coutanceau, C.; Hourch, A. El; Crouigneau, P.; Leger, J. M. & Lamy, C. (1995). Conducting polymer electrodes modified by metal tetrasulfonated phthalocyanines: preparation and electrocatalytic behavior towards dioxygen reduction in acid medium. *Electrochim. Acta*, Vol.40, No.17, pp. 2739-48, ISSN 0013-4686

Cui, L.; Shen, J.; Chen, F.; Tao, Z. & Chen, J. (2011). SnO_2 nanocomposites@polypyrrole nanowires composite as anode materials for rechargeable lithium-ion batteries. *J. Power Sources*, Vol.196, No.2, pp.2195-2201, ISSN 0378-7753

Del Valle, M.A.; Diaz, F.R.; Bodini, M.E.; Pizarro, T.; & Córdova, R. *et al.* (1998) Polythiophene, polyaniline and polypyrrole electrodes modified by electrodeposition of Pt and Pt+Pb for formic acid electrooxidation. *J. Appl. Electrochem.*, Vol.28, No.9, pp. 943-946, ISSN 1572-8838.

Deronzier, A. & Moutet, J.C. (1996). Polypyrrole films containing metal complexes: syntheses and applications. *Coord.Chem. Rev.*, Vol.147, pp. 339-371, ISSN 0010-8545

Diab, N. & Schuhmann W. (2001). Electropolymerized manganese porphyrin/polypyrrole films as catalytic surfaces for the oxidation of nitric oxide. *Electrochim. Acta*, 47, No.1-2, pp. 265-273, ISSN 0013-4686

Diaz, A.F.; Kanazawa, K.K. & Gardini, G.P. (1979). Electrochemical polymerization of pyrrole. *J. Chem. Soc. Chem. Commun.*, No.14, pp. 635-636

Ding, K.; Jia, H.; Suying Wei, & Zhanhu Guo. (2011). Electrocatalysis of Sandwich-Structured Pd/Polypyrrole/Pd Composites toward Formic Acid Oxidation. *Ind. Eng. Chem. Res.*, Vol.50, No.11, pp. 7077-7082

Ding, K.Q. & Cheng, F.M. (2009). Cyclic voltammetrically prepared MnO_2-PPy composite material and its electrocatalysis towards oxygen reduction reaction (ORR). *Synth. Met.*, Vol.159, No.19-20, pp. 2122-2127, ISSN 0379-6779

Dong, S. & Liu, M. (1994). Preparation and properties of polypyrrole film doped with a Dawson-type heteropolyanion. *Electrochim. Acta*, Vol.39, No.7, pp. 947-51, ISSN 0013-4686

Fang, Y.; Liu, J.; Yu, D.J.; Wicksted J.P.; Kalkan, K.; Topal, C.O.; Flanders, B.N.; Wu, J. & Li, J. (2010). Self-supported supercapacitor membranes: Polypyrrole-coated carbon nanotube network enabled by pulsed electrodeposition. *J. Power Sources*, Vol.195, No.2, pp.674-679, ISSN 0378-7753

Ferreira, C.A.; Domenech, S.C. & Lacaze, P.C. (2001). Synthesis and characterization of polypyrrole/TiO_2 composites on mild steel. *J. Appl. Electrochem.*, Vol.31, No.1, pp. 49-56, ISSN 1572-8838

Flamini, D.O. & Saidman, S.B. (2010). Electrodeposition of polypyrrole onto NiTi and the corrosion behaviour of the coated alloy. *Corros. Sci.*, Vol.52, No.1, pp. 229-234, ISSN 0010-938X

Funt, B.L. & Diaz, A.F. (1991). Organic electrochemistry: An introduction and a guide. 1337, Marcel Dekker

Garcia, B.; Lamzoudi, A.; Pillier, F.; Lee, H.N.T.; Deslouis, C. (2002). Oxide/Polypyrrole Composite Films for Corrosion Protection of Iron. *J. Electrochem. Soc.*, Vol.149, No.12, pp. B560-B566, ISSN 1945-7111

Gautier, J.L.; Marco, J.F.; Gracia, M.; Gancedo, J.R.; Garza Guadarrama, V.D.L. & Nguyen Conh, H. & Chartier, P. (2002). $Ni_{0.3}Co_{2.7}O_4$ spinel particles/polypyrrole composite electrode: Study by X-ray photoelectron spectroscopy. *Electrochim. Acta*, Vol.48, No2, pp. 119-125, ISSN 0013-4686

Gemeay, A.H.; Nishiyama, H.; Kuwabata, S.; & Yoneyama, H. (1995).Chemical Preparation of Manganese Dioxide/Polypyrrole Composites and Their Use as Cathode Active Materials for Rechargeable Lithium Batteries. *J. Electrochem. Soc.*, Vol.142, No.12, pp. 4190-4195, ISSN 1945-7111

Genies, E.M. & Pernaut, J.M. (1985). Characterization of the radical cation and the dication species of polypyrrole by spectroelectrochemistry: Kinetics, redox properties, and structural changes upon electrochemical cycling. *J. Electroanal. Chem. Interfacial Electrochem.*, Vol.191, No.1, pp. 111-126

Genies, E.M.; Bidan, G. & Diaz, A.F. (1983). Spectroelectrochemical study of polypyrrole films. *J. Electroanal. Chem.*, Vol.149, No.1-2, pp. 101-113. ISSN 1572-6657

Grgur, B.N.; Gvozdenović, M.M.; Stevanović, J.; Jugović, B.Z. & Marinović, V.M. (2008). Polypyrrole as possible electrode materials for the aqueous-based rechargeable zinc batteries. *Electrochim. Acta*, Vol.53, No.14, pp. 4627-4632, ISSN 0013-4686

Guo, Z.P.; Wang, J.Z.; Liu, H.K. & Dou, S.X. (2005). Study of silicon/polypyrrole composite as anode materials for Li-ion batteries. *J. Power Sources, Vol.*146, No.1-2, pp. 448-451, ISSN 0378-7753

Hacaloglu, J.; Tezal, F. & Kucukyavuz, Z. (2009).The Characterization of Polyaniline and Polypyrrole Composites by Pyrolysis Mass Spectrometry. *J. Appl. Polym. Sci.*,Vol.113, No.5, pp. 3130–3136. ISSN 1097-4628

Hammache, H.; Makhloufi, L. & Saidani, B. (2001). Electrocatalytic oxidation of methanol on PPy electrode modified by gold using the cementation process. *Synth. Met.*, Vol.123, No.3, pp. 515-522, ISSN 0379-6779

Hao, Y.; Yang, M.; Li, W.; Qiao, X. Zhang, L. & Cai, S. (2000). A photoelectrochemical solar cell basedon ZnO/dye/polypyrrole "lm electrode as photoanode. *Sol. Energy Mater. Sol. Cells*, Vol.60, No.4, pp. 349-359, ISSN 0927-0248

Haseko, Y.; Shrestha, N.K.; Teruyama, S. & Saji, T. (2006), Reversal pulsing electrodeposition of Ni/polypyrrole composite film. *Electrochim. Acta*, Vol.51, No.18, pp. 3652-3657, ISSN 0013-4686

Herrasti, P. & Ocon, P. (2001). Polypyrrole layers for steel protection. *Appl. Surf. Sci.*, Vol.172, No.3-4, pp. 276-284, ISSN 0169-4332

Herrasti, P.; del Rio, A.I. & Recio, J. (2007). Electrodeposition of homogeneous and adherent polypyrrole on copper for corrosion protection. *Electrochim. Acta*, Vol.52, No.18, pp. 6496–6501, ISSN 0013-4686

Herrasti, P.; Kulak, A.N.; Bavykin, D.V.; Ponce de Leon, C.; Zkonyte, J. & Walsh, F. C. (2011). Electrodeposition of polypyrrole–titanate nanotube composites coatings and their corrosion resistance. *Electrochim. Acta*, Vol.56, No.3, 1323-1328, ISSN 0013-4686

Hosseini, M.G.; Sabouri, M. & Shahrabi, T. (2007). Corrosion protection of mild steel by polypyrrole phosphate composite coating. *Prog. Org. Coat.*, Vol. 60, No.3, pp. 178-185, ISSN 0300-9440

Ikeda, O.; Okabayashi, K. & Tamura, H. (1983). Electrocatalytic reduction of oxygen on cobalt-doped polypyrrole films. *Chem. Lett.*, Vol.12, pp. 1821-4, ISSN

Iroh, J.O. & Levine, K. (2003). Capacitance of the polypyrrole/polyimide composite by electrochemical impedance spectroscopy. *J. Power Sources*, Vol.117, No.1-2, pp. 267-272, ISSN 0378-7753

Jeyabharathi, C.; Venkateshkumar, P.; Mathiyarasu, J. & Phani, K.L.N. (2010). Carbon-supported palladium-polypyrrole nanocomposite for oxygen reduction and its tolerance to methanol. *J. Electrochem. Soc.*, Vol.157, No.11, pp. B1740-B1745, ISSN 1945-7111

Jin, B.; Kim, J.U. & Gu, H.B. (2003). Electrochemical properties of lithium–sulfur batteries. *J. Power Sources,* Vol.117, No.1-2, pp. 148-152, ISSN 0378-7753

Johanson, U.; Marandi, M.; Sammelselg, V. & Tamm, J. (2005). Electrochemical properties of porphyrin-doped polypyrrole films. *J. Electroanal. Chem.*, Vol. 575, No.2, pp. *267-273,* ISSN 1572-6657

Joseph, S.; McClure, J.C.; Chianelli, R.; Pich, P. & Sebastian. P.J. (2005). Conducting polymer-coated stainless steel bipolar plates for proton exchange membrane fuel cells (PEMFC). *Int. J. Hydrogen Energy*, Vol.30, No.12, pp. 1339 – 1344, ISSN 0360-3199

Jurewicz, K.; Delpeux, S.; Bertagna, V.; Béguin, F. & Frackowiak, E. (2001). Supercapacitors from nanotubes/polypyrrole composites. *Chem. Phys. Lett.*, Vol.347, No.1-3, pp. 36-40, ISSN 0009-3199

Juttner, K.; Mangold, K.-M.; Lange M. & Bouzek K. (2004). Preparation and Properties of Composite Polypyrrole/Pt Catalyst Systems. *Russ. J. Electrochem.*, Vol.40, No. 3, pp. 317-325, ISSN 1023-1935

Kanno, R.; Kubo, H.; Kawamoto, Y.; Kamiyama, T.; Izumi, F.; Takeda, Y. & Takano, M. (1994). Phase Relationship and Lithium Deintercalation in Lithium Nickel Oxides. *J. Solid Chem.*, Vol.110, No.2, pp. 216-225, ISSN 0022-4596

Kaplin, D.A. & Qutubuddin, S. (1995). Electrochemically synthesized polypyrrole films: effects of polymerization potential and electrolyte type. *Polymer*, Vol.36, No.6, pp. 1275-1285, ISSN 0032-3861

Kim, B.C.; Ko, J.M. & Wallace, G.G. (2008b). A novel capacitor material based on Nafion-doped polypyrrole. *J. Power sources*, Vol.177, No.2, pp. 665-668, ISSN 0378-7753

Kim, J.Y.; Kim, K.H. & Kim, K.B. (2008a). Fabrication and electrochemical properties of carbon nanotube/polypyrrole composite film electrodes with controlled pore size. *J. Power Sources*, Vol.176, No.1, pp. 396-402, ISSN 0378-7753

Kim, K.J.; Song, H.S. & Kim, J.D. (1988). Mechanism of Electropolymerization of Pyrrole in Acidic Aqueous Solutions. *Bull. Korean Chem. Soc.*, Vol.9, No.4, pp. 248-251, ISSN 0253-2964

Konwer, S.; Maiti, J. & Dolui, S.K. (2011). Prepration and optical/electrical/electrochemical properties of expanded graphite-filled polypyrrole nanocomposite. *Mater. Chem. Phys.*, In Press, ISSN 0254-0584

Kowalski, D.; Ueda, M.; Ohtsuka, T. (2008). The effect of ultrasonic irradiation during electropolymerization of polypyrrole on corrosion prevention of the coated steel. *Corros. Sci.*, Vol.50, No.1, pp. 286-290, ISSN 0010-938X

Kuwabata, S.; Masui, S. & Yoneyama, H. (1999). Charge–discharge properties of composites of $LiMn_2O_4$ and polypyrrole as positive electrode materials for 4V class of rechargeable Li batteries. *Electrochim. Acta*, Vol.44, No.25, pp. 4593-4600, ISSN 0013-4686

Lang, G. & Inzelt, G. (1999). An advanced model of the impedance of polymer film electrodes. *Electrochim. Acta.* Vol.44, No.12, pp. 2037-2051, ISSN 0013-4686

Lee, H.N.T.; Garcia, B.; Pailleret, A. & Deslouis, C. (2005). Role of doping ions in the corrosion protection of iron by polypyrrole films. *Electrochim. Acta*, Vol.50, No.7-8, pp. 1747–1755, ISSN 0013-4686

Lehr, I.L. & Saidman, S.B. (2006a). Electrodeposition of polypyrrole on aluminium in the presence of sodium bis(2-ethylhexyl) sulfosuccinate. *Mater. Chem. Phys.*, Vol.100, No.2-3, pp. Pages 262-267, ISSN 0254-2584

Lehr, I.L. & Saidman, S.B. (2006b). Characterisation and corrosion protection properties of polypyrrole electropolymerised onto aluminium in the presence of molybdate and nitrate. *Electrochim. Acta,* Vol.51, No.16, pp. 3249–3255, ISSN 0013-4686

Lenz, D. M.; Delamar, M. & Ferreira, C.A. (2003). Application of polypyrrole/TiO_2 composite films as corrosion protection of mild steel. *J. Electroanal. Chem.*, Vol.540, pp. 35-44, ISSN 1572-6657

Lenz, D. M.; Delamar, M. & Ferreira, C.A. (2007). Improvement of the anticorrosion properties of polypyrrole by zinc phosphate pigment incorporation. *Prog. Org. Coat.*, Vol.58, No.1, pp. 64-69, ISSN 0300-9440

Leroux, F.; Koene, B.E. & Nazar, L.F. (1996). Electrochemical Lithium Intercalation into a Polyaniline/V_2O_5 Nanocomposite. *J. Electrochem. Soc.*, Vol.143, No.9, pp. L181-L183, ISSN 1945-7111

Levi, M.D. & Aurabch, D. (1997). Simultaneous Measurements and Modeling of the Electrochemical Impedance and the Cyclic Voltammetric Characteristics of Graphite Electrodes Doped with Lithium. *J. Phys. Chem. B.*, Vol.101, No.23, pp. 4630-4640, ISSN 1520-5207

Levi, M.D. & Aurabch, D. (2004). Impedance of a Single Intercalation Particle and of Non-Homogeneous, Multilayered Porous Composite Electrodes for Li-ion Batteries. *J Phys Chem. B.* Vol.108, No.31, pp. 11693-11703, ISSN 1520-5207

Li, C.M.; Sun, C.Q.; Chen, W. & Pan, L. (2005). Electrochemical thin film deposition of polypyrrole on different substrates. *Surf. Coat. Technol.*, Vol.198, No.1-3, pp. 474–477, ISSN 0257-8972

Liang, X.; Wen, Z.; Liu, Y.; Wang, X.; Zhang, H.; Wu, M. & Huang, L. (2010). Preparation and characterization of sulfur–polypyrrole composites with controlled morphology as high capacity cathode for lithium batteries. *Solid State Ionics,* doi: 10.1016/j.ssi.2010.016, ISSN 0167-2738

Liu, Y.C. & Chuang, T.C. (2003). Synthesis and Characterization of Gold/Polypyrrole Core-Shell Nanocomposites and Elemental Gold Nanoparticles Based on the Gold-Containing Nanocomplexes Prepared by Electrochemical Methods in Aqueous Solutions. *J. Phys. Chem. B*, Vol.107, No. pp. 12383-12386, ISSN 1520-5207

Madani, A.; Nessark, B.; Boukherroub, R. & Chehimi, M.M. (2011). Prepration and electrochemical behavior of PPy-CdS composite films. *J. Electroanal. Chem.*, Vol.650, No.2, pp. 176-181, ISSN 1572-6657

Malviya, M.; Singh, J.P. & Singh, R.N. (2005a). Electrochemical characterization of polypyrrole/cobalt ferrite composite films for oxygen reduction. *Ind. J. Chem.* Vol.44A, pp. 2233-2239, ISSN 0376-4710

Malviya, M.; Singh, J.P.; Lal, B & Singh, R.N. (2005b). Transport behavior of Cl^- in composite films of polypyrrole and $CoFe_2O_4$ obtained for oxygen reduction. *J. New Mater. Electrochem. Syst.*, Vol.8, pp. 223-228

Mangold, K.-M.; Meik, F. & Jüttner, K. (2004) Polypyrrole/palladium composites for the electrocatalyzed Heck reaction. *Synt. Met.*; Vol.144, pp. 221-227, ISSN 0379-6779

Millán, W. M.; Thompson, T.T.; Arriaga, L.G. & Smit, M.A. (2009). Characterization of composite materials of electroconductive polymer and cobalt as electrocatalysts for the oxygen reduction reaction. *Int. J. Hydrogen Energy*, Vol.34, No.2, pp. 694-702, ISSN 0360-3199

Mohamedi, M.; Takahashi, D.; Uchiyama, T.; Itoh, T.; Nishizawa, M. & Uchida, I. (2001). Explicit analysis of impedance spectra related to thin films of spinel $LiMn_2O_4$. *J. Power Sources*, Vol.93, No.1-2, pp. 93-103, ISSN 0378-7753

Mohana Reddy, A. L.; Rajalakshmi, N. & Ramaprabhu, S. (2008). Cobalt-polypyrrole-multiwalled carbon nanotube catalysts for hydrogen and alcohol fuel cells. *Carbon*, Vol.46, No.1, pp. 2-11, ISSN 0008-6223

Nguyen Cong; H.; Abbassi, K.E.; Chartier, P. (2000). Electrically conductive polymer/metal oxidecomposite electrode for oxygen reduction. *Electrochem. Solid State Lett.*, Vol.3, No4, pp. 192-195, ISSN 1944-8775

Nguyen Cong; H.; Abbassi, K.E.; Chartier, P. (2002a). Electrocatalysis of oxygen reduction on polypyrrole/mixed valence spinel oxide nanoparticles. *J Electrochem. Soc.*, Vol.149, No5, pp. A525-A530, ISSN 1945-7111

Nguyen Cong, H.; Garza Guadarrama, V.D.L.; Gautier, J.L. & Chartier, P. (2002b). $Ni_xCo_{3-x}O_4$ mixed valence oxide nanoparticles/polypyrrole composite electrodes for oxygen reduction. *J. New Mater. Electrochem. Syst.*, Vol.5, No.1, pp. 35-40

Nguyen Cong, H.; Garza Guadarrama, V.D.L.; Gautier, J.L. & Chartier, P. (2003). Oxygen reduction on $Ni_xCo_{3-x}O_4$ spinel particles/polypyrrole composite electrodes: hydrogen peroxide formation. *Electrochim. Acta*, Vol.48, No.1, pp. 2389-2395, ISSN 0013-4686

Nguyen Cong; H.; Abbassi, K.E.; Gautier, J.L. & Chartier, P. (2005). Oxygen reduction on oxide/polypyrrole composite electrodes: effect of doping anions. *Electrochim. Acta*, Vol.50, pp. 1369-1376, ISSN 0013-4686

Ohzuku T. & Ueda A. (1994). Solid-State Redox Reactions of $LiCoO_2$ ($R\bar{3}m$) for 4 Volt Secondary Lithium Cells. *J. Electrochem. Soc.*, Vol.141, No.11, pp. 2972-2977, ISSN 1945-7111

Osaka, T.; Naoi, K.; Hirabayashi, T. & Nakamura, S. (1986). Electrocatalytic oxygen reduction at (tetrasulfonatophthalocyaninato)-cobalt incorporated polypyrrole film electrode. *Bull. Chem. Soc. Jpn.*, Vol.59, No.9, pp. 2717-2222, ISSN 1348-0634

Osaka, T.; Nishikawa, M. & Nakamura, S. (1984). Electrocatalytic oxygen reduction with metal porphyrins or metal phthalocyanine incorporated in polypyrrole film electrode. *Denki Kagaku oyobi Kogyo Butsuri Kagaku*, Vol.52, No.6, pp. 370-371

Panah, N.B. & Danaee, I. (2010). Study of the anticorrosive properties of polypyrrole/polyaniline bilayer via electrochemical techniques. *Prog. Org. Coat.*, Vol.68, No.3, pp. 214-218, ISSN 0300-9440

Pasquier, A.D.; Orsini, F.; Gozdz, A.S. & Tarascon J.M. (1999). Electrochemical behaviour of $LiMn_2O_4$-PPy composite cathodes in the 4-V region. *J. Power Sources*, Vol.81, pp. 607-611, ISSN 0378-7753

Passiniemi, P. & Vakiparta, K. (1995). Characterizaton of polyaniline blends with AC impedance measurements. *Synth. Met.* Vol. 69, No.1-3, pp. 237-238, ISSN 0379-6779

Pina, C.D.; Falletta, E.; Rossi, M. (2011). Conductive materials by metal catalyzed polymerization. *Catal. Today*, Vol.160, No.1, pp. 11-27. ISSN 0920-5861

Pournaghi-Azar M. H. & Ojani R. (2000). Electrochemistry and electrocatalytic activity of polypyrrole/ferrocyanide films on a glassy carbon electrode. *J. Solid State Electrochem.*, Vol.4, pp. 75-79, ISSN 1433-0768

Pournaghi-Azar M.H. & Ojani R., (1999). Electrochemistry and electrocatalytic activity of polypyrrole/ferrocyanide films on a glassy caron electrode. *J. Electroanal. Chem.*, Vol.474, No.2, pp. 113-122, ISSN 1572-6657

Qiu, Y.J. & Reynolds, J.R. (1992). Electrochemically initiated chain polymerization of pyrrole in aqueous media. *J. Polym. Sci., Part A: Polym. Chem.*, Vol.30, No.7, pp. 1315–1325. ISSN 1099-0518

Qu, B.; Xu, Y.T.; Lin, S.J.; Zheng, Y.F. & Dai, L.Z. (2010). Fabrication of Pt nanoparticles decorated PPy–MWNTs composites and their electrocatalytic activity for methanol oxidation. *Synth. Met.*, Vol.160, No.7-8, pp. 732-742, ISSN 1379-6779

Rahman, S.U. (2011). Corrosion protection of steel by catalyzed polypyrrole films. *Surf. Coat. Technol.*, Vol.205, No.8-9 pp. 3035–3042, ISSN 0257-8972

Raoof, J.B.; Ojani, R. & Rashid-Nadimi, S. (2004). Preparation of polypyrrole/ferrocyanide films modified carbon paste electrode and its application on the electrocatalytic determination of ascorbic acid. *Electrochim. Acta*, Vol.49, No.2, pp. 271-280, ISSN 0013-4686

Rapecki, T.; Donten, M. & Stojek, Z. (2010). Electrodeposition of polypyrrole–Au nanoparticles composite from one solution containing gold salt and monomer. *Electrochem. Commn.*, Vol.12, No.5, pp. 624-627, ISSN 1388-2481

Redondo, M.I. & Breslin, C.B. (April 2007). Polypyrrole electrodeposited on copper from an aqueous phosphate solution: Corrosion protection properties. *Corros. Sci.*, Vol.49, No.4, pp. 1765-1776, ISSN 0010-938X

Ren, Y.J. & Zeng, C.L. (2008). Effect of conducting composite polypyrrole/polyaniline coatings on the corrosion resistance of type 304 stainless steel for bipolar plates of proton-exchange membrane fuel cells. *J. Power Sources*, Vol.182, No.2, pp. 524–530, ISSN 0378-7753

Riaz, U.; Ashraf, S.M. & Ahmad, S. (2007). High performance corrosion protective DGEBA/polypyrrole composite coatings. *Prog. Org. Coat.*, Vol.59, No.2, pp. 138-145, ISSN 0300-9440

Rodrigues, N.P.; Obirai, J.; Nyokong, T. & Bedioui, F. (2005). Electropolymerized pyrrole-substituted manganese phthalocyanine films for the electroassisted biomimetic catalytic reduction of molecular oxygen. *Electroanalysis*, Vol.17, No.2, pp. 186-190, ISSN 1521-4109

Sabouri, M.; Shahrabi, T.; Faridi, H.R.; Hosseini M.G. (2009). Polypyrrole and polypyrrole-tungstate electropolymerization coatings on carbon steel and evaluating their corrosion protection performance via electrochemical impedance spectroscopy. *Prog. Org. Coat.*, Vol.64, No.4, pp. 429-434, ISSN 0300-9440

Sadki, S.; Schottland, P.; Brodie, N. & Sabouraud, G. (2000). The mechanism of pyrrole electropolymerization. *Chem. Soc. Rev.*, Vol.29, No.5, pp. 283-293. ISSN 0306-0012

Seeliger, W. & Hamnett, A. (1992). Novel electrocatalysts for oxygen reduction. *Electrochim. Acta*, Vol.37, No.4, pp. 763-765, ISSN 0013-4686

Selvaraj, V.; Alagar, M. & Hamerton, I. (2006). Electrocatalytic properties of monometallic and bimetallic nanoparticles-incorporated polypyrrole films for electro-oxidation of methanol. *J. Power Sources*, Vol.160, No.2, pp. 940–948, ISSN 0378-7753

Senadeera, G.K.R.; Kitamura, T.; Wadab, Y. & Yanagida, S. (2006). Enhanced photoresponses of polypyrrole on surface modified TiO_2 with self-assembled monolayers. *J. Photochem. Photobio. A: Chem.*, Vol.184, No.1-2, pp. 234–239, ISSN 1010-6030

Sharifirad, M.; Omrani, A.; Rostami, A.A. & Khoshroo, M. (2010). Electrodeposition and characterization of polypyrrole films on copper. *J. Electroanal. Chem.*,Vol. 645, No.2, pp.149-158, ISSN 1572-6657

Sharma, R.K.; Rastogi, A.C. & Desu, S.B. (2008). Pulse polymerized polypyrrole electrodes for high energy density electrochemical supercapacitor. *Electrochem. Commun.*, Vol.10, No.2, pp. 268-272, ISSN 1388-2481

Singh, R.N. Malviya, M.; & Chartier, P. (2007b). Electrochemical Characterization of composite films of $LaNiO_3$ and polypyrrole for electrocatalysis of O_2 reduction. *J. New Mater. Electrochem. Syst.* Vol.10, pp. 181 - 186

Singh, R.N., Malviya, M. & Anindita. (2007c). Electrochemical characterization of composite films of polypyrrole and $La_{1-x}Sr_xCoO_3$ ($0 \leq x \leq 0.4$) for electrocatalysis of O_2 reduction. *Ind. J. Chem.*, Vol.46A, pp. 1923-1928, ISSN

Singh, R.N.; Lal, B. & Malviya, M. (2004). Electrocatalytic activity of electrodeposited composite films of polypyrrole and $CoFe_2O_4$ nanoparticals towards oxygen reduction reaction. *Electrochim. Acta*, Vol.49, No.26, pp. 4605-4612, ISSN 0013-4686

Singh, R.N.; Malviya, M.; Anindita; Sinha, A.S.K. & Chartier, P. (2007a). Polypyrrole and $La_{1-x}Sr_xMnO_3$ (0 x 0.4) composite electrodes for electroreduction of oxygen. *Electrochim. Acta*, Vol.52, No.13, pp. 4264-4271, ISSN 0013-4686

Snook, G.A.; Kao, P. & Best, A.S. (2011). Review: Conducting-polymer-based supercapacitor devices and electrodes. *J. Power Sources*, Vol.196, No.1, pp. 1-12, ISSN 0378-7753

Song, M.S.; Han, S.C.; Kim, H.S.; Kim, J.H.; Kim, K.T.; Kang, Y.M.; Ahn, H.J.; Dou, S.X. & Lee, J.Y. (2004). Effects of Nanosized Adsorbing Material on Electrochemical Properties of Sulfur Cathodes for Li/S Secondary Batteries. *J. Electrochem. Soc.*, Vol.151, No.6, pp. A791- A795, ISSN 1945-7111

Street, G.B. (1986). Handbook of conducting polymers. In: Skotheim, T.J. (Ed.), 188, 1st ed., Marcel Dekker, New York

Sun, W. & Chen, X. (2009). Preparation and characterization of polypyrrole films for three-dimensional micro supercapacitor. *J. Power Sources*, Vol.193, No.2, pp. 924-929, ISSN 0378-7753

Sun, Z.; Ge, H.; Hu, X. & Peng, Y. (2010). Preparation of foam-nickel composite electrode and its application to 2,4-dichlorophenol dechlorination in aqueous solution. *Sep. Purif. Technol.* Vol.72, No.2, pp. 133-139, ISSN 1383-5866

Tarascon, J.M.; McKinnon, W.R.; Coowar, F.; Bowmer, T.N.; Amatucci, G.; & Guyomard, D. (1994). Synthesis conditions and oxygen stoichiometry effects on Li insertion into the spinel $LiMn_2O_4$. *J. Elctrochem. Soc.*, Vol.141, No.6, pp. 1421-1431, ISSN 1945-7111.

Tüken, T.; Arslan, G.; Yazıcı, B. & Erbil, M. (2004). The corrosion protection of mild steel by polypyrrole/polyphenol multilayer coating. *Corros. Sci.*, Vol.46, No.11, pp. 2743-2754, ISSN 0010-938X

Turhan, M.C.; Weiser, M.; Jha, H. & Virtanen, S. (2011a).Optimization of electrochemical polymerization parameters of polypyrrole on Mg–Al alloy (AZ91D) electrodes and corrosion performance. *Electrochim. Acta*, Vol.50, No.15, pp. 5347-5354, ISSN 0013-4686

Turhan, M.C.; Weiser, M.; Killian, M.S.; Leitner, B. & Virtanen, S. (2011b). Electrochemical polymerization and characterization of polypyrrole on Mg–Al alloy (AZ91D). *Synth. Met.*, Vol.161, No.3-4, pp. 360-364, ISSN 0379-6779

Upadhyay, D.N.; Bharathi, S.; Yegnaraman, V. & Prabhakara Rao, G. (1995). Photoelectrochemical and electrochemical behavior of gold electrode modified with bilayers of polypyrrole and polyaniline. *Sol. Energy Mater. Sol. Cells*, Vol.37, No.3-4, pp. 307-314, ISSN 0927-0248

Veeraraghavan, B.; Paul, J.; Haran, B. & Popov, B. (2002). Study of polypyrrole graphite composite as anode material for secondary lithium-ion batteries. *J. Power Sources*, Vol.109, No.2, pp. 377-387, ISSN 0378-7753

Vishnuvardhan, T.K.; Kulkarni, V.R.; Basavaraja, C. & Raghavendra, S.C. (2006). Synthesis, Charectrization and a.c. conductivity of polypyrrole/Y_2O_3 composites. *Bull. Mater.Sci.*, Vol.29, No.1, pp.77-83, ISSN

Vork, F. & E. Barendrecht. (1989). Application and characterization of polypyrrole-modified electrodes with incorporated Pt particles. *Synth. Met.*, Vol.28, No.1-2, pp.121-126, ISSN 0379-6779

Waltman, R. J. & Bargon, J. Reactivity/structure correlations for the electropolymerization of pyrrole: An INDO/CNDO study of the reactive sites of oligomeric radical cations. *Tetrahedron*. Vol.40, No.20, pp. 3963-3970. ISSN 0040-4020

Waltman, R.J. & J. Bargon. (1986). Electrically conducting polymers: a review of the electropolymerization reaction, of the effects of chemical structure on polymer film properties, and of applications towards technology. *Can. J. Chem.*, Vol.64; No.1, pp. 76-95

Wang, G.X.; Yang, L.; Chen, Y.; Wang, J.Z.; Bewlay, S. & Liu, H.K. (2005). An investigation of polypyrrole-LiFePO4 composite cathode materials for lithium-ion batteries. *Electrochim. Acta*, Vol.50, No.24, pp. 4649-4654, ISSN 0013-4686

Wang, J.; Chen, J.; Konstantinov, K.; Zhao, L.; Ng, S.H.; Wang, G.X.; Guo, Z.P. & Liu, H.K. (2006). Sulphur-polypyrrole composite positive electrode materials for rechargeable lithium batteries. *Electrochim. Acta*, Vol.51, No.22, pp. 4634-4638, ISSN 0013-4686

Wang, J.; Xu, Y.; Chen, X. & Du, X. (2007). Electrochemical supercapacitor electrode material based on poly(3,4-ethylenedioxythiophene)/polypyrrole composite. *J. Power Sources*, Vol.163, No.2, pp. 1120-1125, ISSN 0378-7753

Wang, J.Z.; Chou, S.L.; Chen, J.; Chew, S.Y.; Wang, G.X.; Konstantinov, K.; Wu, J.; Dou, S.X. & Liu, H.K. (2008). Paper-like free-standing polypyrrole and polypyrrole–$LiFePO_4$

composite films for flexible and bendable rechargeable battery. *Electrochem. Commun.*, Vol.10, No.11, pp. 1781-1784, ISSN 1388-2481

Wu, B.; Yang, C.; Wu, H. (1999). Reduction of oxygen on glassy carbon electrode modified by polypyrrole film with ferriporphyrin. *Yingyong Huaxue*. Vol.16, No.5, pp. 17-20, ISSN 1000-0518

Xia, J.; Chen L. & Yanagida, S. (2011). Application of polypyrrole as a counter electrode for a dye-sensitized solar cell. *J. Mater. Chem.*, Vol.21, pp. 4644-4649

Xiaoming; F.; Xianwei; H.; Zhuo; T. & Bin. Z. (2011). Application Research of Polypyrrole/Graphite Composite Counter Electrode for Dye-Sensitized Solar Cells. *Acta Chim. Sinica*, Vol. 69, No.6, pp. 653-658

Yalcinkaya, S.; Demetguil, Timur, M. & Colk, N. (2010). Electrochemical synthesis and characterization of polypyrrole/chitosan composite on platinum electrode: Its electrochemical and thermal behaviors. *Carbohydr. Polym.*, Vol.79, No.4, pp.908-913, ISSN 0144-8617

Yoneyama, H. & Shozi, Y. (1990). Incorporation of WO_3 into Polypyrrole, and Electrochemical Properties of the Resulting Polymer Films. *J. Electrochem. Soc.*, Vol.137, No.12, pp. 3826-3830, ISSN 1945-7111

Yoneyama, H. Kishimoto, A. & Kuwabata, S. (1991). Charge-discharge properties of polypyrrole films containing manganese dioxide particles. *J. Chem. Soc., Chem. Commun.*, Vol.15, pp. 986-987

Yuasa, M.; Yamaguchi, A.; Itsuki, H.; Tanaka, K.; Yamamoto, M. & Oyaizu, K. (2005). Modifying Carbon Particles with Polypyrrole for Adsorption of Cobalt Ions as Electrocatalytic Site for Oxygen Reduction. *Chem. Mater.*, Vol.17, No.17, pp. 4278-4281, ISSN 0897-4756

Zhang, D.; Zhang, X.; Chen, Y.; Yu, P.; Wang, C. & Ma Y. (2011). Enhanced capacitance and rate capability of graphene/polypyrrole composite as electrode material for supercapacitors. *J. Power sources*, Vol.196, No.14, pp. 5990-5996, ISSN 0378-7753

Zhang, G. & Yang, F. (2007). Electrocatalytic reduction of dioxygen at glassy carbon electrodes modified with polypyrrole/anthraquinonedisulphonate composite film in various pH solutions. *Electrochim. Acta*, Vol.52, No.24, pp. 6595-6603, ISSN 0013-4686

Zhang, G.; Yang, F. & Yang, W. (2007). The effect of polypyrrole-bound anthraquinonedisulphonate dianion on cathodic reduction of oxygen. *React. Funct. Polym.*, Vol.67, No.10, pp. 1008-1017, ISSN 1381-5148

Zhang, T. & Zeng, C.L. (2005). Corrosion protection of 1Cr18Ni9Ti stainless steel by polypyrrole coatings in HCl aqueous solution. *Electrochim. Acta*, Vol.50, No.24, pp. 4721–4727, ISSN 0013-4686

Zhang, X.; Wang, J.; Wang, Z. & Wang S. (2005a). Electrocatalytic reduction of nitrate at polypyrrole modified electrode. *Synth. Met*, Vol.155, No.1, pp. 95-99, ISSN 0379-6779

Zhang, Z.; Yuan, Y.; Liang, L.; Cheng, Y.; Xu, H.; Shi, G.; Jin, L. (2008). Preparation and photoelectrochemical properties of a hybrid electrode composed of polypyrrole encapsulated in highly ordered titanium dioxide nanotube array. *Thin Solid Films*, Vol.516, No.23, pp. 8663–8667, ISSN 0040-6779

Zhao, H.; Li, L.; Yang, J. & Zhang, Y. (2008a). Nanostructured polypyrrole/carbon composite as Pt catalyst support for fuel cell applications. *J. Power Sources*, Vol.184, No.2, pp. 375–380, ISSN 0378-7753

Zhao, H.; Li, L.; Yang, J.; Zhang, Y. & Li, H. (2008b). Synthesis and characterization of bimetallic Pt–Fe/polypyrrole–carbon catalyst as DMFC anode catalyst. *Electrochem. Commun.*, Vol.10, No.6, pp. 876–879, ISSN 1388-2481

Zhao, H.; Yang, J.; Li, L.; Li, H.; Wang, J. & Zhang, Y. (2009). Effect of over-oxidation treatment of Pt–Co/polypyrrole-carbon nanotube catalysts on methanol oxidation. *Int. J. Hydrogen Energy*, Vol. 34, No.9, pp. 3908–3914, ISSN 0360-3199

Zhao, S.; Zhang, G.; Fu, L.; Liu, L.; Fang, X. & Yang, F. (2011). Enhanced Electrocatalytic Performance of Anthraquinonemonosulfonate-Doped Polypyrrole Composite: Electroanalysis for the Specific Roles of Anthraquinone Derivative and Polypyrrole Layer on Oxygen Reduction Reaction. *Electroanalysis*, Vol.23, No.2, pp. 355-363, ISSN 1521-4109

Zhou, Q.; Li, C. M.; Li, J.; Cui, X. & Gervasio, D. (2007). Template-Synthesized Cobalt Porphyrin/Polypyrrole Nanocomposite and Its Electrocatalysis for Oxygen Reduction in Neutral Medium. *J. Phys. Chem. C*, Vol.111, No.30, pp. 11216-11222, ISSN 1932-7455

8

Polypyrrole Soft Actuators

Yasushiro Nishioka
College of Science and Technology/Nihon University
Japan

1. Introduction

Organic soft actuators attract strong attentions because they have many advantages compared to conventional mechanical actuators. Organic soft actuators are generally light and flexibly deformed. In addition, they operate under low voltage ranges as low as 1 volt or so, and generate no sound during deformation. Amongst those soft actuators, bending soft actuators are of special interest because small volume change in organic materials can cause a large bending displacement. For example, soft actuators consisting of Ionic Conducting Polymer Films (ICPF) have been widely used as bending actuators (Guo et al., 1996). However, the fabrication processes for ICPF actuators seem complicated. In contrast, polypyrrole films synthesized using electropolymerization as a new material for organic soft actuators have been extensively studied (Hara et al., 2004a, 2004b, Hatchison et al., 2000, Jager et al., 1999). In addition, the amount of the volume change can be modified by altering the electropolymerization conditions. Hara et. al. recently reported that the expansion and contraction ratio of their polypyrrole actuators exceeded 40%, which is very encouraging (Hara et al., 2005). Bending actuators can be easily fabricated by forming a bimorph structure consisting of a polypyrrole film and other film material.

In this section, simple bimorph actuators using an electropolymerized polypyrrole (PPy) film and a both-side adhesive tape were fabricated. It turned out that those actuators nonuniformly bent depending on the distance between the actuator and the counter electrode. We fabricated a structure consisting of polypyrrole/both-side adhesive tape/polypyrrole whose two polypyrrole films have different extension/contraction ratios, and found that the actuator exhibited more uniform bending deformation which was nearly independent of the distance between the electrodes.

1.1 Principle of PPy actuator

Figure 1.1. describes the principle of the PPy actuator functions. A PPy actuator and a counter electrode are placed in an electrolysis solution. When a negative voltage is applied to the actuator, negative ions in the solution will be driven out from the actuator, the actuator shrinks. This is called dedoping process. When positive voltage is applied to the actuator, the negative ions are absorbed into the PPy actuator (doping process), and the actuator expands. The conductive PPy polymer networks are loosely dangled, and the spacing between the polymer chains can expand and shrink during the doping and dedoping processes.

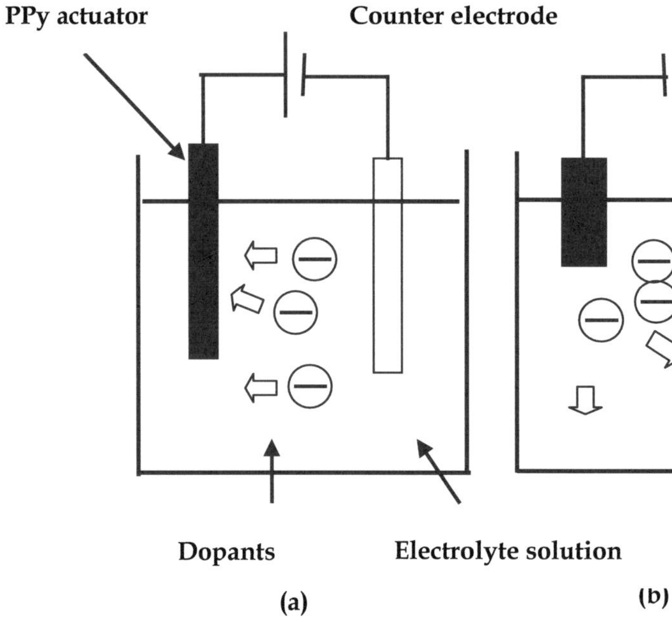

Fig. 1.1. The conceptual description of principle for the expansion and contraction processes of PPy actuators.

1.2 Experimentals

1.2.1 Galvanostatic electropolymerization of pyrrole

Figure 1.2. describes the experimental setup for the Galvanostatic electropolymerization to form PPy films. A counter electrode (Ti), a reference electrode (Ag/AgCl), and a working electrode (Ti) were immersed into a solvent containing pyrrole monomers and an electrolyte, and the bias voltage was controlled to keep constant current between the counter electrode and the working electrode during the PPy polymerization. The PPy film deposited on the working electrode was peeled off, and was used as an actuator material. The characteristics of PPy polymers had very different characteristics influenced by different synthetic conditions of the galvanostatic electropolymerization. For example, if different electrolytes were used, the expansion contraction ratios were very different (Han et al., 2004). The polymerization was done using a computer controlled potentiostat (HZ-5000, Hokuto Denko Corp.).

Here, PPy films with different expansion/contraction behaviors were prepared under different fabrication conditions (PPy1 and PPy2) listed below.

a. PPy1; Electrolyte: N.N-Diethyl-N-methyl-N-(2-methoxyethyl) ammoniumbis (trifluoro methane slfonyl) imide, Solvent: Methyl Benzoate, Room temperature, Current: 0.55 mA/cm^2, Deposition time: 4hrs)
b. PPy2; Electrolyte: tetra-*n*-butylammonium trifluoromethanesulfonate ($TBACF_3SO_3$), Solvent: Methyl Benzoate, Room temperature, Current: 0.50 mA/cm^2, Deposition time: 2hrs)

Here, two films with the thickness of 68.8 μm and 138.8 μm were prepared under the PPy1 conditions, and a film with the thickness of 28.6 μm was prepared under the PPy2 conditions. The thicknesses of the PPy films were measured using a micrometer. Three samples of PPy1(68.8 μm)/adhesive tape, PPy1(138.8 μm)/adhesive tape, and PPy1(138.8 μm)/adhesive tape/PPy2(28.6 μm) were fabricated. The PPy film and the adhesive tape film were stacked together. The stacked structure was cut to form actuators with the appropriate size. The area of the samples investigated here was 5×20 mm^2. The adhesive tape used here is a both-side adhesive tape (NW-10S) produced by Nichiban Inc.

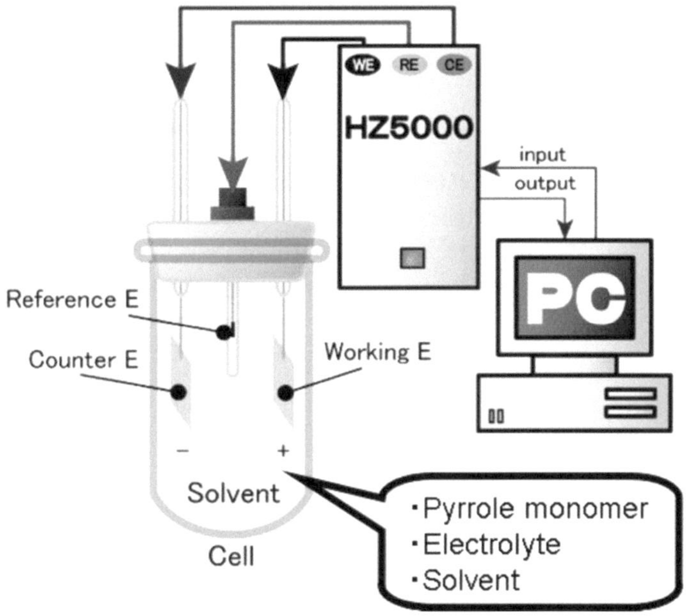

Fig. 1.2. Setup for Galvanostatic electropolymerization.

1.2.2 Actuator characterizations

The bending experiments for the fabricated devices were performed using the same experimental setup for the PPy polymerization as shown in Fig. 1.2. Here, the potentiostat HZ-5000, Hokuto Denko Corp. was also used to perform the actuation experiments. The top part of the PPy actuator was connected to the working electrode using a metal clip to make the electrical contact. Then, the counter electrode, the reference electrode, and the the actuator were immersed in a water solution of an electrolyte, lithium bis-trifluoromethane sulphonyl imide (LiTFSI). A bias voltage was given at the working electrode during the bending experiments. The bias ranges was -1.0~+1.0 V, and the bias sweep rate was 10 mV/s. Please note that only the lower part of the actuator was immersed in the LiTFSI solution, and that the conducting PPy actuator acts as the working electrode.

1.2.3 Results and discussions

Figure 1.3a and figure 1.3b show the photographs during deformation of the PP1(68.8 μm)/adhesion tape actuator. Obviously, the bi-directional bending of the actuator was observed. However, as seen in the Fig. 1.3a, the actuators edge area close to the counter electrode shows a curl, and the actuator bending was not uniform.

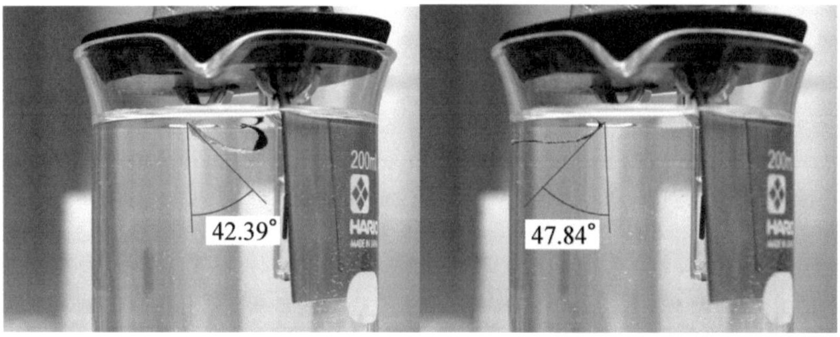

(a) (b)

Fig. 1.3. Bi-directional bending organic actuator of PPy1(68.8 μm)/adhesive tape.

This phenomenon may be due to the fact that the electric field between the actuator and the counter electrode increases resulting in the increase of the ionic current when the actuator comes closer to the counter electrode. Therefore, it was considered that some techniques to prevent this phenomenon were necessary. It was considered that if PPy1/adhesive tape/PPy2 with different extension/contraction ratios or different thickness may prevent non-uniform bending of the actuators. Figure 1.4. exhibits the bending characteristics of the PPy1 (138.8 μm)/both-side adhesive tape/PPy2 (38.8 μm). Note that the actuator shown in Fig. 1.4. shows fairly uniform bending independent of the relative positions from the counter electrodes with some expense of the bending angle.

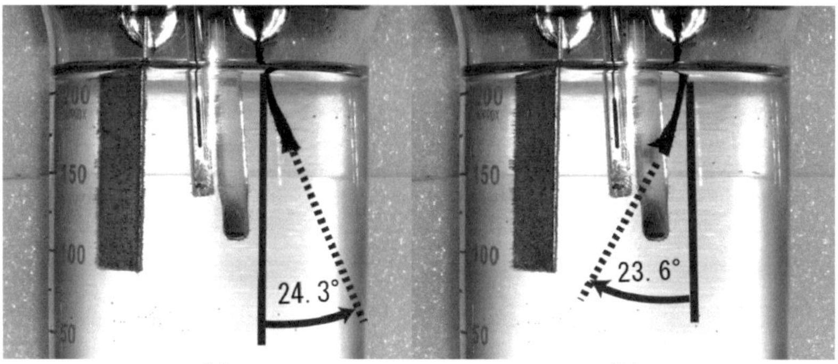

(a) (b)

Fig. 1.4. Bi-directional bending organic actuator of PPy1(138.8 μm)/adhesive tape/PPy2(28.6 μm).

The experimental results are summarized in Table.1. As seen in Table.1, the bending angle increases when the PPy film thickness increases, and the bending angle can be adjusted by changing the thicknesses of the PPy films on both sides.
The reason for the uniform bending has still not yet been clarified. However, it might be speculated that when a part of the actuator comes closer to the counter electrode, the amount of dopants penetrating into the part will be larger, which may cause the nonuniform bending of the PPy/adhesive actuator. On the other hand, this nonuniform bending may be canceled in the PPy1/adhesive tape/PPy2 structure, because the amounts of the dopants of the PPy1 and the PPy2 films near the counter electrode become larger simultaneously.

Thickness (μm)		Bending angle (deg)		Total (deg)
PPy1 (counter electrode side)	PPy2 (Rear side)			
68.8	None	+42.4	-47.8	90.2
138.8	None	+11.3	-85.4	96.7
138.8	28.6	+24.3	-23.6	47.9

Table 1. Results of bending angles of the PPy actuators ※ Positive angle means that actuators bend towards direction opposite to the counter electrode.

1.3 Conclusion

Two kinds of polypyrrole films were galvanostatically fabricated, and the polypyrrole/adhesive bending actuators were fabricated. The actuators exhibited a large bending angle range of nearly 90°. However, when the actuators approached the counter electrode, the bending angle increased which resulted in non-uniform bending of the actuators. Then, the three-layered actuator of PPy1/adhesive tape/PPy2, whose PPy films have different expansion/contraction ratios, was fabricated and characterized. The three layered actuator showed pretty uniform bending deformation with some expense of the bending angle.

2. Polypyrrole soft actuator having corrugated structures

2.1 Introduction

PPy actuators have been regarded as a possible candidate for an artificial muscle because the generation of stress is nearly ten times larger than human muscles. However, the expansion and contraction ratio of a PPy actuator was only 1-3 % which was far smaller than the value of human muscle, 25% (Hara et al., 2005, Baughman, 1996). However, Hara et. al. recently reported that an expansion and contraction ratio of their PPy actuators exceeded 40%, which is very encouraging (Hara et al., 2004). There are also several reports to improve the expansion and contraction ratios by modifying PPy film deposition conditions such as temperatures and modifying supporting electrolyte solutions (Hara et al., 2005a, 2005b, 2005c, Zama et al., 2005a, 2005b, Ogasawara et al., 1986).

In this section, a new structure of PPy actuators to improve the expansion and contraction ratio is proposed. Here, the details of fabrication procedures and the characterization results of the corrugated PPy soft actuator are reported.

2.2 Experimentals

2.2.1 Fabrication processes of PPy soft actuator

The PPy thin film was fabricated by electropolymerization. The electropolymerization of PPy was done in a methyl benzoate solution with a volume of 50 ml in which pyrrole monomers with a concentration of 0.25 mol dm^{-3} and the electrolyte (N, N- Diethyl- N-methyl-N-(2-methoxyethyl) ammonium bis (trifluoromethanesulfonyl) imide with a concentration of 0.2 mol dm^{-3} are dissolved.

The working electrode was a Au thin film sputtered on an acrylic board and the counter electrode was a Ti plate. The area of the electrodes immersed in the solvent was 30×25 mm^2. The polymerization was done at a constant current of 0.2 mA cm^{-2} for 4 hrs at the room temperature. The PPy thin film was polymerized with the doped anions in the supporting electrolyte. The PPy thin film was then peeled off from the electrode, and it was used as an actuator.

A working electrode with a corrugated structure and a plane structure were used for fabricating PPy thin film having the corrugated structure and the plane structure. Figure 2.1. describes the process to fabricate the corrugated PPy actuator. The corrugated working electrode was made by the following methods. Firstly, an acrylic board was processed to form a corrugated structure which had the ditches of 1 mm in depth, and spacing between the ditches was 5 mm as seen in Fig. 2.1. Next, the Au thin film was coated on the processed surface of the acrylic board by sputter deposition. PPy was polymerized on the acrylic board covered by the Au thin film which worked as a working electrode. Then, the corrugated PPy thin film was peeled off from the electrode by dissolving the acrylic board in acetone. The PPy thin film was cut into slices with the width of 6 mm, and these were used as actuators. PPy films without the corrugated structure were also fabricated on the plane acrylic board with the Au film for comparison. The thicknesses of the PPy films were measured using a micrometer, and these are approximately 14.5 μm for both of the corrugated PPy film and the plane PPy film.

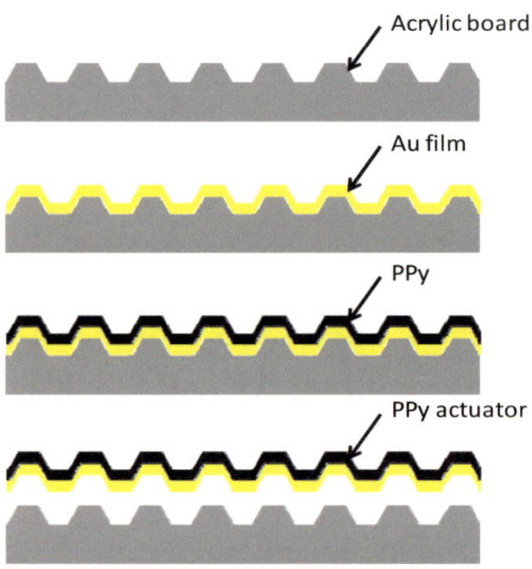

Fig. 2.1. Conceptual description of the corrugated PPy actuator fabrication process.

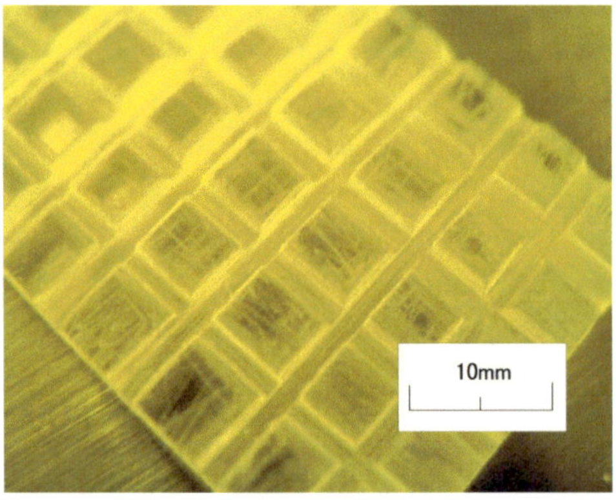

Fig. 2.2. Optical microscope image of the surface of the corrugated acrylic board.

Figure 2.3a. and 2.3b. show the scanning electron microscope (SEM) images for the PPy films formed on the corrugated working electrode and the plane working electrode, respectively. The surfaces of the both PPy films look like a sponge surface. However, no notable differences were observed between the two films.

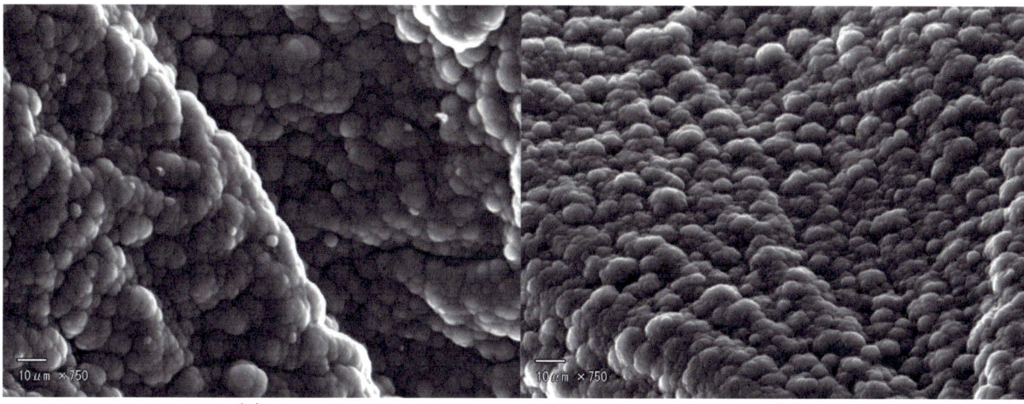

(a) (b)

Fig. 2.3. Scanning electron microscope images of the PPy surfaces formed on the corrugated working electrode (a), and the PPy surface formed on the plane working electrode (b).

2.2.2 Characterization of PPy soft actuator

Figure 2.4. shows the actuator characterization system which utilizes a balance to measure the expansion and contraction ratios. The PPy actuator was used as a working electrode in the LiTFSI solution of 1 mol dm^{-3}, and the both of PPy actuator edges were suspended by

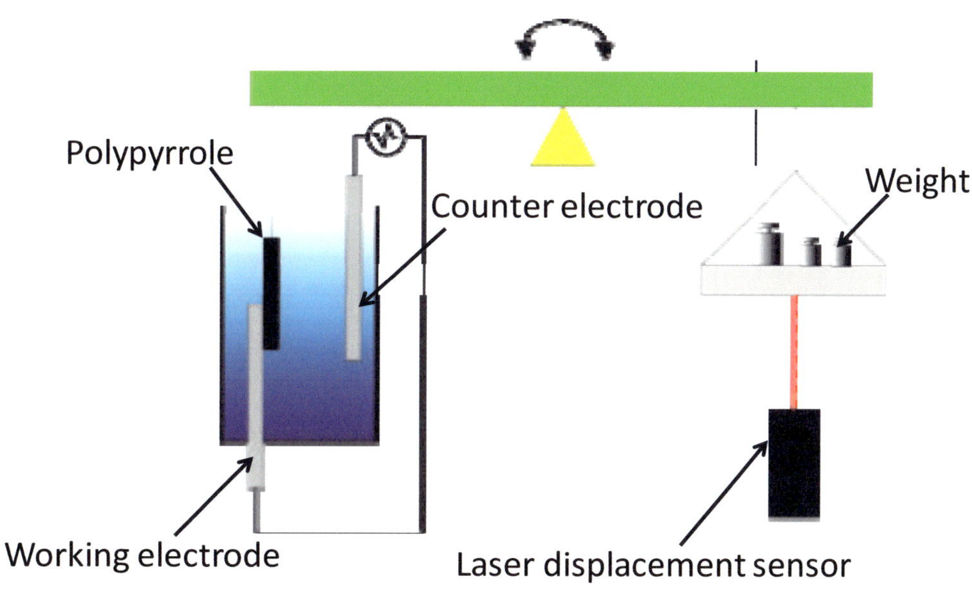

Fig. 2.4. Conceptual description of experimental setup for measuring the expansion and contraction ratio of PPy actuators.

clips. The size of the moving part of the PPy actuators was 6 mm in width, 23.8 mm in length, and 14.5 μm in thickness. The PPy actuator exhibited the expansion and contraction motions under the altering bias with the triangular wave shape applied between the PPy actuator and the counter electrode. The peak values of the bias voltage were -1 V and +1 V, and the bias voltage sweep rate was 20 mVs^{-1}. The extension and contraction of the PPy actuator was measured by monitoring the displacement of the weight position using a laser displacement sensor. Moreover, an arbitrary load stress was applied on the PPy actuator by putting weights on the saucer of the balance. The weight on saucer was adjusted so that the stress to the cross section of the PPy actuator was 0.3 MPa. The reason for selecting the stress of 0.3 MPa was that the generated stress of human muscle is 0.3 MPa. Hereafter, the expansion and contraction ratio is defined as the actuator length change divided by the initial length.

2.3 Result and discussion

The expansion and contraction ratios of the corrugated PPy actuator and the plane PPy actuator were compared during a bias cycle between -1 V and +1 V at the sweep rate, 20 mVs^{-1} as shown in Fig. 2.5. The expansion ratio of the corrugated actuator, 6.6% was larger than that of the plane actuator, 3.1%. However, the contraction ratios of these actuators are nearly the same. The full swing of the expansion and contraction ratio of the corrugated PPy actuator was 11.6%, while that of the plane (normal) PPy actuator was 8.1%. In addition, the response time of the corrugated PPy actuator is notably shorter than that of the plane PPy actuator.

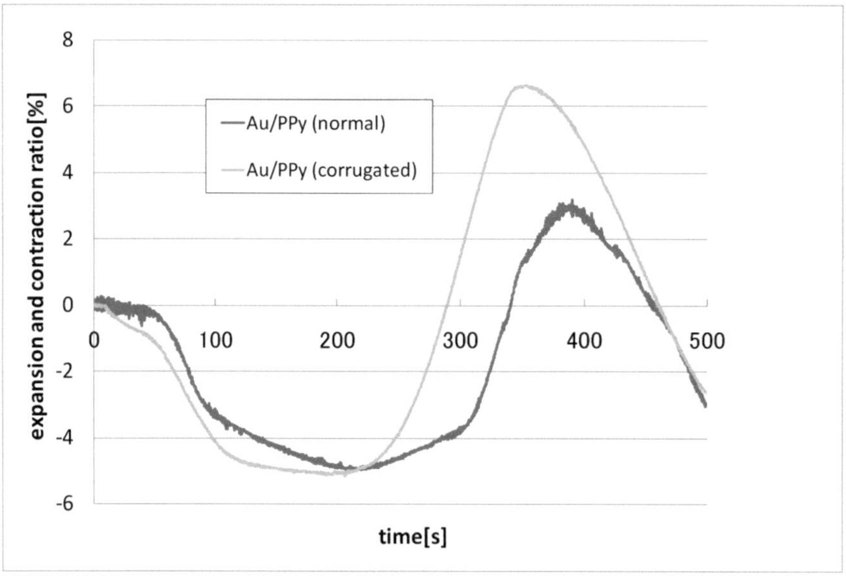

Fig. 2.5. Comparison of the expansion and contraction ratios of the corrugated PPy actuator and the plane PPy actuator during one bias cycle (sweep rate 20 mVs-1 between -1 V and +1 V).

The reason for the increase of the expansion ratio of the corrugated PPy actuator has not yet been clarified. One possible cause for that is that the corner parts of the corrugated PPy actuators expands and become more flat, and another cause is that the corrugated PPy film surface area is larger than that of the plane PPy surface. This is because the amount of the absorbed dopants in the corrugated actuator was larger than that of the plane PPy.
The behaviors of the expansion and contraction for the repeated 15 bias cycles are also shown in Fig. 2.6. The peak and bottom values of the expansion and contraction ratios of both of the PPy actuators continued to increase as the cycle numbers increases. It is clear that the creep or memory effect is observed. In other word, the total length of the both actuators continued to increase. It should be also noted that the difference between the peak value and the bottom values becomes smaller in the plane actuator, while that of the corrugated actuator remains the same level. The mechanisms for the different behaviors of those PPy actuators are not clear. However, the corrugated PPy actuator seems to have better performances than those of the plane PPy actuator.

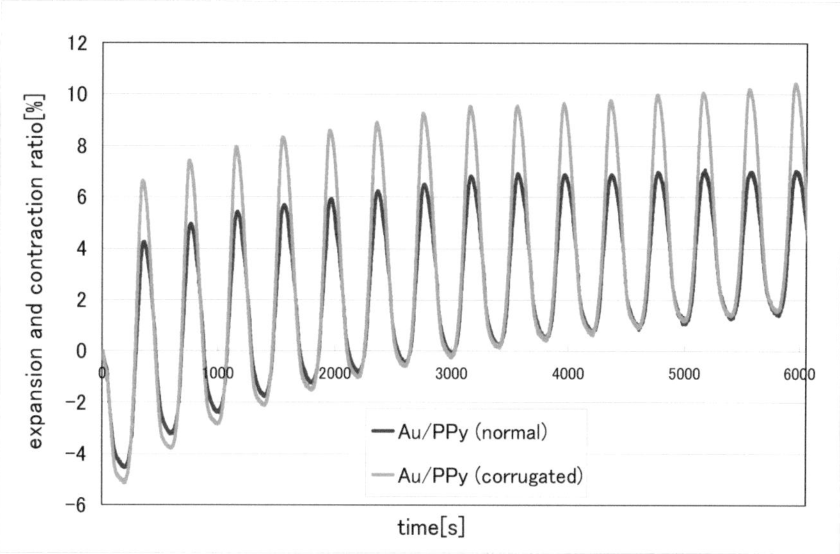

Fig. 2.6. Expansion and contraction of the PPy actuators during 15 biasing cycles. The bias voltage range was between -1 V and +1 V, and the voltage sweep rate was 20 mVs-1.

2.4 Conclusion

The corrugated PPy actuator was fabricated using the electropolymerization method. The PPy film was deposited on the Au film deposited corrugated acrylic board which worked as a working electrode during the electropolymerization. The acrylic board was then etched off to release the actuator structure. The fabricated corrugated PPy actuator exhibited the larger total swing of the expansion and contraction ratio of 11.6 %, while that of the plane (normal) PPy actuator exhibited 8.1%. Although the creep effects was observed in the both type of the PPy actuators, the total swing of the expansion and contraction ratio remained at the similar level in the corrugated PPy actuator, and that of the plane (normal) PPy actuator continued to decrease within the voltage swing (-1 V to + 1 V) for 15 cycles. Thus, it may possible to conclude that the corrugated PPy actuator have better performances than those of the plane PPy actuator.

3. Effect of 2-propanol concentration in electrolyte solution on polypyrrole actuator performance

3.1 Introduction

Recently, it has been reported that some PPy actuators exhibit strains of more than 10%, and that some of those even achieved strains of up to 40% (Hara et al., 2004). The improved strain has been mostly achieved using an electrolyte of tetra-n-butylammonium bis(trifluoromethansulfonyl)imide (TBATFSI) during PPy electropolymerization. These actuators generally function under a low potential voltage range less than 1 V.

Hara et al. reported that their TFSI-doped porous PPy films exhibited increased deformation when their aqueous lithium bis(trifluoromethansulfonyl)imide (LiTFSI) electrolyte solutions contained propylene carbonate (Hara et al., 2004). They attributed those effects to the swelling of the PPy film caused by the penetration of propylene carbonate. The swelled PPy film could more easily pass TFSI anions. We also immersed TFSI-doped PPy films into several organic chemicals, and found that the PPy films showed notable swelling in 2-propanol (Hoshino et al., 2011). Therefore, it was interesting to investigate whether the PPy actuators show improved deformation behaviors in aqueous LiTFSI solutions containing 2-propanol. In this section, the electrochemical deformation characteristics of TFSI-doped PPy soft actuators in aqueous LiTFSI solutions with different 2-propanol concentrations are reported.

3.2 Experimental procedure

The polymerization of PPy films was carried out using a computer-controlled potentio-galvanostat (Hokuto Denko HZ-5000). A counter electrode (Ti), a reference electrode (Ag/AgCl), and a working electrode (Ti) were immersed into methyl benzoate solutions of 0.25 M pyrrole and 0.2 M N,N-diethyl-N-methyl-N-(2-methoxyethyl)ammonium bis(trifluoromethanesulfonyl)imide, and the potential voltage was controlled to keep a constant current of 0.2 $mAcm^{-2}$ for 4 h at 20 ^{o}C between the counter electrode and the working electrode. The thickness of the PPy films was measured to be approximately 150 μm using a micrometer. The obtained films were peeled off from the electrode, rinsed with acetone, and dried in air. The PPy films were cut into 20 x 5 mm^2 strips to form the PPy actuators.

The actuator characterization system that utilizes a balance to measure the expansion and contraction ratios under load stress was described in Fig. 2.4. (Morita et al., 2010, Chida et al., 2010). The PPy actuator was used as the working electrode in the 1 M LiTFSI aqueous electrolyte solutions containing 2-propanol at various concentrations of 0, 20, 30, 40, 60, 80, and 100%. Both of the PPy actuator ends were clipped with two metal plates. The PPy actuator exhibited the expansion and contraction motions under the alternating potential with the triangular wave shape applied between the PPy actuator and the counter electrode. The potential voltage difference between the PPy actuator and the electrolyte solution was monitored using the Ag/AgCl reference electrode. The peak values of the potential voltage were -1 and +1 V, and the potential sweep rate was 10 mVs^{-1}. The extension and contraction of the PPy actuator was measured by monitoring the displacement of the weight position using a laser displacement sensor (Keyence LE-4000). Moreover, a load stress of 0.3 MPa was applied on the PPy actuator by placing corresponding weights on the saucer of the balance.

3.3 Results and discussion

Figure 3.1a shows the time dependences of the strain of the actuators, as measured by the displacement of the weight as a function of time under a load stress of 0.3 MPa. The strain in Fig. 3.1a is defined as the length change of the PPy actuators divided by the length prior to deformation. No potential voltage was given for the first 30 s, and after that repeated voltage sweeps with a period of 400 s were applied to the PPy actuators, as shown in Fig. 3.1b. The characteristics of the actuators were measured in aqueous solutions of LiTFSI with different 2-propanol concentrations of 0, 20, 80, and 100%. During the initial 130 s of the first cycle, the PPy actuators immersed in the electrolyte solutions with 2-propanol concentrations of 0 and 20% showed slight reduction of the PPy length. Notable actuator elongation of approximately 3.5% due to swelling was observed in the electrolyte solutions with the 2-propanol concentrations of 80 and 100% even when the PPy potential was negative for the first 130 s. In this case, there should be no penetration of $TFSI^-$ anions into the PPy actuators. This may be due to the fact that neutral 2-propanol molecules penetrate into the PPy porous structures resulting in the elongation of the actuators (swelling) under stress.

Here, it should be noted that the strain of the measured PPy actuator consists of the electrochemical strain caused by anion motions and the swelling (creeping) strain. Therefore, the electrochemical strain is defined by the difference in the peak value minus the lowest value of the strain for each potential cycle. The electrochemical strain of the PPy actuator driven in the 0% 2-propanol electrolyte for the second cycle was nearly 3%. In contrast, the electrochemical strain of the PPy actuator driven in the aqueous solution of LiTFSI containing 20% 2-propanol was increased up to 12%. However, the electrochemical strain of the actuator driven in the LiTFSI solution of 80% 2-propanol became nearly 5%, and it continued to decrease as time elapsed. In this case, the maximum strain of the PPy actuator stayed at approximately 12.5%, but the minimum strain of the actuators continuously increased. The electrochemical strain of the actuator driven in the LiTFSI solution of 100% 2-propanol decreased to 2%. Here, the creeping effect seems to dominate, but the electrochemical strain is suppressed. This may be due to the fact that the ionic conductivity of the LiTFSI electrolyte solution is much smaller than that of the aqueous solution. The minimum strain of the actuator continues to increase as time elapses. This behavior appears to be similar to the creeping effect in metal deformation processes. Sendai et al. recently reported their detailed study on the creeping effect of PPy actuators, and concluded that this elongation could be recovered by releasing the stress during the deformation [15]. Hence, they called this phenomenon the memory effect.

Figure 3.2. shows the 2-propanol concentration dependence of the electrochemical strain. Note that the electrochemical strain of the PPy actuators shows the maximum at a 2-propanol concentration between 20 and 40%. Since this 2-propanol concentration range is fairly wide, slight change in the 2-propanol concentration may not affect the performance of the PPy actuators.

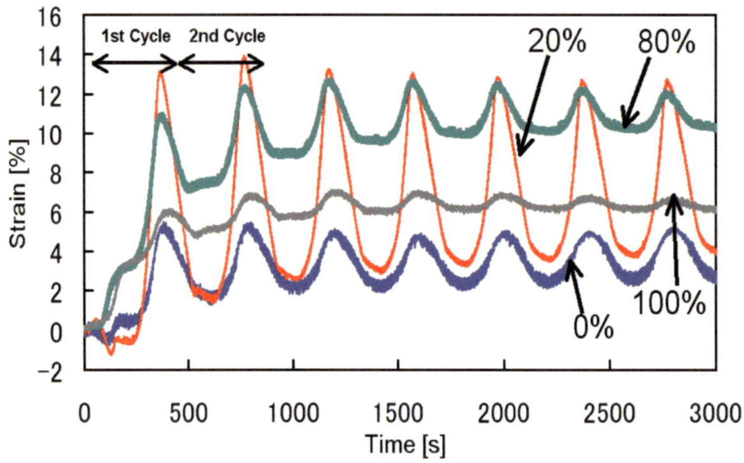

(a)

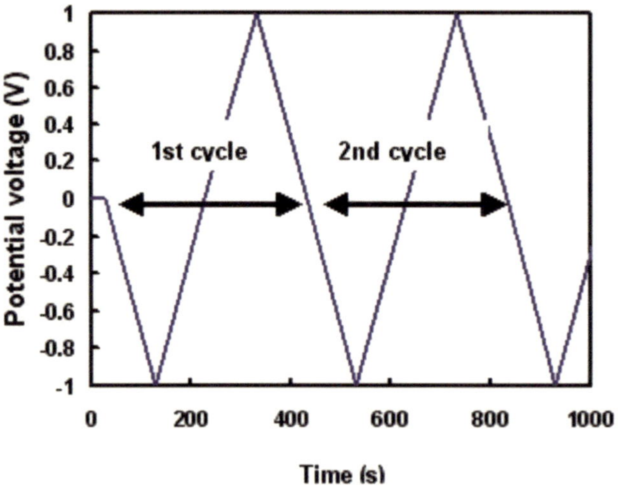

(b)

Fig. 3.1. (a) Relationship between the strain of the actuator and time under the repeated operation of the actuator for 8 times. The actuators were electrochemically deformed in aqueous solutions of 1 M LiTFSI mixed with 2-propanol concentrations of 0, 20, 80, and 100%. The load stress during the deformation was 0.3 MPa. (b) Potential voltage shape shown as a function of time. The potential voltage during the first 30 s was maintained at 0 V, and no stress was applied. A stress of 0.3 MPa was applied after 30 s.

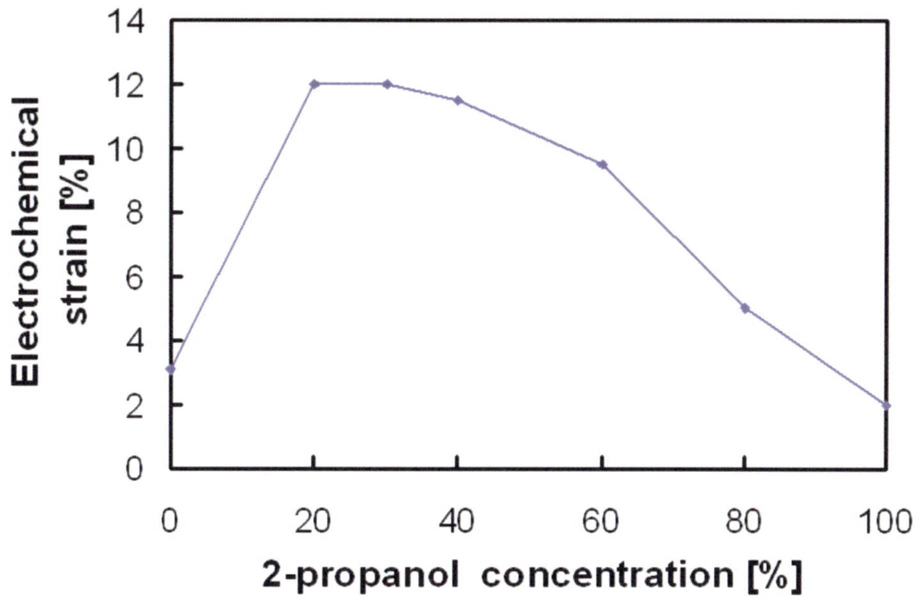

Fig. 3.2. Relationship between the electrochemical strain and the 2-propanol concentration of LiTFSI electrolyte solutions.

Figure 3.3. shows comparisons of (a) the potential voltage dependences of the strain of the second cycle for the PPy actuators deformed in the electrolyte solutions with 0 and 20% 2-propanol concentrations, and (b) the potential voltage dependence of the ionic current flowing into the PPy actuators for the second cycles. The curve (b) is called a "cyclic voltammogram (CV)". When the current increases, the PPy actuator expands. This corresponds to the anion ($TFSI^-$) penetration into the positively biased PPy actuator. In contrast, Li^+ cations will penetrate into the PPy film when the actuator is negatively biased. However, the diameter of the cations is much smaller than that of $TFSI^-$ anions. Thus, the elongation during the negative bias should be negligibly small. It is clear that both CV characteristics corresponding to the 0 and 20% 2-propanol concentrations suggest that the deformation of the PPy actuator is due to the penetration of the $TFSI^-$ anions. The electrochemical strain is much larger for the PPy actuator driven in the 20% 2-propanol electrolyte solution than that in the 0% 2-propanol electrolyte solution. As shown in Fig. 3.3.b, the hysteresis of the PPy actuator in the 20% 2-propanol electrolyte is much larger than that in the 0% 2-propanol electrolyte. This means that the number of $TFSI^-$ ions penetrating into the PPy actuator and out diffusing from that in the 20% 2-propanol electrolyte solution is much larger than that in the 0% 2-propanol electrolyte solution, which explains the increased deformation of the PPy actuator in the 20% 2-propanol electrolyte solution.
Figure 3.4. shows a comparison of the CV characteristics of the PPy actuators deformed in the LiTFSI aqueous solutions with the 2-propanol concentrations of 30, 40, 60, 80, and 100%. The large hysteresis was maintained for the CV curves for 20 [Fig. 3.3.b], 30, and 40%. These CV curves became smaller as the concentration of 2-propanol increases above 60%. Both cation and anion currents are suppressed, possibly due to the reduced ionization ratio of LiTFSI in the 60, 80 and 100% 2-propanol solutions, which explains the reduced deformation range above 60%.

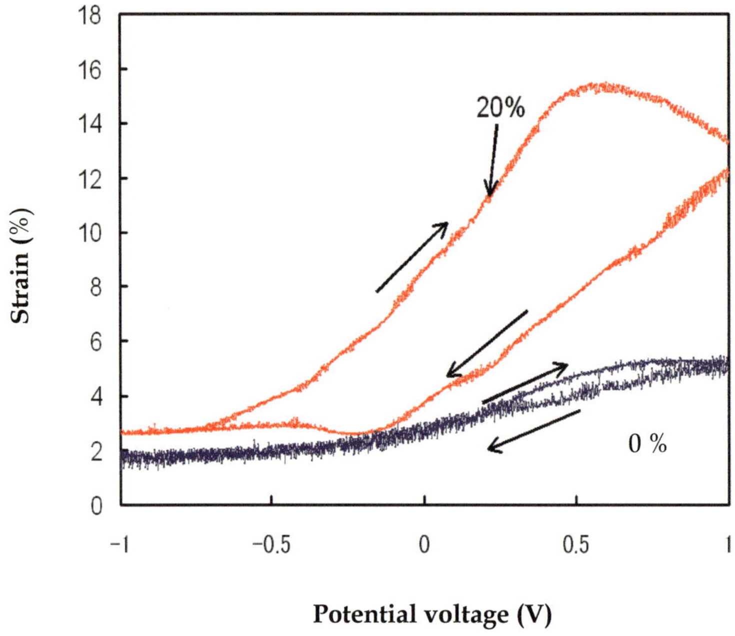

(a)

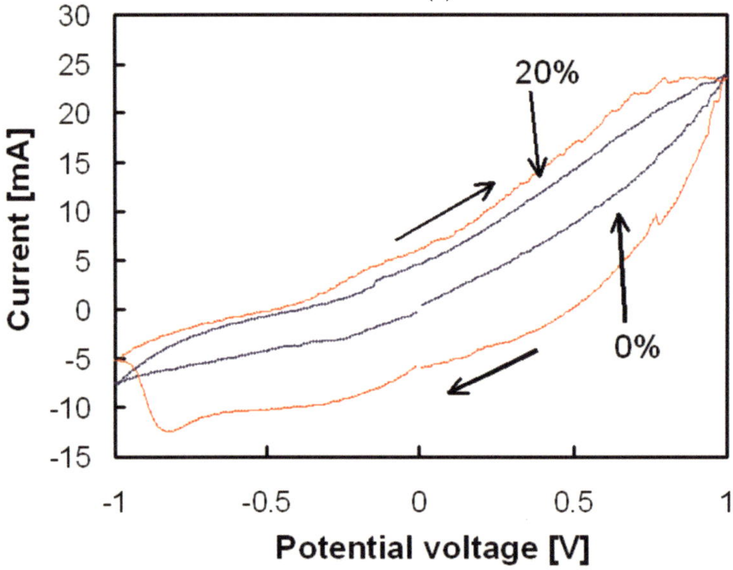

(b)

Fig. 3.3. (a) Relationships between potential voltage applied on PPy actuators and the strain in the LiTFSI solutions with 2-propanol concentrations of 0 and 20 %, and (b) the corresponding cyclic voltammograms of the PPy actuators and the strain in the LiTFSI solutions with 2-propanol concentrations of 0 and 20 %.

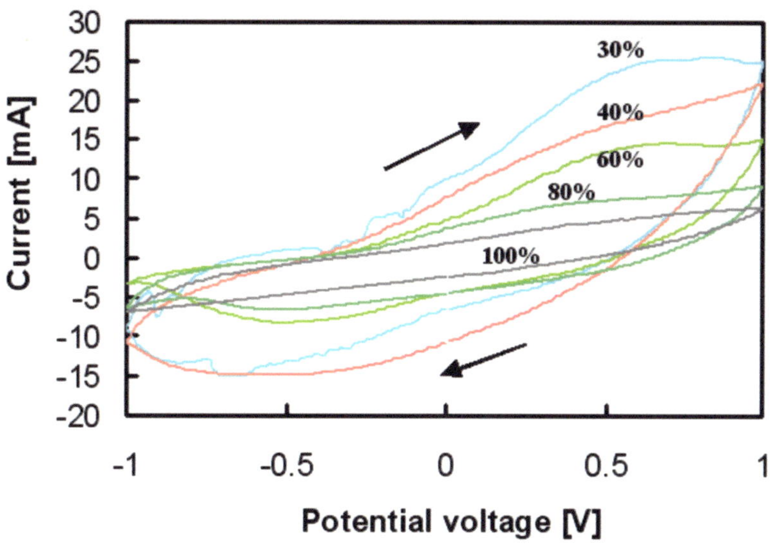

Fig. 3.4. Cyclic voltammograms for deformations in electrolyte solutions of LiTFSI mixed with 2-propanol concentrations of 30, 40, 60, 80, and 100%.

Hara et al. reported that their TFSI-doped PPy actuators exhibited the maximum deformation when their aqueous LiTFSI electrolyte solution contained 20-40% of propylene carbonate (Hara et al., 2005). They pointed out that these phenomena were due to the swelling of the PPy film with propylene carbonate. The PPy film was reported to have porous and sponge-like structures, and the swelling of the film increased the porous spacing. This will make the anions more easily penetrate into the PPy film, which results in the increase in the deformation of the PPy actuators.

It should be pointed out that the PPy actuators investigated in this research showed smaller levels of the creep effects in the electrolyte solutions containing 20-40% 2-propanol while they showed the maximum deformation range. This could not be always explained only by the swelling of the PPy films. It is speculated that the surface tension and viscosity of water and 2-propanol affect the electrochemical deformation of the PPy actuators. Table 2 compares the surface tension and viscosity of pure water and 2-propanol at 20 °C. The data were taken from the web page of the National Institute of Standards and Technology (NIST). The surface tension of 2-propanol is 21.7 dyn.cm^{-1}, which is nearly 30% that of water. Therefore, when the PPy actuator is positively biased, TFSI$^-$ anions along with 2-propanol molecules might more easily penetrate into the porous structure of PPy. Thus, the increased deformation was observed for the 2-propanol concentration range between 20 and 40%. On the other hand, the contraction of the PPy actuator was disturbed in the electrolyte solutions with the 2-propanol concentrations larger that 60%. The TFSI$^-$ ions diffused into the PPy porous structure in the positive potential region could be disturbed to escape from the PPy structures due to 2-propanol's high viscosity nature in the negative potential region.

The mechanisms for these behaviors still need to be more carefully investigated. However, the introduction of optimised amounts of 2-propanol into the LiTFSI electrolyte solution significantly improves the deformation ranges of the PPy actuators.

	Surface tension (dyn.cm^{-1})	Viscosity (mPa.s)
Water	72.8	1.01
2-propanol	21.7	2.37

Table 2. Comparison of surface tension and viscocity for water and 2-propanol at 20 ^{o}C (data taken from the web page of the National Institute of Standards and Technology (http://webbook.nist.gov/chemistry/fluid/).

3.4 Conclusions

Soft actuators were fabricated by galvanostatic electropolymerization of the polypyrrole (PPy) thin film using a methyl benzoate electrolyte solution of N,N-diethyl-N-methyl-N-(2-methoxyethyl)ammonium bis(trifluoromethanesulfonyl)imide. The electrochemical deformation behaviors of the PPy actuators were investigated in the aqueous solutions of an electrolyte, lithium bis(trifluoromethanesulphonyl)imide (LiTFSI), mixed with various concentrations of 2-propanol. The potential voltage range given to the actuators was between -1 and 1 V with a sweeping rate of 10 mVs^{-1}, and a stress of 0.3 MPa was applied. The actuator exhibited nearly 3% of the electrochemical strain in the electrolyte solution without 2-propanol, and exhibited an electrochemical strain of up to 12% in the electrolyte solutions with the 2-propanol concentration between 20 and 40%. However, the reduction of the electrochemical strain and the acceleration of the creeping effect were observed when the actuators were deformed in the electrolyte solution with 2-propanol concentration of more than 60%.

4. Dynamic behaviors of polypyrrole actuators in electrolyte solution mixed with 2-propanol

4.1 Introduction

Hoshino et al. tried to immerse TFSI-doped PPy films into several organic chemicals, and found that the PPy films showed notable swelling in 2-propanol. They reported that the PPy actuators showed the increased electrochemical strains in aqueous LiTFSI solutions containing 20 to 40% of 2-propanol as described in the previous section (Hoshino et al., 2011).

In this section, the dynamic electrochemical deformation characteristics of TFSI-doped PPy soft actuators under potential sweep rates between 10 and 25 mVs^{-1} in aqueous LiTFSI solutions with different 2-propanol concentrations are reported.

4.2 Experimental procedure

4.2.1 The experimental procedures were already described in the previous section. 4.3 Results and discussion

Figure 4.1. shows the time dependences of the strain of the actuators, as measured by the displacement of the weight as a function of time under the load stress of 0.3 MPa. The strain in Fig. 4.1. was defined as the length change of the PPy actuators divided by the length prior to deformation. No potential voltage was given for the first 30 s, and after that repeated voltage sweeps with a period of 400 seconds at the potential sweep rate of 10 mVs^{-1} were applied to the PPy actuators.

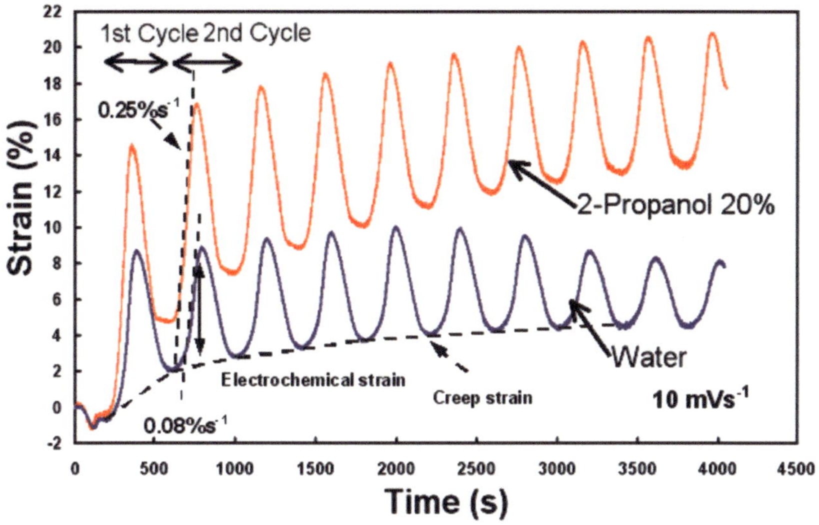

Fig. 4.1. Relationships between the strain of the actuator and time under the repeated operation of the actuator for 10 times with the potential sweep rate of 10 mVs^{-1}. The actuators were electrochemically deformed in aqueous solutions of 1 M LiTFSI mixed with 2-propanol concentrations of 0 (water) and 20%. The load stress during the deformation was 0.3 MPa.

The characteristics of the actuators were measured in aqueous solutions of LiTFSI with the different 2-propanol concentrations of 0 (water) and 20%. During the initial 130 s of the first cycle, the PPy actuators immersed in the electrolyte solutions with the 2-propanol concentrations of 0 and 20% showed slight reduction of the PPy length. Here, it should be noted that the strain of the measured PPy actuator consists of the electrochemical strain caused by anion motions and the swelling (creep) strain. Therefore, the electrochemical strain was defined by the difference of the peak value minus the lowest values of the strain for each potential cycle. The electrochemical strain of the PPy actuator driven in the 0% 2-propanol electrolyte for the second cycle exhibited nearly 7%. In contrast, the electrochemical strain of the PPy actuator driven in the aqueous solution of LiTFSI containing 20% of 2-propanol was 12%. In this case, the electrochemical strain of the PPy actuator gradually decreased after the repeated potential cycles but it stayed around 10%, while the creep strain of the actuators continuously increased. This creep strain of the PPy actuator in the 20% 2-propanol electrolyte was larger than that in the 0% 2-propanol (water) electrolyte. This seems to suggest that the creep strain is larger in the former case due to the swelling of the PPy film in 2-propanol. These creep behaviors look similar to creep effects in metal deformation processes. Sendai et al. recently reported in their detailed study on creep effects of PPy actuators, and concluded that this elongation could be recovered by releasing the stress during the deformation (Chida et al., 2010). Hence, they called this phenomenon the memory effect. The dotted straight lines correspond to the tangents of the second peaks, and the tangential slopes correspond to the electrochemical strain rate of 0.25%s^{-1} for the actuator in the 20% 2-propanol electrolyte and 0.08%s^{-1} for the actuator in the 0% 2-propanol (water) electrolyte, respectively.

It is interesting to investigate whether the response time of the PPy actuator improves when the potential sweep rate is increased. Figure 4.2. shows the potential sweep rate dependences of the electrochemical strain of the PPy actuators in the electrolyte solutions with 0 and 20% 2-propanol. The electrochemical strains continuously decreased with the sweep rate. The electrochemical strain of the PPy actuator in the electrolyte solution with 20% 2-propanol is larger than those in the electrolyte solution with 0% 2-propanol in the sweep rate range between 10 and 25 mVs^{-1}.

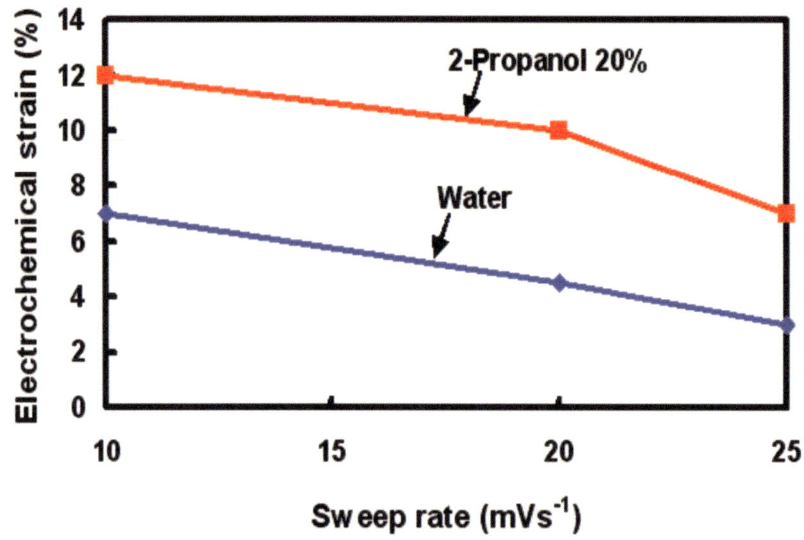

Fig. 4.2. Sweep rate dependences of electrochemical strains for PPy actuators in LiTFSI solutions with 2-propanol concentrations of 0 (water) and 20%.

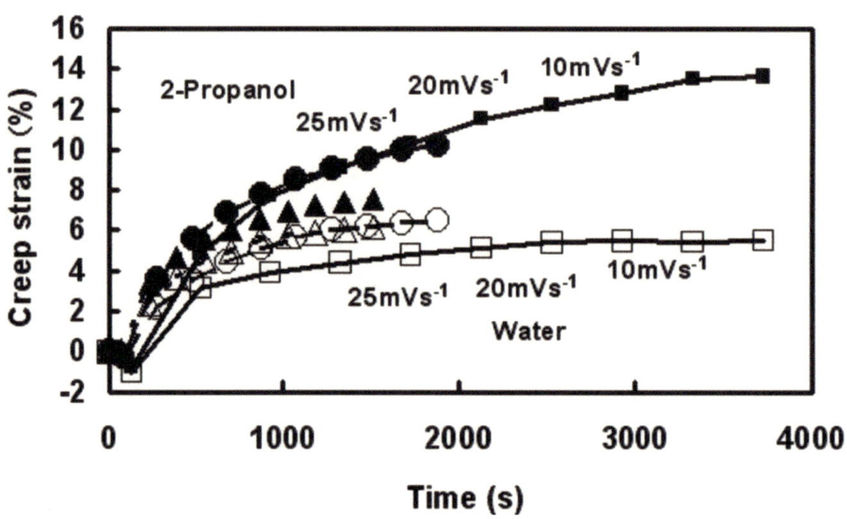

Fig. 4.3. Comparison of creep strains for PPy actuators driven in LiTFSI solutions of 0 (water) and 20% of 2-propanol with different potential sweep rates.

Figure 4.3. shows the relationships between the creep strains and time of the PPy actuators driven in the electrolyte solutions with the different potential sweep rates. The time dependences look similar, which possibly suggests that the creep strain of the PPy actuators mostly depends on time but not on the potential sweep rate. This may mean that the PPy films swell both in the water and 2-propanol water solutions, and that the swelling of the PPy is larger in the 2-propanol electrolyte solution. This situation is more clearly described in Fig. 4.4. Figure 4.4. shows the potential sweep rate dependence of the creep strain measured at the time of 1000 s. The creep strains are mostly determined by the time under the 0.3 MPa stress.

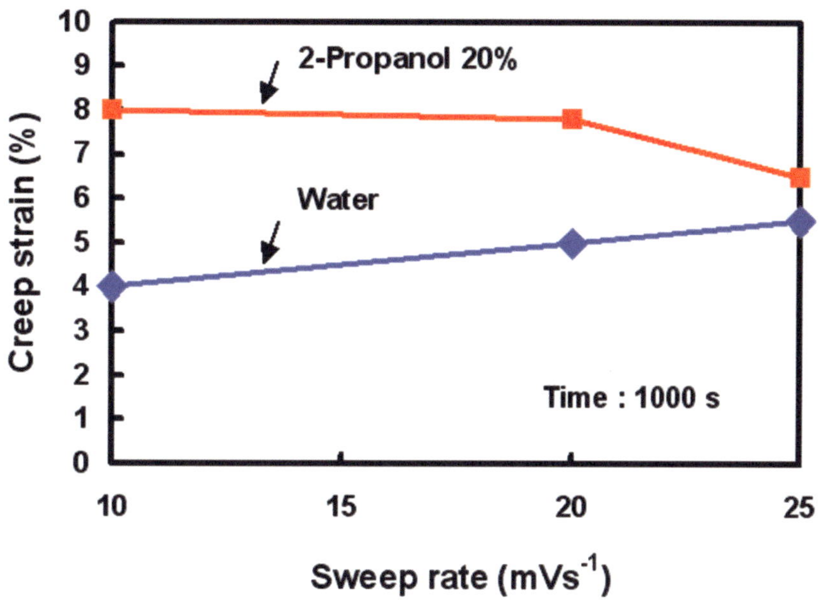

Fig. 4.4. Sweep rate dependences of creep strains for PPy actuators in LiTFSI solutions with 2-propanol concentrations of 0 (water) and 20%.

Figure 4.5. shows the potential sweep rate dependences of electrochemical strain rate of the PPy actuator in the electrolyte solution containing 0 and 20% of 2-propanol. The electrochemical strain evidently is independent of the potential sweep rate, and those for the electrolyte solution with 20% 2-propanol are always larger than those for the electrolyte solution with 0% 2-propanol. If the ionic flow rate in the electrolyte solution influences the electrochemical strain, the electrochemical strain rate should be dependent on the potential sweep rate. This is because the potential sweep rate should modify the ionic flow rate in the electrolyte solution. Since this is not the case, another mechanism should be considered in the PPy actuator functions. Hara et al. reported that their TFSI-doped PPy actuators exhibited the maximum deformation when their aqueous LiTFSI electrolyte solution contains 20 to 40% of propylene carbonate (Hara et al., 2005). They pointed out that these phenomena were due to the swelling of the PPy film with propylene carbonate. The PPy film was reported to have porous and sponge like structures, and the swelling of the film increased the porous spacing. This will make the anions more easily penetrate into the PPy film, which resulted in the increase of the deformation of the PPy actuators.

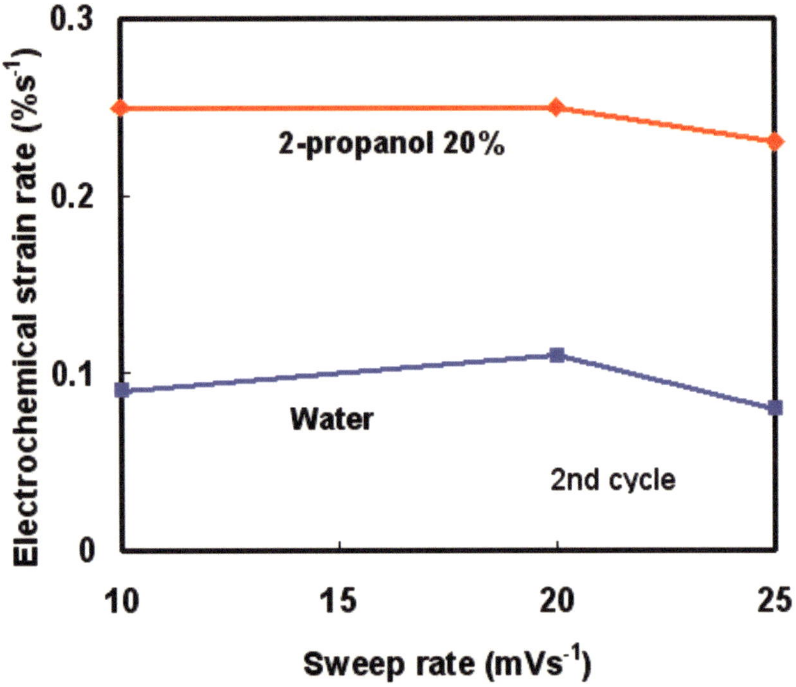

Fig. 4.5. Sweep rate dependences of electrochemical strain rates for PPy actuators in LiTFSI solutions with 2-propanol concentrations of 0 (water) and 20%.

4.3 Conclusion

The influences of the potential sweep rate on electrochemical and creep strains PPy actuators in aqueous LiTFSI electrolyte solutions with 2-propanol concentrations of 0 and 20% were compared under the load stress of 0.3 MPa. The electrochemical strain rates in the 0 and 20% 2-propanol solution were found to be approximately $0.1\%s^{-1}$ and $0.25\%s^{-1}$, respectively, and they are nearly independent of the sweep rate between 10 and 25 mVs^{-1}. These results suggest that the TFSI anion penetration into the PPy films might be limited by the interactions between the electrolyte and the PPy surface, and the reduced viscosity in the LiTFSI electrolyte solution containing 2-propanol possibly enhances the doping and dedoping of TFST anions along with the swelling effect of the PPy film by 2-propanol. The creep rate was more rapidly increased as time elapsed in the electrolyte solution containing 20% 2-propanol, which was due to the swelling of PPy film in the electrolyte.

5. Comparison of polypyrrole organic thin film actuators with or without silicon microspring

5.1 Introduction

Recently, Ding et al. reported a new type of PPy actuator that had a tubular geometry and a helical wire interconnect (Ding et al., 2003). The actuator was fabricated by forming a PPy

film on a platinum wire (125 μm in diameter) that was wrapped around with thinner (25 μm in diameter) platinum wire. A maximum strain of 5% and the response of 10%s^{-1} were achieved. A similar structure to minimize the response time of PPy actuators was also reported by Hara et al. (Hara et al., 2003, 2004a, 2004b). They deposited a PPy film on a tungsten helical coil (250 μm in diameter) made of a tungsten wire with a diameter of 30 μm. This fibrous PPy actuator exhibited a strain of 11.6 % under the load of 0.2 N. The tungsten wire also helped to reduce the potential drop within the PPy for the improved performance. Another trial to increase the extension and contraction ratios employing a corrugated PPy structure has been reported (Morita et al., 2010), and a bimorph structure of a PPy actuator for more uniform bending has also been reported (Chida et al., 2010).
In contrast, we considered applying these PPy soft actuator techniques to the actuation of the very small mechanisms used in silicon microelectromechanical systems (MEMS). To the best of our knowledge, there have been very few applications of these techniques to the actuation of small MEMS mechanisms (Chida et al., 2010, Guo et al., 1996, Hutchison et al., 2000). In this section, two kinds of PPy thin film actuators with or without a silicon MEMS microspring were fabricated and compared.

5.2 Linear actuator design

Figure 1 describes the designed actuator. The silicon microspring has a width of 0.5 mm, length of 15 mm, and thickness of 60 μm. The microspring consists of silicon wires having a cross-section of 10 x 60 μm^2. The surface of this microspring is covered by a 91 μm-thick PPy

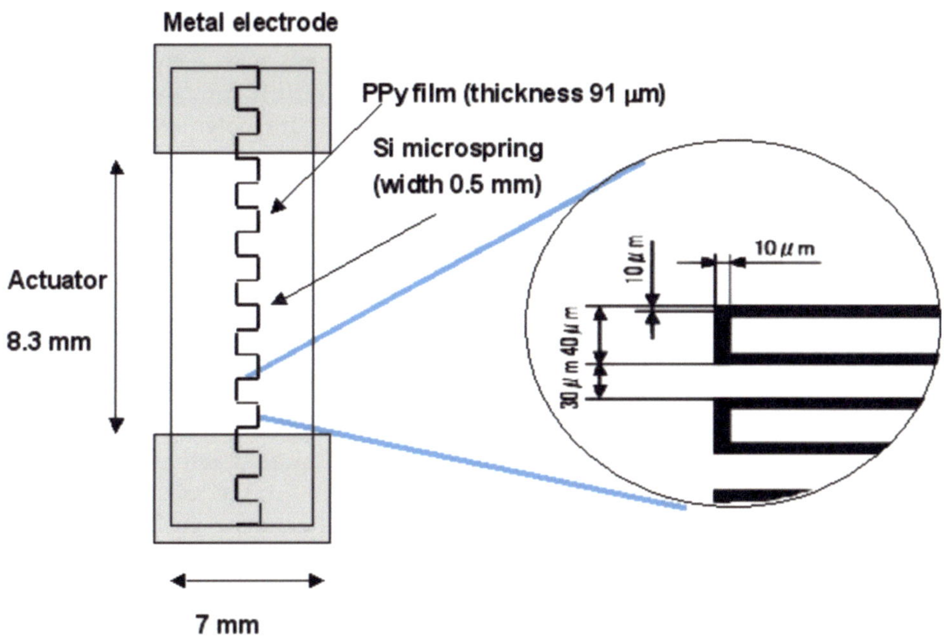

Fig. 5.1. Design of the PPy-driven silicon linear actuator; (a) corresponds to designed silicon micro spring, and (b) corresponds to the designed linear actuator driven by PPy expansion and contraction motions.

film of length and width 15 mm and 7mm, respectively. The PPy film covering the microspring shrinks and expands along with the PPy film beside the microspring, which causes the actuation of the microspring. The top and bottom parts of the actuator were clipped with metal electrodes. As a result, the working area of the actuator was 8.3 mm in length and 7 mm in width. A PPy actuator with the same dimensions without the silicon MEMS micro spring was also fabricated for comparison.

5.3 Actuator fabrication processes

Figure 5.2. describes the fabrication processes of the microactuator. First, an extremely thin (60 μm) silicon film was anodically bonded to a glass substrate of 1.5 mm in thickness (a). Next, the silicon microspring pattern was photolithographycally defined on the silicon film, followed by silicon etching using an inductively coupled plasma (ICP) dry etcher (b). The silicon microspring was released by immersing the structure in a buffered HF solution (c). The microspring was placed on an acrylic board, and a 100 nm-thick Au film was sputter deposited on the whole surface of the microspring and the acrylic board (d). Finally, the PPy film with a thickness of approximately 91 μm was electrochemically deposited on the whole surface of the sputtered Au film, and the structure was peeled off from the acrylic board in acetone (e).

The polymerization was done using a computer-controlled potentio-galvanostat (Hokuto Denko HZ-5000). A counter electrode (Ti), a reference electrode (Ag/AgCl), and a working electrode (Au) were immersed into a solvent containing pyrrole monomers and an electrolyte, and the potential voltage was controlled to keep a constant current between the counter electrode and the working electrode of the Au surface covering the silicon microspring during the PPy polymerisation. The electropolymerization of PPy was done in a methyl benzoate solution with a volume of 50 ml in which pyrrole monomers with a concentration of 0.25 mol.dm^{-3} and the electrolyte tetra-n-butylammonium bis(trifluoromethansulfonyl)imide (TBATFSI) with a concentration of 0.2 mol.dm^{-3} are dissolved. The polymerization was done at a constant current density of 0.2 mA.cm^{-2} for 4 h at room temperature. The PPy film deposited on the Au surface was peeled off, and cut into the actuator dimension of 7 x 15 mm^2 as shown in Fig. 5.3b. The PPy actuator without the Si microspring of 7 x 15 mm^2 was also cut from the same PPy film. The thickness of the plane part of the PPy film was measured to be approximately 91 μm using a micrometer. The measured electro conductivity of the PPy film was approximately 12 Scm^{-1}. In the previous publications, fairly large strain and fast response of PPy actuators, that were polymerized electrochemically with TBATFSI at the temperature of -10 ^{o}C, were reported (Hara et al., 2005a, 2004b). Therefore, the PPy actuators fabricated using (TBATFSI) were focused in this research.

Figure 5.3. shows the optical microscope image of the fabricated actuator taken from the backside (substrate side) of the actuator. The Au film in the plane PPy area was unintentionally peeled off from the PPy film during the actuator peeling off process in acetone, while the Au film covering the silicon microspring was not peeled off. Figure 5.4a. and 5.4b. show scanning electron microscope (SEM) images of the PPy actuator observed from the surface of the micro spring (electrolyte solution side) and the backside of the micro spring, respectively. It was indicated that the PPy film almost covered the silicon spring surface, and it had a rugged surface structure. Similar rugged structures were observed in the image taken from the backside. It is believed that this sponge-like structure is the origin of the large expansion and contraction ratios during the ion doping and dedoping processes.

The space between the microsprings was not filled with the PPy film. This may mean that the polypyrrole film on the microspring does not contribute the extension and contraction of the silicon actuator. Therefore, it was determined that the actuation was realized utilizing the extension and contraction of the plane PPy film deposited beside the microspring.

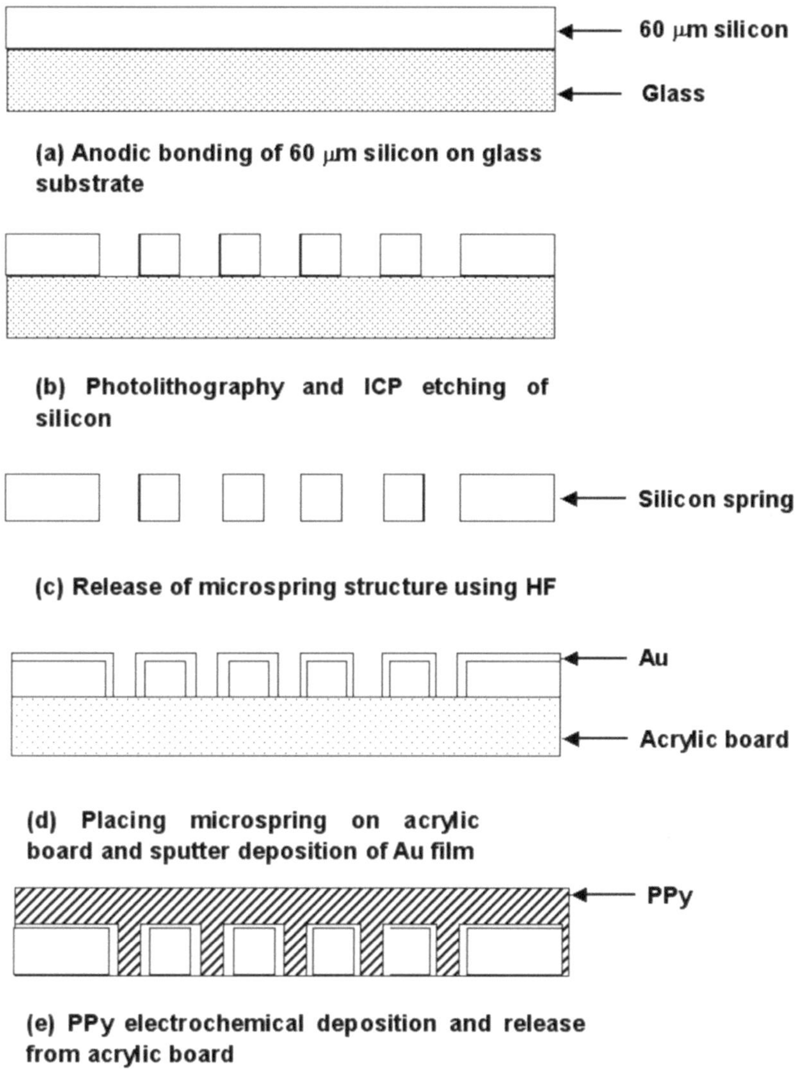

Fig. 5.2. Cross- sectionally described linear actuator fabrication processes

5.4 Characterizations

The actuator characterization system that utilizes a balance to measure the expansion and contraction ratios under the load stress has already been described in Fig. 2.4. The PPy actuator was used as the working electrode in the lithium bis(trifluoromethansulfonyl)imide (LiTFSI) electrolyte solution of 1 mol.dm^{-3}, and both of the PPy actuator ends were

Fig. 5.3. Optical microscope image of the fabricated actuator taken from the backside (substrate side) of the actuator. The Au film in the plane PPy area was peeled off, while the Au film covering the silicon micro spring was not peeled off.

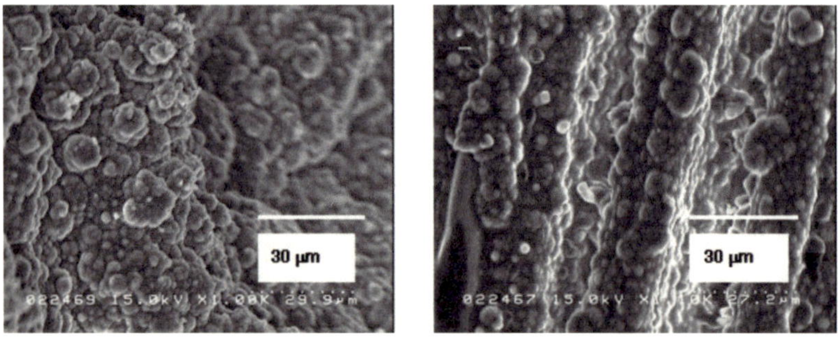

Fig. 5.4. Electron microscope (SEM) images of the PPy actuator observed from the surface (electrolyte solution side) of the micro spring (a) and backside (substrate side) of the micro spring (b).

suspended by metal clips. The size of the moving part of the PPy actuators was 7 mm in width, 8.3 mm in length, and 91 µm in thickness. The PPy actuator exhibited the expansion and contraction motions under the alternating potential with the triangular wave shape applied between the PPy actuator and the counter electrode. The potential voltage difference between the PPy actuator and the electrolyte solution was monitored using an Ag/AgCl reference electrode. The peak values of the potential voltage were -1 and +1 V, and the potential sweep rate was 10 mVs^{-1}. The extension and contraction of the PPy actuator was measured by monitoring the displacement of the weight position using the laser displacement sensor. An arbitrary load stress was applied on the PPy actuator by putting weights on the saucer of the balance. The weights used here were 0.05, 0.2, and 0.5 N. The load stresses for these weights correspond to 0.07, 0.3, and 0.76 MPa, respectively.

Figure 5.5. shows the time dependences of the displacement (length change) of the actuators as measured with different weights of (a) 0.05 N, (b) 0.2 N, and (c) 0.5 N for the initial period of 800 seconds. Those weights correspond to 0.07, 0.3 and 0.76 MPa, respectively. The displacement vs. time curve for the eight of 0.05 N [Fig. 5.5a] seems to consist of two components of the electrochemical strain caused by the anion doping/dedoping processes and the creeping strain possibly due to the swelling of the PPy film. Here, the peak height was subtracted by the creeping strain and as shown in Fig. 5.5a, and the subtracted peak

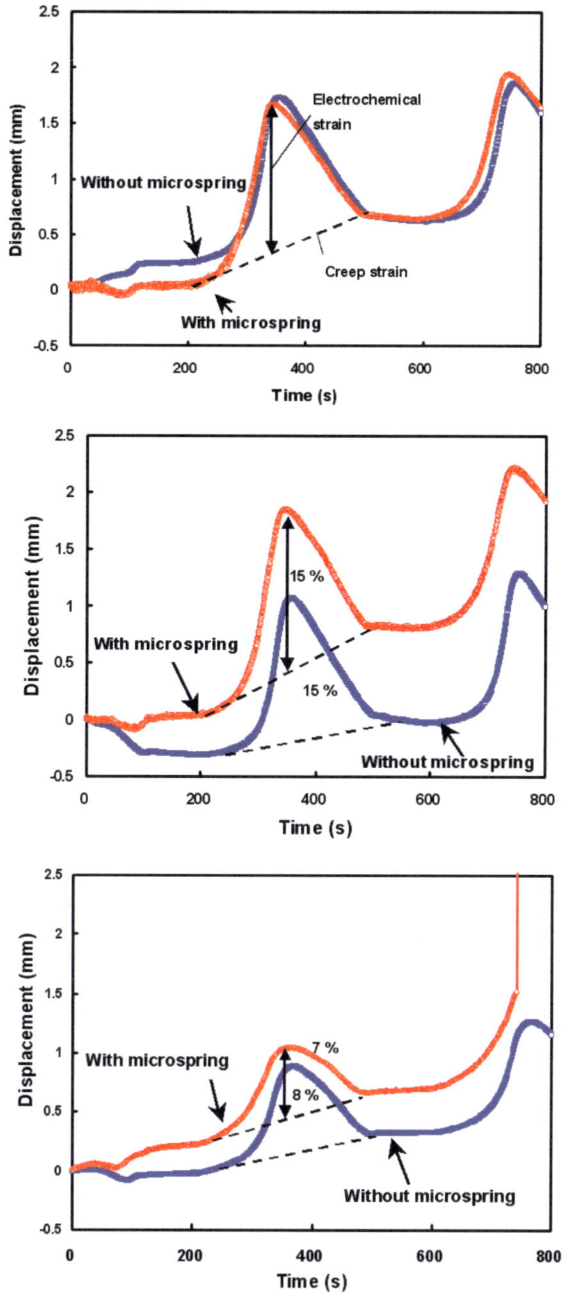

Fig. 5.5. Time dependences of the displacement (length change) of the actuators as measured with different weights of (a) 0.05 N, (b) 0.2 N, and (c) 0.5 N for the initial period of 800 seconds. Those weights correspond to 0.07, 0.3 and 0.76 MPa, respectively.

height value was divided by the initial length of the actuator to define the electrochemical strain. In this case of the weight of 0.05 N (0.07 MPa), the electrochemical strains of the PPy actuators with or without the silicon microspring are nearly identical. In contrast, when the weight increased up to 0.2 N (0.3 MPa), the creeping strain of the PPy actuator with the silicon microspring increased notably, while the electrochemical strains of two actuators are nearly identical as seen in Fig. 5.5b. When the 0.5 N (0.76 MPa) was applied, both of the actuators showed the smaller electrochemical strains of approximately 7-8%. In addition, the PPy actuator with the microspring torn off while the PPy actuator without the microspring did not tear off. The optical microscope observation of the torn off actuator showed cracks at the spring/PPy interface. The reason for the reduced electrochemical strain in the high load stress is not clear. The stressed polymer networks might possibly block the volume expansion caused by penetration of anions.

Figure 5.6. shows the repeated operation of the actuator for 15 times. The displacement of the actuators gradually increased due to creeping, while the electrochemical strains of them slightly decreased. The actuator with the microspring continued to have the electrochemical strain of 13%, while the actuator without the microspring continued to have that of 15%. The creeping effect of the PPy actuator with the silicon microspring was larger than the PPy actuator without the silicon microspring. Although the PPy actuator with the silicon microspring had slightly degraded performances compared to the PPy actuators without the silicon microspring, it will be beneficial for MEMS applications because of its large stress and strain.

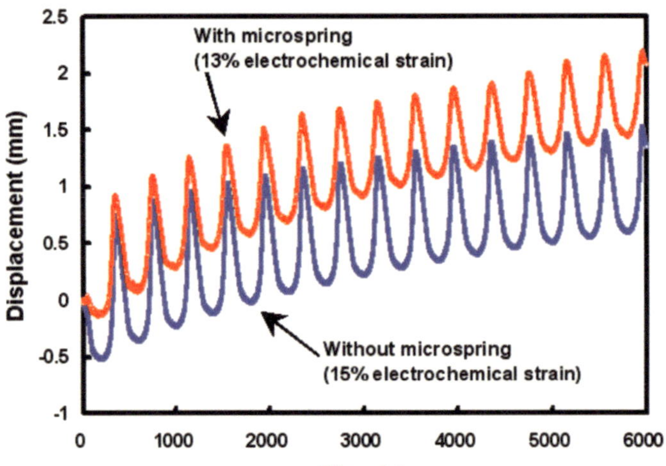

Fig. 5.6. Relationships between the displacement of the actuators and time under the repeated operation of the actuator for 15 times. The load weight was 0.2 N that corresponds to 0.3 MPa stress.

The length of the actuator continues to increase as time elapses. This behavior looks similar to the creeping effect in metal deformation processes. Zama and others recently reported on their detailed study for this creeping effect of PPy actuators, and concluded that this elongation could be recovered by releasing the stress during the deformation (Zama et al.,

2005a, 2005b, Sendai et al., 2009). Although the actuator investigated here exhibited a slow response, the generating stress of the order of 0.3 MPa is relatively large as an actuating mechanism for MEMS actuation. The PPy actuator requires an electrolyte solution during actuation. Therefore, some protecting film such as an artificial skin to cover the electrolyte solution surrounding the PPy actuator may be needed for actual MEMS applications.

5.5 Conclusion

Two kinds of PPy thin film actuators with or without the silicon MEMS microspring were fabricated and compared. The polypyrrole thin films with the thickness of 91 m were deposited by galvanostatic electropolymerization of a polypyrrole thin film using a methyl benzoate electrolyte solution of tetra-n-butylammonium bis(trifluoromethansulfonyl)imide (TBATFSI). One of the actuators was inserted with the silicon MEMS microspring with the length of 15 mm, the width of 0.5 mm, and the thickness of 60 μm. The MEMS PPy actuator exhibited nearly 12% of the electrochemical strain under the load of 0.2 N in a water solution of an electrolyte, lithium bis-trifluoromethane sulphonyl imide (LiTFSI) at the bias sweep rate of 10 mVs^{-1} in the voltage range between -1 and 1V. The load stress was approximately 0.3 MPa. Although the performances of the MEMS actuators showed some degradation compared to the PPy actuator without the MEMS microspring, the MEMS PPy actuator may be beneficial to drive MEMS structures, which require a large strain and a large stress with a low voltage actuation.

6. Acknowledgements

The author express heartfelt applications to his collaborators, Mr. Hiroyuki Katsumata, Mr. Takayuki Fujiya, Mr. Daiki Hoshino, Mr. Tsuyoshi Morita, Mr. Yutaka Chida, Mr. Zongfan Duan, Mr. Yutaro Suzuki, Mr. Shou Ogihara, Mr. Syota Kaihatsu, Mr. Masahiro Higashi, and other students. The staffs of the Micro Functional Device Research Center are also appreciated.

7. References

Baughman, R.H. (1996). *Synth. Met., 78,* 339.

Chida, Y., Morita, T., Machida, R., Hoshino, D., & Nishioka, Y. (2010). *Mol. Cryst. Liq. Cryst., 519,* 115.

Guo, S., Fukuda, T., Kosuge, K., Arai, F., Oguro, K., & Negura, M. (1996). *Transactions of the Japan Society of Mechanical Engineers, C62(596),* 1384.

Han G., & Shi, G. (2004). *Sensors and Actuators, B 99,* 525.

Hara, S., Zama, T., Takashima, W., & Kaneto, K. (2004). *J. Mater. Chem.,* 14, 1516.

Hara, S., Zama, T., Ametani, A., Takashima, W., & Kaneto, K. (2004). *J. Mater. Chem., 14,* 2724.

Hara, S., Zama, T., Takashima, W., & Kaneto K. (2004). *Polym. J., 36,* 933.

Hara, S., Zama, T., Takashima, W., & Kaneto, K. (2005). *Synthetic Metals, 149,* 199.

Hara, S., Zama, T., Takashima, W., & KANETO, K. (2005). *Smart Mater. Struct., 14,* 1501.

Hara, S., Zama, T., Takashima, W., & Kaneto, K. (2006). *Synth. Met., 156,* 351.

Hoshino, D., Morita, T., Chida, Y., Duan, Z., Ogihara, S., Suzuki, Y., & Nishioka, Y. (2011). *Jpn. J. Appl. Phys. 50 (2011)* 01BG10 .
Hutchison, A. S., Lewis, T. W., Moulton, S. E., Spinks G. M., & Wallace, G. G. (2000). *Synthetic Metals 113,* 121.
Jager, E. W. H., Smela, E., Inganas O., & Lundstrom, I. (1999). *Synthetic Metals 102,* 1309.
Morita, T., Chida, Y., Hoshino, D., Fujiya, T., & Nishioka, Y. (2010). *Mol. Cryst. Liq. Cryst. 519,* 121.
Ogasawara, M., Funahashi, K., Demura, T., Hagiwara T., & Iwata, K. (1986). *Synth. Met., 14,* 61.
Sendai, T., Suematsu, H., & Kaneto, K. (2009). *Jpn. J. Appl. Phys., 48,* 51506.
Zama, T., Hara, S., Takashima, W., & Kaneto, K. (2005). *Bull. Chem. Soc. Jpn., 78,* 506.
Zama, T., Hara, S., Takashima, W. & Kaneto, K. (2005). *Jpn. J. Appl. Phys., 44,* 8153.
Zama, T., Hara, S., Takashima, W., & Kaneto, K. (2005). *Synth. Met., 149,* 199.

9

Electrochemical Sensors Based on Electropolymerized Films

Xu Qin, Hu Xiao-Ya and Hao Shi-Rong
College of Chemistry and Chemical Engineering, Yangzhou University
China

1. Introduction

Chemical sensors are the devices that provide a certain type of responses related to the quantity of a specific species. All chemical sensors consist of a transducer and a chemically selective layer. The transducers transform the response into a detectable signal on modern instrumentation, and the chemically selective layer isolates the response of the analyte from its immediate environment. According to the properties to be determined, chemical sensors can be classified as electrical, optical, mass or thermal sensors. Among all of them electrochemical sensors obtained more attention because they are sensitive and selective, fast and accurate, portable and in expensive.

The modification of electrodes surfaces by some special layers has been the major growth area in electrochemical sensors in recent years. Compared to conventional electrodes, greater control of electrode characteristics and reactivity is achieved by surface modification, since the immobilization transfers the physicochemical properties of the modifier to the electrode surface. This process could impart a high degree of selectivity or sensitivity to the electrochemical transducers. Different procedures such as chemical reaction, chemisorption, composite formation or polymer coating have been used to modify electrodes. Several papers have reviewed the application of the modified electrodes in electrochemical sensors areas [1]

Electrochemical polymerization (ECP) refers to the application of electrochemical methods in the cathode or anode during the polymerization reaction process. The polymerization method provides a new current or potential control factor, so some important advantages were recognized at the beginning of its development such as: short process and low cost. The film thickness and composition can be achieved easily by controlling the electrochemical parameters during the electrochemical process. In addition, electrochemical polymerization can make raw monomer aggregate directly on the substrate film to avoid the use of a large number of volatile organic solvents to achieve the aim of clean production. Because of these excellent features, electropolymerization method holds a promising future for construct simple design, high stable, rapid response and enhanced selectivity sensors. This review will briefly present the development and application of the sensors prepared by electropolymerization method in the recent ten years. The main goal of this contribution is not to collect all papers published recently, but to discuss the development and advantages of the electropolymerized films for analytical purpose. For application, we specifically focused on the most recent and promising applications of those sensors in environmental and clinical monitoring.

2. Classification of films obtained by electrochemical polymerization method

Polymerization is a reaction in which large molecules are created from many small monomers. Normally it is a process which must be controlled carefully under strict conditions. Recently electropolymerization method has been successfully used for the controllable preparation of films because of the advantages we referred in the introduction section. Three types of electrochemical methods are generally employed for the polymerization of different monomers: (1) a constant current (galvanostatic); (b) a constant potential (potentiostatic) and (3) a potential scanning/cycling or sweeping. Different types of polymer films were applied for the construction of sensors. They are classified as conductive and nonconductive films.

2.1 Conducting films prepared by electropolymerization method

Conducting polymers have the polyconjugated structures with electronic properties similar to metals, but retaining the properties of conventional organic polymers. They have gained much attention in sensor areas [2,3] in recent years because of their unique characters [4]. A wide variety of organic molecules have been used as the monomers for the preparation of conducting polymers, such as polycyclic benzenoid, nonbenzenoid hydrocarbons, acetylene, polyaromatic and heterocyclic compounds. **Fig. 1** gives some examples of the conductive

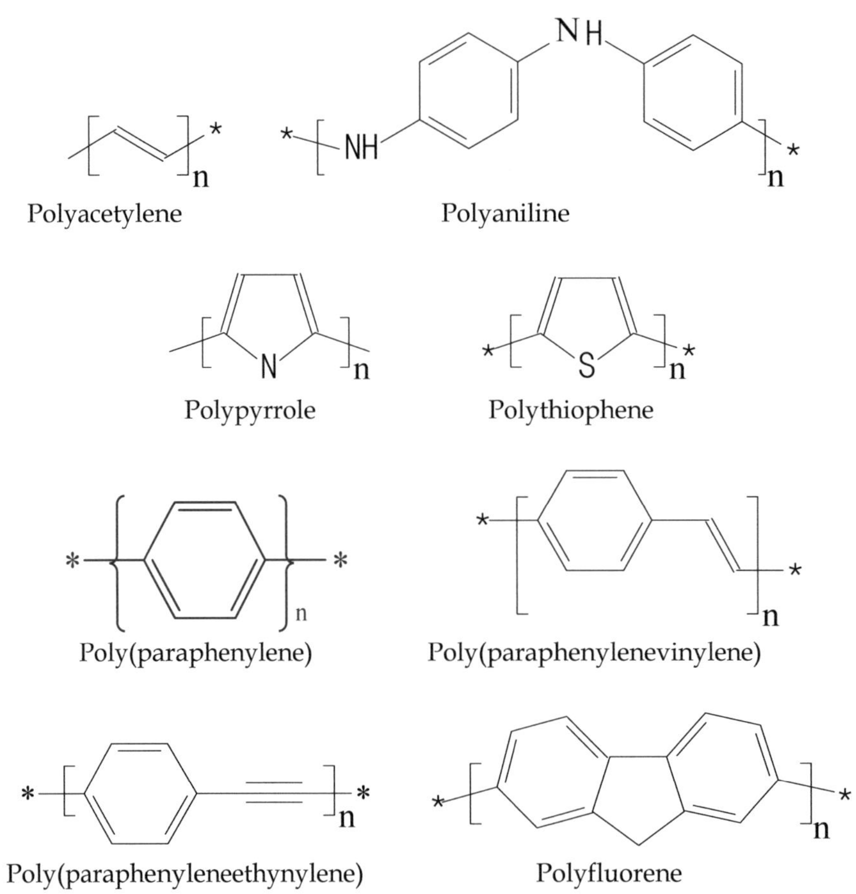

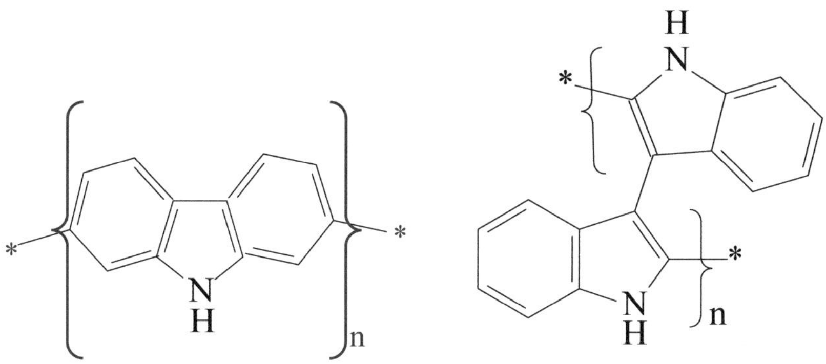

Fig. 1. Main chasses of conductive polymers

polymer films. Their unusual electrochemical properties are caused by the conjugated π-electron backbones. A large number of reviews devoted to the fabrication and description of the properties of conducting polymers have been published. Numbers of reviews focus on their use as electrochemical sensors. However there is still considerable interest in the development of new conductive polymers by the electropolymerization method, and new application of the films continually appear. Novel work in the literature from 2000 up to present will be reviewed.

2.2 Non-conducting films prepared by electropolymerization method

Non-conducting films prepared by electropolymerization method are also important. The resulting non-conducting film usually has a small thickness and is self-controlled by the increase in electrical resistance during its growth on the electrode. Because non-conducting polymers are always thin (10-100 nm), substrates and products can diffuse rapidly to and from the film modified electrodes. Therefore, fast response time and high selectivity could be expected for non-conducting polymer based electrochemical sensors. In most cases, phenol and its derivates are always used for the synthesis of non-conductive films by electrochemical methods [5]. **Fig. 2** illustrated the process for the preparing of the phenol related films. Phenylate will be oxidized to generate phenolate radicals which would couple together by ortho- or para- coupling way. Subsequent reactions produce oligomers and, finally, poly(phenylene oxide) films are polymerized on the surface of the electrode. Mahmoudian MR et al prepared poly (pyrrole-co-phenol) (co-PyPh) film by using cyclic voltammetry in the mixture electrolytes of dodecyl benzene sulphonic acid (DBSA) and oxalic acid solution on steel electrodes [6] . It can be used to protect the corrosion of steel. Tahar NB and Savall A[7] have studied the electrochemical oxidation of phenol at different temperatures in basic aqueous solution on a vitreous carbon electrode at different temperatures by cyclic voltammetry and chronoamperometry techniques. Other phenol derivatives have also been prepared by the electropolymerization method. Matsushita et al reported the electropolymerization of coniferyl alcohol in an aqueous system (0.2 M NaOH) and in an organic solvent system [CH_2Cl_2/methanol (4:1 v/v) in the presence of 0.2 M $LiClO_4$][8]. Ciriello R et al. investigated the electrosynthesis mechanism of 2-naphthol (2-NAP) in phosphate buffer at pH 7 on Pt electrodes. The voltammetric behavior suggested the formation of a non-conducting polymer (poly(2-NAP)) through an irreversible electrochemical process complicated by 2-NAP adsorption and fast electrode passivation [9].

Non-conductive polymers obtained from amino acid or their derivates have also obtained particular interest because they bears specific groups which can interact with some electroactive species through the formation of covalent bonds between either the amino and aldehyde or amino and carboxyl groups. We have used cyclic voltammetric method to form the L-cysteine modified electrode for the detection of sinomenine [10] , dopamine [11], terbinafine [12] and adenine [13]. Cystein was electropolymerized on a glassy carbon electrode in 0.04 M HCl solution in the scan range from -1.20 to 2.60 V at the scan rate of 100 mV/s [13].

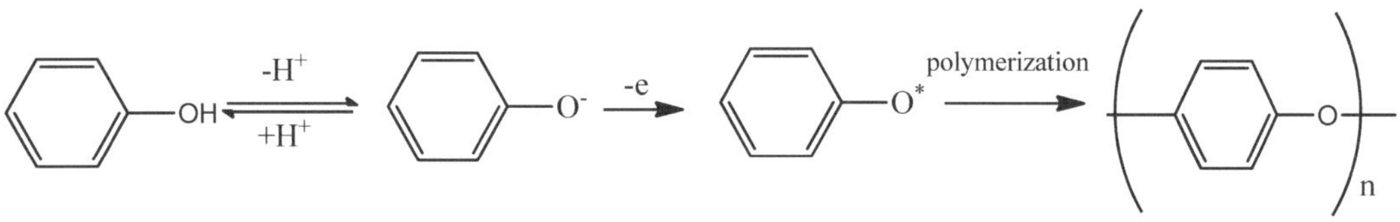

Fig. 2. The process for the preparing of the Phenol film by electrochemical method.

Fig. 3 illustrated the structures of some of the non-conductive films obtained by electropolymerization method. They included the already discussed polyphenol, polymers of phenylendiamines and the overoxidized polypyrrole (PPy). These polymers can be used as a novel support matrix for the immobilization of biomolecules to construct different electrochemical sensors. We would discuss the preparation and application of these films in the following section.

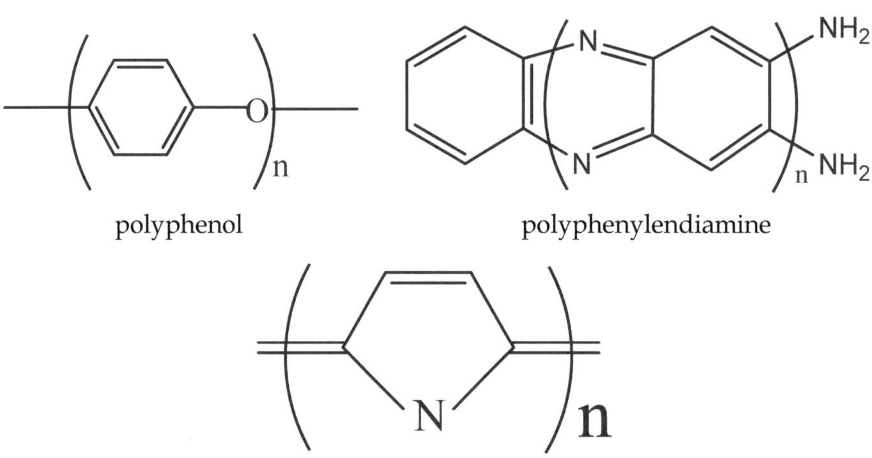

Fig. 3. Structures of some non-conductive polymerfilms

3. Application of the electropolymerized film in analytical areas

Significant advances in electropolymerization areas during the 1990's are certain to facilitate the application of the sensors to different analytical areas. To date, electropolymerized films have been widely used for clinical and environmental detection purpose.

3.1 Electropolymerized films for clinical monitoring

Electrochemical sensors in clinical assay was developed from 1962 by Clark and Lyons who used the glucose oxidase (GOx) enzyme to construct an amperometric electrode for dissolved oxygen detection [14]. From that time, the application of electrochemical sensors to determine the concentration of substances and other parameters of biological interest has represented a rapidly expanding field of instrument. The electrochemical sensors have been widely used in clinical analysis because of their high sensitivity and selectivity, portable field-based size, rapid response time and low-cost. Some of these sensor devices have been routinely used in clinical, industrial, environmental, and agricultural areas. Many works and reviews related on this area have been reported. Lakshmi D et al has reviewed the application of electrochemical sensors for uric acid detection in mixed and clinical samples[15]. Ronkainen et al reviewed the application of electrochemical biosensors from two points: biocatalyst and affinity [16].

Since the original work reported by Diaz et al. [17], the films prepared by electropolymreization method have attracted considerable interest due to their versatility. Polymers have gain considerable interest in the clinical analysis area because of their unique and biochemical properties. In 1992, Davies et al have extensively reviewed the application of the polymer membranes in clinical sensor application [18]. In this work, the authors were concerned with the relationship between the polymer design and the proposed application. They highlighted the permeability, permselectivity and transmembrane potential of the polymer membranes and the role of polymer membranes as matrices for the immobilization of reactive chemical and biological agents. Cosnier reviewed the application of the electropolymerized films on the construction of affinity sensor[19]. He compared the different strategies for the immobilization of biomolecules on electropolymerized films to construct affinity sensors which can be used as clinical sensors. Table 1 [20-52] summarized the numerous of recent applications in clinical areas based on the electropolymerized films. The information on the analytes, the polymer films, and the characters of the sensors has been listed.

Non-conductive polymer from polyphenylenediamines has been used as a matrix for the entrapment of enzymes. Glucose oxidase has been caged into the microtubule structures of polycarbonate membrane by using poly (1, 3- phenylenediamine) to fill the pores. The sensitivity of the sensor increased 60 times [23]. Different techniques for the electropolymerization of 1,2-, 1,3- and 1,4-phenylenediamine, such as cyclic voltammetry and chronoamperometry, were compared by Currulli et al [53]. When heparin was co-immobilized with glucose oxidase during the electropolymerization of a non-conductive poly (1,2- phenylenediamine) film, an implantable glucose biosensor could be constructed [54]. This sensor could prevent the fibrin formation and clotting when the glucose sensor was exposed to blood.

Phenol and its derivative have also been widely used for clinical analysis. The electropolymerization of phenol derivatives is similar to that of phenol. We reported the polymerization of the acid chrome blue K on a glassy carbon electrode by cyclic viltammetric method in 0.05 M pH 7.0 phosphate buffer solution in the potential range from -0.4 V to 1.5 V at the scan range of 100 mV/s by 25 cycles [28]. This film can be used to separate the electrochemical response of dopamine (DA), ascorbic acid (AA) and uric acid (UA). Under the optimum conditions, the calibration curves for DA, AA and UA were obtained in the range of 1.0–200.0, 50.0–1000.0 and 1.0–120.0μM, respectively. Both poly

analyte	monomer	linearity or detection limit	Ref
glucose	toluidine blue O	0.1-1.2 mM	20
	preoxidized catecholamines	0.3 µM	21
	thioaniline functionalized gold nanoparticle	0-200 mM	22
	1,3-phenylenediamine	0.25 µM - 18 mM	23
	thioaniline-modified glucose oxidase		24
dopamine	N-methylpyrrole	0.1- 10 µM	25
	aniline/gold nanoparticles	3-115 µM	26
	1-aminoanthracene	0.56-100 µM	27
	acid chrome blue K	1.0 – 200.0 µM	28
prostate specific antigen	poly(1,2-diaminobenzene) as the template for the electropolymerization of polyaniline	1-100 pg/mL	29
uric acid	3-aminophenol		30
	2-aminophenol	0.5 – 0.9 mM	31
DNA	pyrrole		32
		3.7-370 nM	33
		0.16 -3.5 fmol	34
	ferrocene-functionalized pyrrole	0.1-200 nM	35
	gold nanoparticles/p-aminobenzoic acid/carbon nanotubes	1.0 fmol -50 nM	36
	gold nanoparticles/L-lysine	0.1 - 10 fmol	37
	silver nanoparticles/3-(3-pyridyl) acrylic acid/carbon nanotubes	9.0 fmol-9.0 nM	38
leptin	Au-pyrrole propylic acid-pyrrole nanocomposite	10-100000 ng/mL	39
Human IgG	pyrrole-3-carboxylic acid		40
urea	styrene sulphonate-aniline	0-75 mM	41
hemoglobin	pyrrole-gold nanoparticles	60-180 µg/mL	42
Myeloperoxidase	o-phenylenediamine/multi-wall carbon nanotubes -ionic liquid/gold nanoparticles	0.25-350 ng/mL	43
17−β estradiol	3,4-ethylenedioxylthiopene/gold Nanocomposite		44
nitric oxide	eugenol or o-phenylenediamine		45
	poly(toluidine blue)	0.18-86µM	46
	m-phenylenediamine, 2 3-diaminonaphthalene, and 5-amino-1-naphthol polymers	From nM to µM	47
	meldola blue/chitosan	10 nM-600µM	48
cholesterol	2-mercaptobenzimidazole	5-30µM	49
nicotinamide adenine dinucleotide	phenothiazine	70 nM	50
	pyronin B	1.0 - 500 µM	51
interleukin 5	pyrrole-pyrrolepropylic acid-gold nanocomposite	10 fg/mL	52

Table 1. Examples of electropolymerized films for clinical analysis

(3-aminophenol) film [30] and poly (2-aminophenol) [31] film have been used for the selectively detection of uric acid. The poly (2-aminophenol) film was electrochemically prepared on Pt electrodes at a constant potential of 0.3 V from a deoxygenated aqueous solution of monomer dissolved in 0.1 M KCl. This film modified electrode allows the penetration of large amounts of uric acid while blocking the electrochemical activity of ascorbic acid in the potential region examined.

3.2 Electropolymerized film for environmental monitoring

Electrochemical sensors play very important roles in the protection of our environment. They can monitor the pollutant on-site and address some other environmental needs. Several electrochemical devices, such as pH- or oxygen electrodes based on the polymerized films, have been used routinely for years in environmental analysis. The electropolymerized pyrrole [55,56], aniline, thiophene, benzene derivatives and others [57,58] have been used for the preparation of pH chemical sensors. Herlem, G et al prepared the polyglycine-like thin film on platinum electrode by anodic oxidation. The film can be used as a pH sensor in the pH range 2-12 because of the proton affinity towards amino groups of polyglycine [59].

Analyte	monomer	linearity or detection limit	Ref
Hg^{2+}	2-mercaptobenzothiazole	1.0-160.0 nM	62
	2,6-diaminopyridine	10 μ M -0. 1 M	63
	3-methylthiophene	1.4 μg/ L	64
nitrite	methylene blue-carbon nanotubes-ionic liquid	0.5-67.9 μ M	65
	carbon nanotube-anillin	0.2μM -3.1 mM	66
	pyrrole	10μM -1 mM	65
	functionalized thiadiazole	0.05 -16μM	68
ammonia	pyrrole	10- 200 μM	69
Ca^{2+}	melatonin	6.2×10^{-7}- 1.0×10^{-4} M	70
Cu^{2+}	2-aminothiazole)-multi-walled carbon nanotubes	0.1 - 20 μM	71
	pyrrole	50 nM- 0.01 M	72
sulfite	copper salen (salen=N,N'-ethylenebis(salicylideneiminato))	4.0 -69μM	73
	aniline	0.006-5 mM	74
methyl-parathion para-nitrophenol	para-phenylenediamine	0.01 to 10 mg/L	75
microbial	4-(2,5-di(thiophen-2-yl)-1H-pyrrol-1-yl)benzenamine (SNS-NH2) polymer	0.1-2.5 mM	76
chloride	3-octylthiophene	10^{-8} – 10^{-1} M	77
4-nitrophenol	carmine	50 nM – 10μM	78

Table 2. Examples of electropolymerized films for environmental analysis

The scope of the electropolymerized films have also been expanded towards a wide range of organic and inorganic contaminants including pesticides, polychlorinated biphenyls, and heavy metals. A new kind of polymer monomer, bis(terthiophene)-appended uranyl-salophen complex, comprising N,N'-bis[4-(5,2':5',2''-terthiophen-3'-yl)salicylidene]-1,2-ethanediamine-uranyl complexes (TUS), has been modified on a glassy carbon electrode by electrochemical polymerizations. This polymer film has both the functionality of ion-to-electron transducers (solid contact) and Lewis-acidic binding sites to construct a monohydrogen phosphate (MHP) ion-selective electrode (ISE). The detection limit was down to 10^{-5} M and the response time is less than 5s [60]. In some cases, the polymers acted as the matrix for the immobilization of enzyme for environmental analytical application. Chen et al reported the entrapment of glucose oxidase into poly-(L-noradrenalin) films (PNA). They studied the inhibition effects of Hg^{2+}, Cu^{2+}, and Co^{2+} on the activity of glucose oxidase. The electrosynthesized PNA matrix to entrap GOX for an inhibitive assay of Ag+ shows the lowest competitive affinity to heavy metal ions and gives the highest sensitivity, so it can be used for Ag^{+} detection61. The applications of the electropolymerized films applied for environmental monitoring were listed in Table 2 [62-78].

4. Conclusion and future perspective

There has been an enormous increase of the preparation and application of the polymer films in analytical areas. Some of the advances and fields of the electropolymerized films have been outlined in the review. The electropolymerization methods have important advantages over the conventional techniques for the modification and preparation of microelectrodes, permitting the regulation of the spatial location and selective control of the film properties. Selective immobilization of biomoleculars in array of microelectrodes can be implanted in biological tissues for the simultaneous detection of several compounds. Minimization arrays, fast responding electrochemical sensors and on-line detection are the developing tendency of the electropolymerized films in sensor areas. Some progress will be necessary to achieve the appearance of the commercial electrochemical sensors based on the electropolymerized films. A lot of future research into the development of new polymer films by electrochemical method can be expected.

5. Acknowledgements

We greatly appreciate the support of the National Natural Science Foundation of China (Nos. 20705030, 20875081, 21075107), 863 Program Foundation (2009AA03Z331), the Foundation of Jiangsu Key Laboratory of Environmental Material and Engineering (K08021) and the Postdoctoral Science Foundation of China (20090461161).

6. References

[1] B.J. Privett, J.H. Shin, M.H. Schoenfisch, Electrochemical Sensors, Analytical Chemistry 82 (2010) 4723-4741.

[2] A. Ramanavicius, A. Ramanaviciene, A. Malinauskas, Electrochemical sensors based on conducting polymer- polypyrrole, Electrochimica Acta 51 (2006) 6025-6037.
[3] H. Peng, L.J. Zhang, C. Soeller, J. Travas-Sejdic, Biomaterials 30 (2009) 2132-2148.
[4] C.Y. Wang, J. Guan, Q.S. Qu, G.J. Yang, X.Y. Hu, Combinatorial Chemistry & High Throughput Screening 10 (2007) 595-603.
[5] Z. Ežerskis, Z. Jusys, Journal of Applied Electrochemistry 31 (2001) 1117-1124.
[6] M.R. Mahmoudian, Y. Alias, W.J. Basirun, Materials Chemistry and Physics 124 (2010) 1022-1028.
[7] N.B. Tahar, A. Savall, Electrochimica Acta 55 (2009) 465-469.
[8] Y. Matsushita, T. Sekiguchi, R. Ichino, K. Fukushima, Journal of Wood Science 55 (2009) 344-349.
[9] R. Ciriello, A. Guerrieri, F. Pavese, A. M. Salvi, Analytical and Bioanalytical Chemistry 392 (2008) 913-926.
[10] J. Guan, Z.X. Wang, C.Y. Wang, Q.S. Qu, G.J. Yang, X.Y. Hu, International Journal of Electrochemical Science 2 (2007) 572-582.
[11] C.Y. Wang, Q.X. Liu, X.Q. Shao, X.Y. Hu, Analytical Letters 40 (2007) 689-704.
[12] C.Y. Wang, Y.D. Mao, D.Y. Wang, G.J. Yang, Q.H. Qu, X.Y. Hu, Bioelectrochemistry 72 (2008) 107-115.
[13] Q. Xu, M. Sun, Q.X. Du, X.J. Bian, D. Chen, X.Y. Hu, Current Pharmaceutical Analysis 5 (2009) 190-196.
[14] L.C. Clark, C. Lyons, Annals of the New York Academy of Sciences 102 (1962) 29-&.
[15] D. Lakshmi, M.J. Whitcombe, F. Davis, P.S. Sharma, B.B. Prasad, Electroanalysis 23 (2011) 305-320.
[16] N.J. Ronkainen, H.B. Halsall, W.R. Heineman, Chemical Society Reviews 39 (2010) 1747-1763.
[17] A F. Diaz, K. K. Kanazawa, G. P. Gardini, Journal of the Chemical Society, Chemical Communications (1979) 635-636
[18] M.L. Davies, C.J. Hamilton, S.M. Murphy, B.J. Tighe, Biomaterials 13 (1992) 971-978.
[19] S. Cosnier, Electroanalysis 17 (2005) 1701-1715.
[20] W.J. Wang, F. Wang, Y.L. Yao, S.S. Hu, K.K. Shiu, Electrochimica Acta 55 (2010) 7055-7060.
[21] C. Chen, Y.C. Fu, C.H. Xiang, Q.J. Xie, Q.F. Zhang, Y.H. Su, L.H. Wang, S.Z. Yao, Biosensors & Bioelectronics 24 (2009) 2726-2729.
[22] O. Yehezkeli, R. Tel-Vered, S. Reichlin, I. Willner, Acs Nano 5 (2011) 2385-2391.
[23] M. Lee, Y. Son, J. Park, Y. Lee, Molecular Crystals and Liquid Crystals 492 (2008) 155-164.
[24] O. Yehezkeli, Y.M. Yan, I. Baravik, R. Tel-Vered, I. Willner, Chemistry-a European Journal 15 (2009) 2674-2679.
[25] N.F. Atta, M.F. El-Kady, A. Galal, Analytical Biochemistry 400 (2010) 78-88.
[26] A.J. Wang, J.J. Feng, Y.F. Li, J.L. Xi, W.J. Dong, Microchimica Acta 171 (2010) 431-436.

[27] E.D. Troiani, R.C. Faria, Electroanalysis 22 (2010) 2284-2289.
[28] R. Zhang, G.-D. Jin, D. Chen, X.-Y. Hu, Sensors and Actuators B 138 (2009) 174–181.
[29] A.C. Barton, F. Davis, S.P.J. Higson, Analytical Chemistry 80 (2008) 6198-6205.
[30] S. Kursun, A. Pasahan, B.Z. Ekinci, E. Ekinci, International Journal of Polymeric Materials 60 (2011) 365-373.
[31] S. Kursun, B.Z. Ekinci, A. Pasahan, E. Ekinci, Journal of Applied Polymer Science 120 (2011) 406-410.
[32] J.H. Jin, E.C. Alocilja, D.L. Grooms, Journal of Porous Materials 17 (2010) 169-176.
[33] H. Peng, C. Soeller, M.B. Cannell, G.A. Bowmaker, R.P. Cooney, J. Travas-Sejdic, Biosensors & Bioelectronics 21 (2006) 1727-1736.
[34] C.D. Riccardi, H. Yamanaka, M. Josowicz, J. Kowalik, B. Mizaikoff, C. Kranz, Analytical Chemistry 78 (2006) 1139-1145.
[35] H.Q.A. Le, S. Chebil, B. Makrouf, H. Sauriat-Dorizon, B. Mandrand, H. Korri-Youssoufi, Talanta 81 (2010) 1250-1257.
[36] Y.Z. Zhang, J. Wang, M.L. Xu, Colloids and Surfaces B-Biointerfaces 75 (2010) 179-185.
[37] J. Wang, S.J. Zhang, Y.Z. Zhang, Analytical Biochemistry 396 (2010) 304-309.
[38] Y.Z. Zhang, K.Y. Zhang, H.Y. Ma, Analytical Biochemistry 387 (2009) 13-19.
[39] W. Chen, Y. Lei, C.M. Li, Electroanalysis 22 (2010) 1078-1083.
[40] R. Janmanee, A. Baba, S. Phanichphant, S. Sriwichai, K. Shinbo, K. Kato, F. Kaneko, Japanese Journal of Applied Physics 50 (2011) Part 3 Sp. Iss. SI, 01BK02.
[41] S.K. Jha, M. Kanungo, A. Nath, S.F. D'Souza, Biosensors & Bioelectronics 24 (2009) 2637-2642.
[42] L. Qu, S.H. Xia, C. Bian, J.Z. Sun, J.H. Han, Biosensors & Bioelectronics 24 (2009) 3419-3424.
[43] B. Liu, L.S. Lu, C.G. Liu, G.M. Xie, Acta Chimica Sinica 69 (2011) 438-444.
[44] R.A. Olowu, O. Arotiba, S.N. Mailu, T.T. Waryo, P. Baker, E. Iwuoha, Sensors 10 (2010) 9872-9890.
[45] B.A. Patel, M. Arundell, K.H. Parker, M.S. Yeoman, D. O'Hare, Analytical Chemistry 78 (2006) 7643-7648.
[46] Y.Z. Wang, S.S. Hu, Biosensors & Bioelectronics 22 (2006) 10-17.
[47] J.H. Shim, H. Do, Y. Lee, Electroanalysis 22 (2010) 359-366.
[48] J. Njagi, J.S. Erlichman, J.W. Aston, J.C. Leiter, S. Andreescu, Sensors and Actuators B-Chemical 143 (2010) 673-680.
[49] A. Aghaei, M.R.M. Hosseini, M. Najafi, Electrochimica Acta 55 (2010) 1503-1508.
[50] Q. Gao, M. Sun, P. Peng, H.L. Qi, C.X. Zhang, Microchimica Acta 168 (2010) 299-307.
[51] S.A. Kumar, S.L. Chen, S.M. Chen, Electroanalysis 21 (2009) 1379-1386.
[52] W. Chen, Z.S. Lu, C.M. Li, Analytical Chemistry 80 (2008) 8485-8492.
[53] A. Curulli, G. Palleschi, In Proceedings of The 2nd Workshop on Chemical Sensors and Biosensors (Mazzei, F. and Pilloton, R., eds), (2000) 439–444

[54] J.Wang, L. Chen, S.B. Hocevar, B. Ogorevc, Analyst 125 (2000)1431–1434
[55] W. Prissanaroon-Ouajai, P.J. Pigram, R. Jones, A. Sirivat, Sensors and Actuators B-Chemical 138 (2009) 504-511.
[56] W. Prissanaroon-Ouajai, P.J. Pigram, R. Jones, A. Sirivat, Sensors and Actuators B-Chemical 135 (2008) 366-374.
[57] G. Herlem, B. Lakard, M. Herlem, B. Fahys, Journal of The Electrochemical Society 148 (2001) E435-E43
[58] R. Aoun, A. Yassin, M. El Jamal, A. Kanj, J. Rault-Berthelot, C. Poriel, Synthetic Metals 158 (2008) 790-795.
[59] G. Herlem, R. Zeggari, J.Y. Rauch, S. Monney, F.T. Anzola, Y. Guillaume, C. Andre, T. Gharbi, Talanta 82 (2010) 417-421.
[60] J. Kim, D.M. Kang, S.C. Shin, M.Y. Choi, S.S. Lee, J.S. Kim, Analytica Chimica Acta 614 (2008) 85-92.
[61] C. Chen, Q.J. Xie, L.H. Wang, C. Qin, F.Y. Xie, S.Z. Yao, J.H. Chen, Analytical Chemistry 83 (2011) 2660-2666.
[62] X.C. Fu, X. Chen, Z. Guo, C.G. Xie, L.T. Kong, J.H. Liu, X.J. Huang, Analytica Chimica Acta 685 (2011) 21-28.
[63] F. Bakhtiarzadeh, S. Ab Ghani, Electroanalysis 22 (2010) 549-555.
[64] H. Zejli, P. Sharrock, J. de Cisneros, I. Naranjo-Rodriguez, K.R. Temsamani, Talanta 68 (2005) 79-85.
[65] Y.H. Li, X.L. Liu, X.D. Zeng, X.Y. Liu, L. Tao, W.Z. Wei, S.L. Luo, Sensor Letters 8 (2010) 584-590.
[66] D.Y. Zheng, C.G. Hu, Y.F. Peng, S.S. Hu, Electrochimica Acta 54 (2009) 4910-4915.
[67] S. Aravamudhan, S. Bhansali, Sensors and Actuators B-Chemical 132 (2008) 623-630.
[68] P. Kalimuthu, S.A. John, Electrochemistry Communications 11 (2009) 1065-1068.
[69] L. Zhang, F.L. Meng, Y. Chen, J.Y. Liu, Y.F. Sun, T. Luo, M.Q. Li, J.H. Liu, Sensors and Actuators B-Chemical 142 (2009) 204-209.
[70] X.P. Wu, W. Liu, H. Dai, G.N. Chen, Electrochemistry Communications 11 (2009) 393-396.
[17] H. Zhao, Z.J. Wu, Y. Xue, Q.A. Cao, Y.J. He, X.J. Li, Z.B. Yuan, Journal of Nanoscience and Nanotechnology 11 (2011) 3381-3384.
[72] A.R. Zanganeh, M.K. Amini, Sensors and Actuators B-Chemical 135 (2008) 358-365.
[73] T.R.L. Dadamos, M.F.S. Teixeira, Electrochimica Acta 54 (2009) 4552-4558.
[74] B. Bahmani, F. Moztarzadeh, M. Hossini, M. Rabiee, M. Tahriri, M. Rezvannia, M. Alizadeh, Asian Journal of Chemistry 21 (2009) 923-930.
[75] I. Tapsoba, S. Bourhis, T. Feng, M. Pontie, Electroanalysis 21 (2009) 1167-1176.
[76] S. Tuncagil, D. Odaci, E. Yidiz, S. Timur, L. Toppare, Sensors and Actuators B-Chemical 137 (2009) 42-47.
[77] P. Sjoberg-Eerola, J. Nylund, J. Bobacka, A. Lewenstam, A. Ivaska, Sensors and Actuators B-Chemical 134 (2008) 878-886.

[78] C.Y. Li, Journal of Applied Polymer Science 103 (2007) 3271-3277.

10

Stability of Peptide in Microarrays: A Challenge for High-Throughput Screening

Marie-Bernadette Villiers[1,2], Carine Brakha[1,2], Arnaud Buhot[3], Christophe Marquette[4] and Patrice N. Marche[1,2]
[1]*INSERM, U823, Grenoble,*
[2]*Université J. Fourrier, UMR-5823, Grenoble,*
[3]*UMR-5819(CEA-CNRS-UJF), INAC/SPrAM, CEA-Grenoble,*
[4]*Université Lyon 1, CNRS 5246 ICBMS, Villeurbanne,*
France

1. Introduction

Microarrays are becoming a common tool in biology for screening large numbers of samples. However, the relevance of such an approach depends on the reproducibility of measurements that is directly linked to the stability of the probes grafted on the chip especially when many cycles of regeneration are performed. Indeed, regeneration of microarray chips is of great interest in improving the throughput and reducing the costs. The impact of treatments performed to remove bound ligands in order to reuse the chip depends partially on the grafted probe and on the characteristics of the probe-ligand interaction. Thus DNA microarrays are considered to be stable as oligonucleotides are highly stable molecules and hybridization reaction depends very little on the conformation of the partners; so, multiple regeneration/rehybridization procedures can be carried out without major loss of signal intensity (Benters et al., 2002, Donhauser et al., 2009). Stability of the probes is much more difficult to achieve when proteins are used. Indeed, such molecules are very complex and heterogeneous, thus there are no general rules to account for their behaviour upon different regeneration steps. Moreover, protein-protein interactions are highly susceptible to partner conformation. This is particularly true in the case of antigen (Ag) - antibody (Ab) binding, and several papers mention a loss of signal after the second or third regeneration step (Barton et al., 2008, Yakovleva et al., 2003). Peptide microarrays are a good alternative, as peptides are shorter and their stability less dependant on their tri-dimensional structure. Thus they are often used for antibody profiling (Cherif et al., 2006, Halperin et al., 2011, Neuman de Vegvar et al., 2003). However, there is no study dealing with the stability of such microarrays during a large samples screening (>20). Furthermore, a conformational change is not the only parameter which can impact on ligand binding. In this chapter, our aim was to analyze the evolution of the signals during samples screening and to determine which parameters are involved in the decay of the chip efficiency: grafting method, saturation step, probe itself or probe-ligand interactions, presence of protease activity in the sample. We use a microarray system based on pyrrole electropolymerization for probe immobilization and surface plasmon resonance imaging (SPRi) for ligand detection. The biological model consists in antigen-antibody interactions, where probes are peptides used as antigens and ligands are antibodies (Ab) contained in serum samples. Our data suggest that modification of the peptide conformation is the main parameter involved in the decrease of the signal observed upon successive uses of the chip. This conformational change leads to both a progressive reduction of the signal due to a decrease of the peptide reactivity with Ab, and the selection of Ab with the highest affinity. This phenomenon can be evaluated and must be taken into account in the analysis of the data resulting from samples screening.

2. Materials and methods

2.1 Reagents

Polyvinylpyrrolidone (PVP, MW 360 kDa) was obtained from Sigma-Aldrich (St Quentin Fallavier, France). Poly(L-Lysine)-PEG (PLL(20KDa) grafted with PEG(2 Kda) having 3.5 Lys units/PEG chains = PLL-PEG) was purchased from SurfaceSolutionS (Zurich, Switzerland).

Peptides were synthesized by Altergen (Bischheim, France) with a pyrrole-modified NH2 terminus as previously described (Villiers et al., 2009). Two peptides are derived from the structural protein core of hepatitis C virus of the genotype 1b (HCV), one from ovalbumine, one from hen egg lysozyme, nine from hepatitis D virus (HDV) of various genotypes. Full sequences can be found on the ExPASy server (http://ca.expasy.org/) under different accession numbers (Table 1).

Peptide	Original protein	Accession number	Amino acids number
C131	HCV	P26663	131-150
C20	HCV	P26663	20-40
Ova75	Ovalbumine	P01012	75-96
HEL101	Lysozyme	P00698	101-120
HDV1	HDV	P0C6M9	155-172
HDV2	HDV	P0C6M9	174-195
HDV3	HDV	P0C6M9	189-211
HDV4	HDV	P0C6M9	65-80
HDV5	HDV	P0C6M9	1-18
HDV6	HDV	Q70E23	155-172
HDV7	HDV	A1IVP7	155-172
HDV8	HDV	Q70E23	174-195
HDV9	HDV	A1IVP7	174-195
HDV10	HDV	Q70E23	189-211 (aa 195: X=W)
HDV11	HDV	A1IVP7	189-211

Table 1. Peptides used in the study

Rabbit immune serums against different peptides were prepared by NeoMPS (Strasbourg, France). Human serums from healthy donors as non immune serum (NIS) were purchased from the Etablissement Français du Sang (La Tronche, France). Human serums from HDV infected patients were provided by Drs E. Gordien and S. Brichler (Hôpital Avicenne, Paris,

France) and Dr P. Morand (Centre Hospitalo-Universitaire, Grenoble, France). Serums were stored at -20°C.

2.2 Materials

Glass prisms coated with a 50 nm gold layer were obtained from Genoptics-HORIBA Scientific (Chilly-Mazarin, France). Electrodeposition was performed using an Omnigrid Micro robotic arrayer (Genoptics-HORIBA Scientific). Surface Plasmon Resonance (SPR) signals were monitored using a surface plasmon resonance imager (SPRi-Plex from Genoptics-HORIBA Scientific). Measurements were performed using SPRi dedicated software (Genoptics-HORIBA Scientific). Sample injections were ensured by a 231XL sampling injector coupled to a 832 temperature regulator (Gilson, Roissy, France).

2.3 Peptide immobilization on gold

Peptides (100μM) were grafted at least in triplicat on the gold surface of the biochip by electrochemical copolymerization of pyrrole-peptide conjugates using a solution containing 20 mmol/L pyrrole, 100 μmol/L of pyrrolated peptides and 10% glycerol in phosphate buffer (50 mmol/L). The polymerisation step was performed by a short 100 ms electrical pulse (2 V) between the counter electrode located in the needle of the microarrayer and the prism gold layer (Cherif et al., 2006, Villiers et al., 2009). Another grafting method based on electro-deposition of diazonium-peptide adducts (Corgier et al., 2009) was also used, as indicated in the text. The prism was rinsed with distilled water and saturated at room temperature for 2h using various mediums as indicated in the text. After washing with distilled water, the prism was positioned in the SPRi-Plex and used immediately.

2.4 SPRi interaction monitoring

All reactions were carried out at room temperature, in phosphate-buffered saline (PBS)/0.01%Tween 20. The flow rate in the chamber was 37 μL/min. Reflectivity was measured at 810 nm, at a fixed incidence angle ($55° < \theta < 56°$). After injection of serum (500 μL, 1/50 or 1/200 as indicated in the text), the biochip surface was rinsed with running buffer (10 min) to remove unbound ligands and specific binding was quantified by measuring the change in reflectivity (ΔR) obtained after 10 min washing. The chip was regenerated using 0.1 M HCl-Glycine (pH 2.3) solution for 10 min and stabilized in the running buffer (10 min). Every twelve injections, a cleaning step was performed by injection of 1% SDS (sodium dodecyl sulphate) in water for 10 min followed by running buffer for 20 min. Saturation with NIS (1/25 in PBS), PLL-PEG or PVP as indicated in the text was realized after each cleaning step.

3. Results and discussion

3.1 Is a grafted peptide sufficiently stable to allow a multiple re-use of the chip?

To assess the stability of the peptide chip during samples screening, we immobilized C131 peptide using pyrrole electropolymerization and performed multiple injection/regeneration cycles using non immune serum (NIS) with periodical injections of anti-C131 serum to monitor the reactivity with the grafted probe. SDS cleaning was realized every twelve injections, as described in §2.4. The SPR (Surface Plasmon Resonance) signal obtained for each injection was monitored. Results from eight independent experiments are presented in Fig. 1.

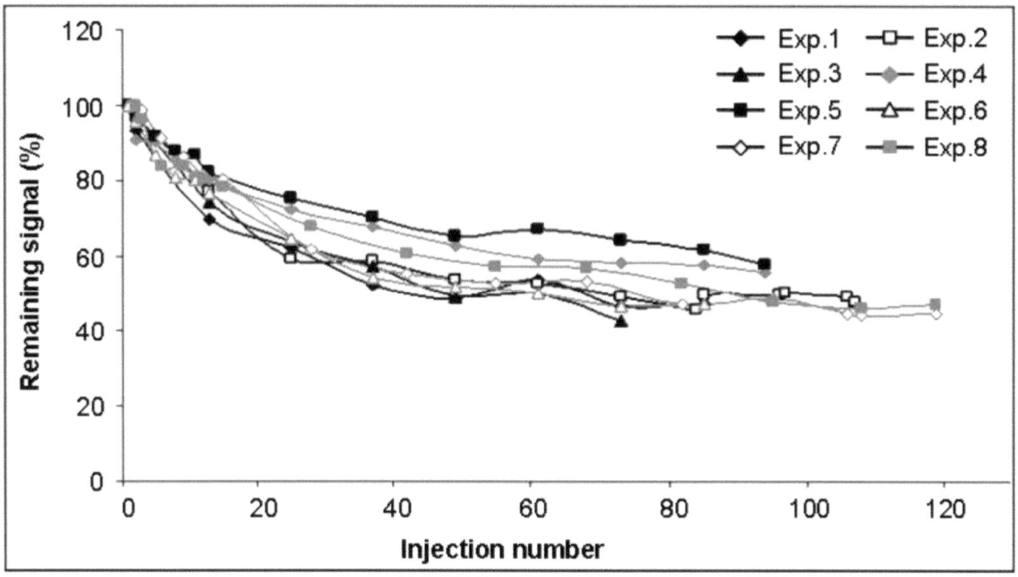

Fig. 1. Peptide chip stability. Eight independent experiments were realized as follows: C131 was immobilized in triplicate on the gold chip surface via pyrrole electropolymerization. Successive injections of non immune serum (1/50) were performed, followed by HCl-Glycine regeneration. Anti-C131 serum (1/200) was periodically injected (every 12 injections) and SPR signal was quantified. Results were standardized according to the signal obtained for the first injection (100%).

We observed a progressive decrease in the SPR signal during the experiments. This decay is very reproducible from one experiment to another and presents a biphasic profile with a rapid drop during the first 12 injections followed by a slighter decrease upon the subsequent injections. Such a loss of efficiency could limit the use of peptide chip for samples screening, therefore it is important to identify the parameters involved in this phenomenon to limit its impact.

3.2 What can influence grafted peptide stability?

SPR signal is directly related to probe-ligand binding which depends on many parameters at both probe and ligand levels. Various hypotheses are summarized in Fig. 2.

SPR signal depends on the quantity and the quality of both probe and ligand. In our experiments, ligands were Ab from a rabbit anti-serum which were aliquoted and kept at 8°C in the autoinjector rack. Thus it is unlikely that any changes in the Ab content of the samples occur.

The problem is more complex when we consider the probe. First, the decrease of the binding capacity could be due to a reduction in the number of available binding sites (epitopes) for the injected Ab during the set of injection/regeneration cycles. This could be due to either a gradual release of the grafted peptide from the chip surface or a proteolysis of the peptide due to the presence of proteolytic activity in the injected samples. The quality of the probe is also an important parameter for Ab binding as it must have a proper conformation and a good accessibility. Thus, if the regeneration step is partially inefficient, remaining Ab may prevent de novo binding, either directly or indirectly by steric hindrance. Various sets of experiments were performed in order to determine which parameters are involved in the chip efficiency decay.

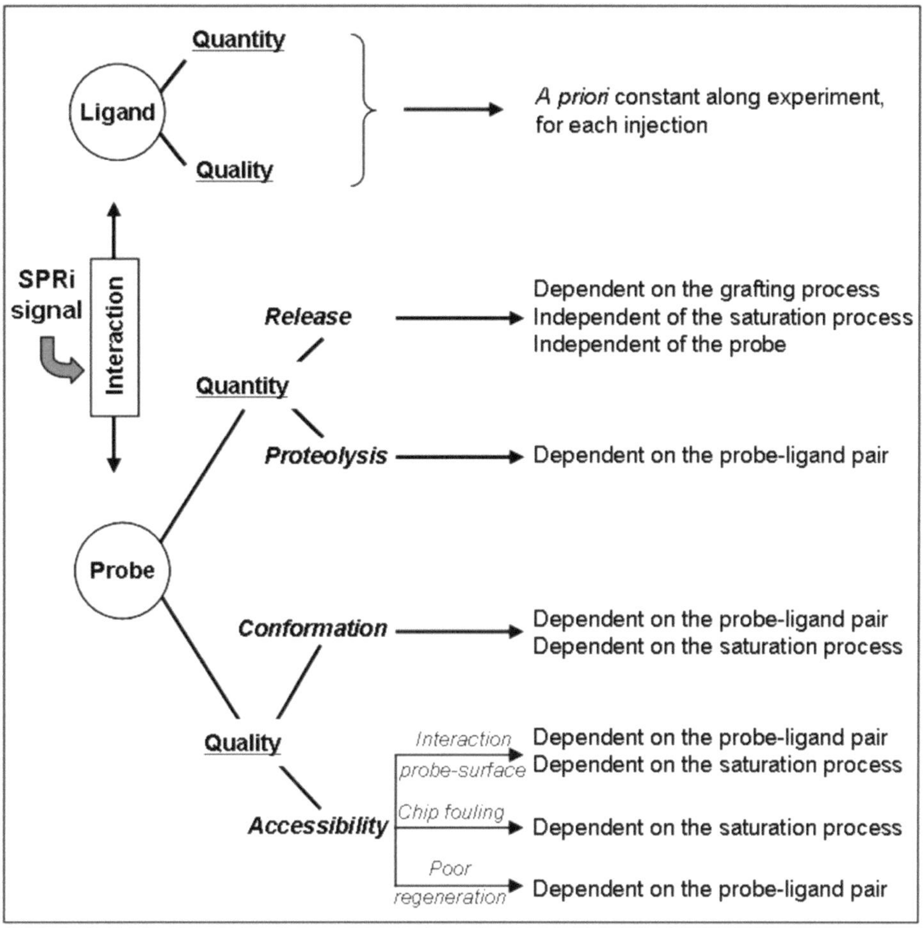

Fig. 2. Parameters involved in the signal loss: tree of possibilities

3.3 Influence of the grafting process

If probe release occurs, this may be related to the nature of the link between the probe and the chip surface, i.e. to the grafting process. To evaluate this hypothesis, C131 peptide was immobilized using either electropolymerization of pyrrole-peptide conjugates or electrodeposition, as a monolayer, of diazonium-peptide adducts and the signals obtained upon successive anti-C131 serum injections were recorded in both cases.

As shown in Fig. 3, the stability of the SPR signal was much better when C131 was immobilized via diazonium, but the change in reflectivity is much lower (likely due to lower probe density on the surface), impairing the sensitivity of the system. At first sight, this observation fits with a probe release in the case of pyrrole electropolymerization protocol, but this eventuality will be discussed later on in light of the other results (§ 4).

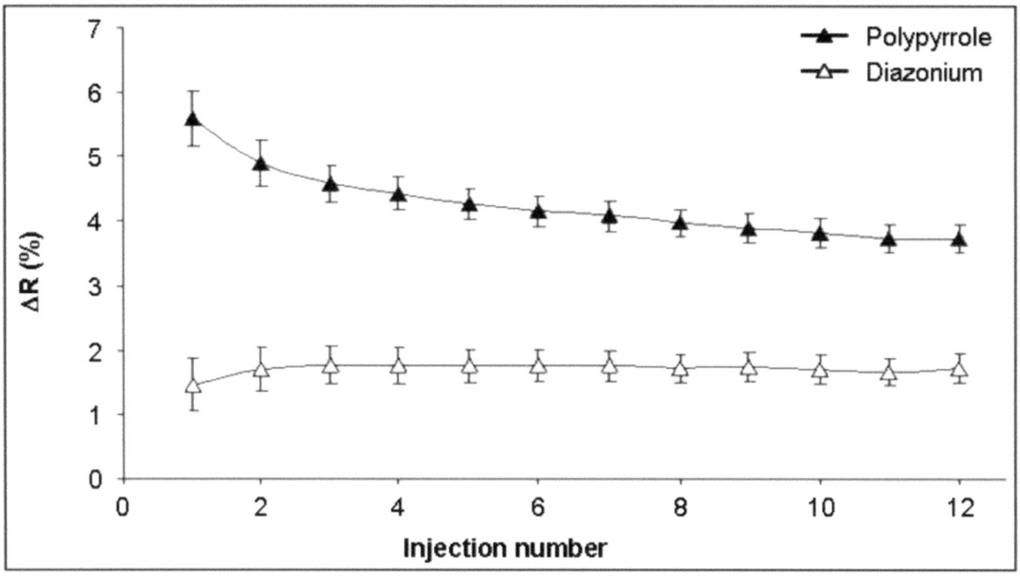

Fig. 3. Influence of the grafting process on the evolution of SPR signal (change in reflectivity = ΔR). C131 peptide was immobilized in triplicate on the gold chip surface via either pyrrole electropolymerization or diazonium electrodeposition. Successive injections of NIS (1/50) with periodical injections of anti-C131 serum (1/200) were performed, followed by HCl-Glycine regeneration.

3.4 Efficiency of the regeneration step

As mentioned above, a potential decrease in the number of binding sites available on the probe could be due to a partially inefficient regeneration process. In these conditions, still bound Ab would lead to a progressive increase in baseline level. To assess this point, we analyzed SPR signal after each regeneration step.

Results presented in Fig. 4 demonstrate that there is no significant increase in the SPR signals measured after each regeneration, attesting that no Ab stays behind the washing step. Therefore, the loss of signal observed for the specific peptide/Ab binding cannot be attributed to incomplete regeneration of the chip.

3.5 Influence of the saturation process

Probe accessibility is a key parameter to ensure a good interaction with injected ligands. This point is especially important when the probe is a small molecule (peptide). Indeed, the molecules used to saturate the chip surface to avoid non-specific interactions could impact on epitope accessibility, either by masking the binding sites or by interfering with peptide conformation through molecular interactions. Furthermore, the conformation of the peptide can evolve during the experiment, affecting the epitope reactivity. Moreover, an incomplete saturation could lead to a progressive fouling of the surface and, thus, to a gradual loss of available binding sites. To saturate chip surface, we usually use non-immune serum (NIS, 1/25 dilution) which leads to non-specific adsorption of many proteins on chip surface, probably predominantly albumin (MW = 65800 Da) as it represents about 60% of total

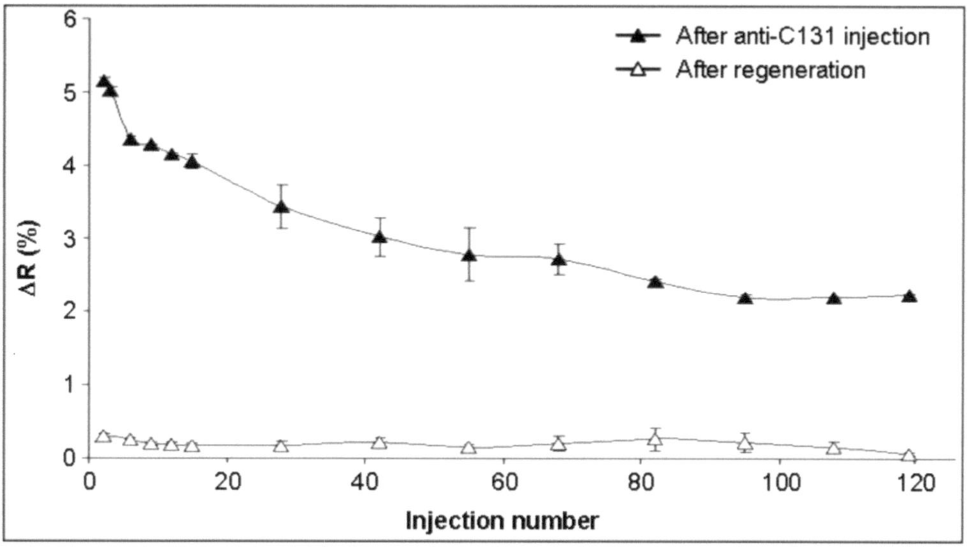

Fig. 4. Efficiency of the regeneration process. C131 peptide was immobilized in triplicate on the gold chip surface via pyrrole electropolymerization. Successive injections of NIS (1/50) with periodical injections of anti-C131 serum (1/200) were performed, followed by HCl-Glycine regeneration. SPR signal was quantified after anti-C131 injection and after the regeneration step.

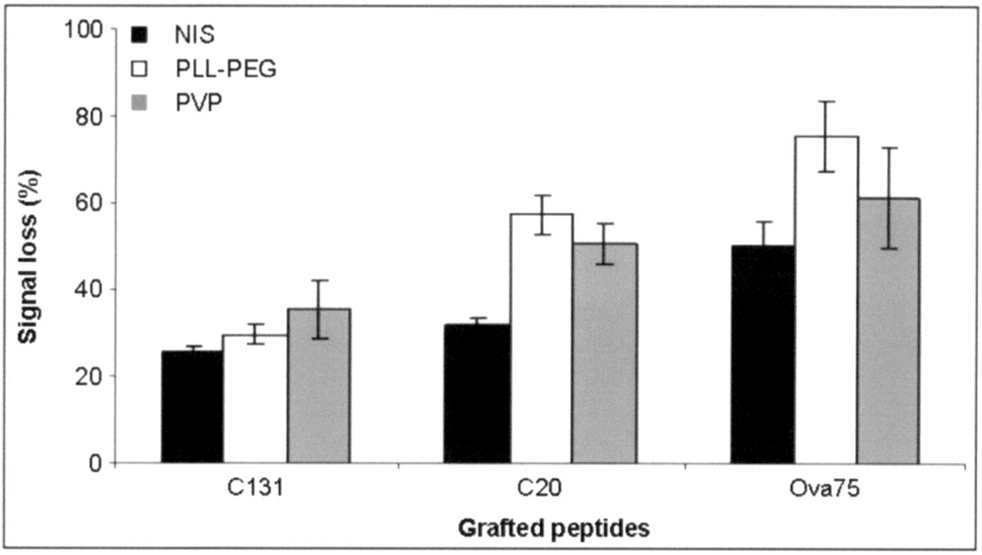

Fig. 5. Influence of the saturation process and of the peptide sequence on SPR signal loss. C131, C20 and Ova75 peptides were immobilized in triplicate on the gold chip surface via pyrrole electropolymerization. Saturation of the chip surface was ensured using either NIS or PLL-PEG or PVP. Successive injections of anti-C131, anti-C20 and anti-Ova75 serums (1/50) were performed, followed by HCl-Glycine regeneration. SPR signal loss after 74 injections.

serum proteins. To determine whether the saturation process is involved in the loss of chip efficiency, we performed the same type of experiments with two anti-fouling molecules: Poly (L-Lysine)-PolyEthyleneGlycol (PLL-PEG) and Polyvinylpyrrolidone (PVP) instead of NIS (0.5mg/mL and 1% p/v respectively). Signal evolution was analyzed in each case. As shown in Fig. 5, signal loss on C131 spot is similar, whatever the saturation process used. As interactions between peptide and anti-fouling molecules could also depend on the physico-chemical characteristics of the peptide, we wondered whether signal loss depends on the grafted probe. Two others peptides (C20 and Ova75) were immobilized on the chip and their corresponding rabbit anti-serums were injected in the same conditions than anti-C131.
As shown in Fig. 5, signal loss depends both upon saturation process and grafted peptide sequence. However, the general shape of the signal decay curve seems to be related with the peptide (Fig. 6).

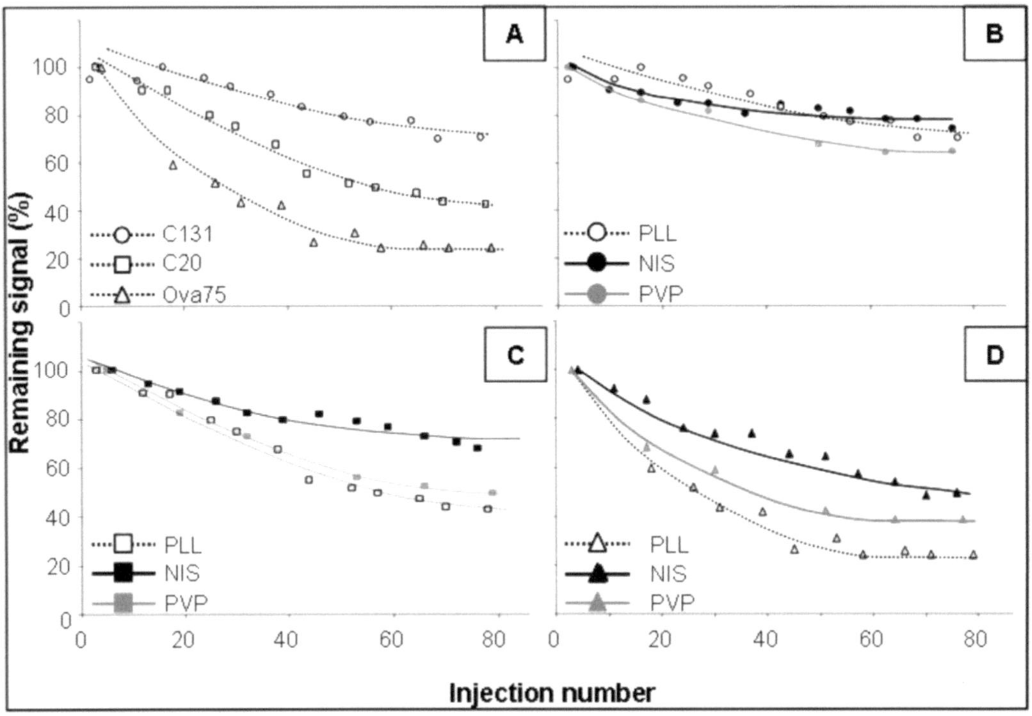

Fig. 6. Evolution of the SPR signal along the experiment. C131, C20 and Ova75 peptides were immobilized in triplicate on the gold chip surface via pyrrole electropolymerization. Saturation of the chip surface was ensured using either NIS or PLL-PEG or PVP. Successive injections of anti-C131, anti-C20 and anti-Ova75 serums (1/50) were performed, followed by HCl-Glycine regeneration. Remaining SPR signal obtained A) on the different spots upon anti-C131, anti-C20 and anti-Ova injections on a chip saturated with PLL-PEG B) on C131 spots after anti-C131 injection on chip saturated with NIS or PLL-PEG or PVP C) on C20 spots after anti-C20 injection on chip saturated with NIS or PLL-PEG or PVP D) on Ova75 spots after anti-Ova injection on chip saturated with NIS or PLL-PEG or PVP.

Altogether, these results indicate that the saturation process impacts on the amplitude of the signal loss while the shape of the signal decay curve is more likely dependant on the peptide. The impact of the saturation process may be related to the physico-chemical properties (hydrophibicity, charge, etc) of both the saturating molecules and the peptides.

3.6 Influence of the ligand

Until now, our analyses were always performed using the same Ab for a given Ag, which did not allow checking the influence of the peptide from that of the Ab. In the next experiment, we compare the signal obtained on C131 and C20 spots after injections of anti-C131 and anti-C20 serums. In each case (anti-C131 and anti-C20) two different serums issued from the same rabbit, but collected at two different days (D=39 and D=66) were tested. As observed in Fig.7, the second sample led to a more stable signal.

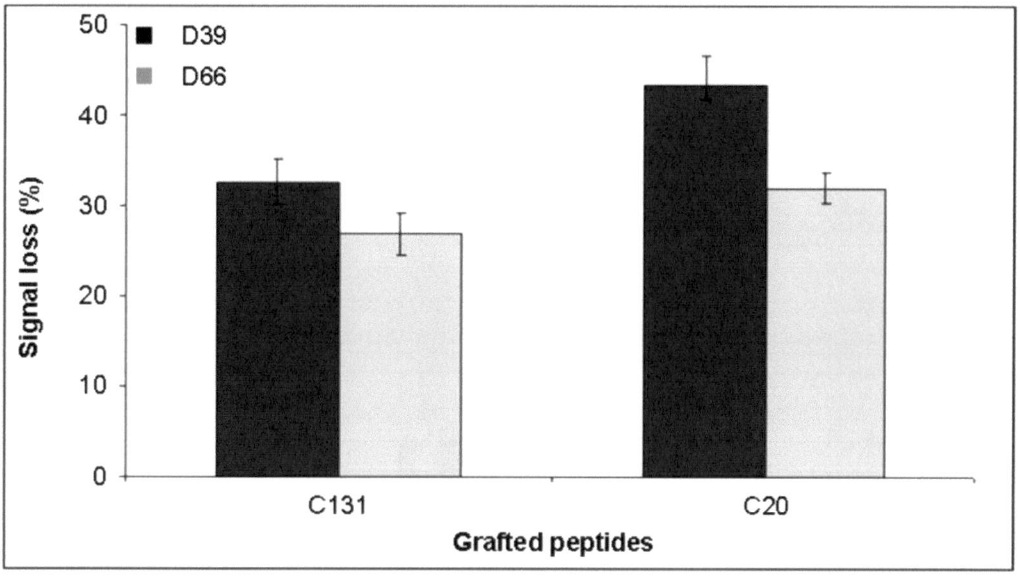

Fig. 7. Influence of the ligand on SPR signal loss. C131 and C20 peptides were immobilized in triplicate on the gold chip surface via pyrrole electropolymerization. Saturation of the chip surface was ensured using NIS. Successive injections of anti-C20 and anti-C131 serums (1/50) were performed, followed by HCl-Glycine regeneration and signal loss after 74 injections was quantified. In each case (anti-C131 and anti-C20) two different serums issued from the same rabbit, but collected at two different days (D=39 and D=66) were injected.

These results could be related to a difference in the affinity of the ligands for the probe. Indeed, it is well known that Ab affinity usually increases gradually during the immune response, which is referred to affinity maturation process (Berek & Ziegner, 1993). Thus, it is likely that D66 sample contains anti-C20 Ab with higher affinity for C20 than D39. But we cannot exclude that, despite the small size of the antigen (20aa), the epitopes recognized by D66 Ab differ from those recognized by D39 Ab, which would impact on the overall characteristics of peptide/Ab interaction.

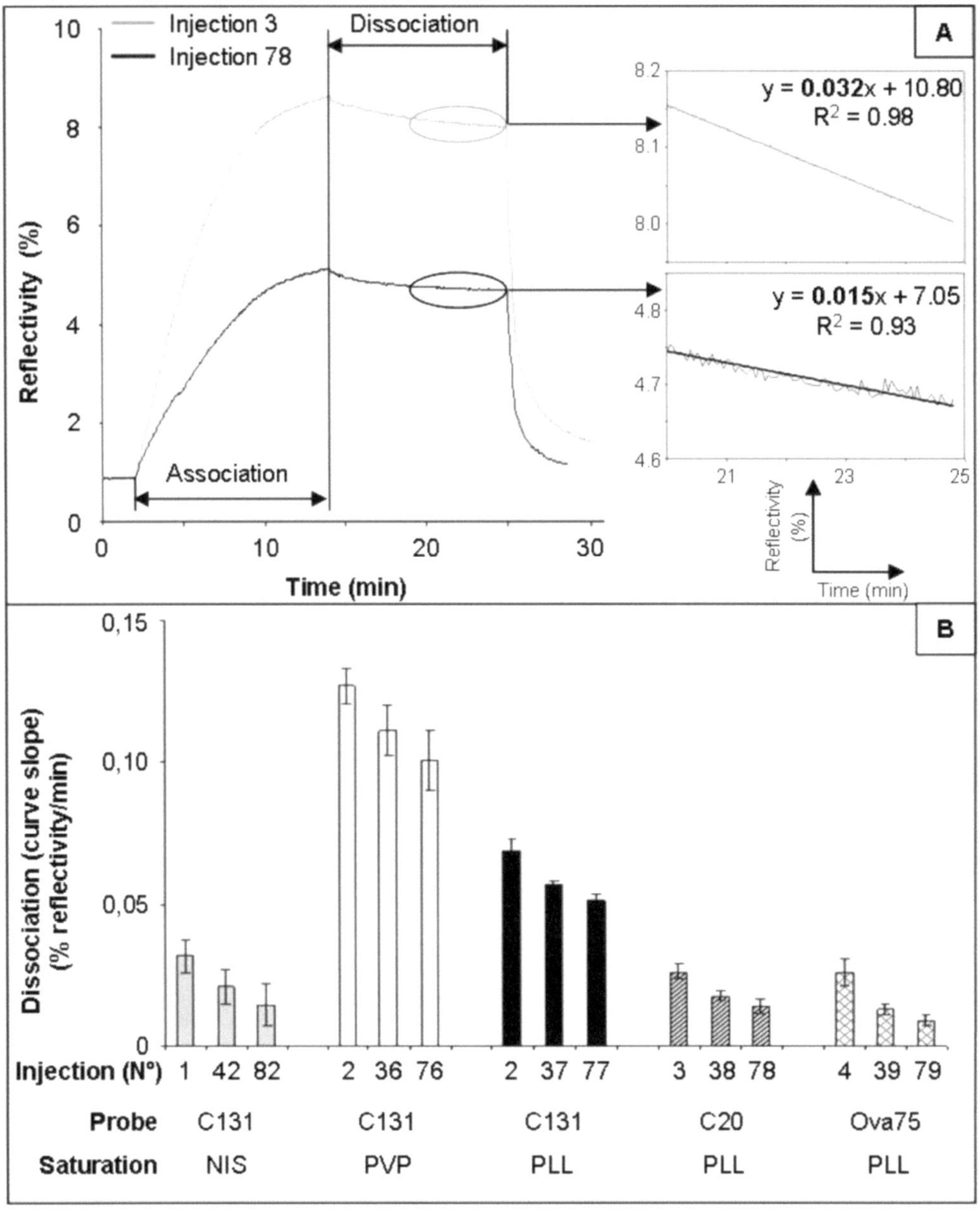

Fig. 8. Ab-peptide binding affinity. C131, C20 and Ova75 peptides were immobilized in triplicate on the gold chip surface via pyrrole electropolymerization. Saturation of the chip surface was ensured using either NIS or PVP or PLL-PEG. Successive injections of anti-C131, anti-C20 and anti-Ova serums (1/50) were performed, followed by HCl-Glycine regeneration. A) Sensorgramms obtained on C20 spots after the third and the 78th injection of anti-C20 serum (PLL-PEG saturation). Insert: detail of the dissociation curves. B) Quantification of antibody dissociation for the different peptides, from the slope of the dissociation curves.

3.7 Is there a change in the affinity of the ligand (Ab) for the probe (peptide) between the first and the last injection?

Affinity is a key parameter in Ab-Ag interaction which could impact on the signal observed upon Ab binding. Actually, Ab binding on the grafted peptides was quantified by measuring the change in reflectivity obtained after washing, a step during which ligands can dissociate, accordingly to their affinity for the probes. Moreover, it is well known that Ag conformation influences Ab binding (Fieser et al., 1987). Thus, a modification of peptide conformation during the experiment could lead to a change in the Ag-Ab affinity. As SPRi technique allows label-free and real-time detection of biomolecular interactions, it gives access to affinity parameters. But, as ligands in our experiments consisted in polyclonal antibodies, the dissociation curves obtained during the washing step after serum injection correspond to a mean value resulting from all the individual dissociation constants of the various Ab. In this case, it is not possible to determine classical affinity parameters, but the slope of the dissociation curve is representative of the overall affinity. In order to analyze the evolution of peptide/Ab affinity during an experiment, the slopes of the dissociation curves were calculated using the last 6 min of washing (Fig. 8A).

As shown in Fig. 8B, the slope of the curve corresponding to the dissociation of Ab-peptide complexes decreased during the experiment. Thus the diminution of the quantity of Ab bound to the peptides (signal loss) is associated with an increase in the affinity of the interaction, whatever the probe/ligand pair used or the saturation process.

These results suggest that successive injection/regeneration cycles lead to a slight conformational change in the probe. At the beginning of the experiment, peptide conformation is suitable for the binding of different Ab with a large range of affinity. As the experiment progress, Ab having lower affinity for the peptide can no longer bind, due to modification in peptide conformation.

3.8 Effect of glycerol and protease inhibitors

It is well known that the stability of protein is enhanced by various molecules among which polyols (Lee & Kim, 2002, Vagenende et al., 2009). In the aim of reducing conformational change in the grafted probes, glycerol (0.1%) was added to both the running buffer and the regeneration solution. As shown in Fig. 9, the presence of glycerol in the running solutions improved the signal stability in the case of Ova75, but not in the case of C131. It seems that the protective effect of glycerol occurs only on the less stable peptides.

As injected samples are complex biological mediums (serums), we wondered if some peptide degradation could occur, due to protease activities. So, we performed the same experiments after addition of protease inhibitors in the samples (cocktail set VII, CalbBiochem, 1/100). We did not observe any improvement of chip stability (Fig. 9), suggesting that probes proteolysis was not responsible for the loss of signal during the experiments.

3.9 Improvement possibilities of the system

Various strategies can be implemented to improve the quality of the data resulting from samples screening on a peptide chip. First, addition of glycerol in the solutions can improve peptide stability, depending on the peptide (§3.8). Second, we observed that signal loss was usually biphasic, with a rapid drop during the first injections followed by a slighter decrease upon the subsequent injections. Thus we suggest performing 10 - 12 blank injections/regeneration cycles before sample analysis to limit the conformational change between the first and the last sample of interest. Finally, we recommend to include a spot with a control peptide on the chip and to perform periodical injections of a control sample. The reduction of signal measured on this control spot can be modelled using a polynomial curve (Fig. 10A), which can be used to determine a correction factor for the others injections.

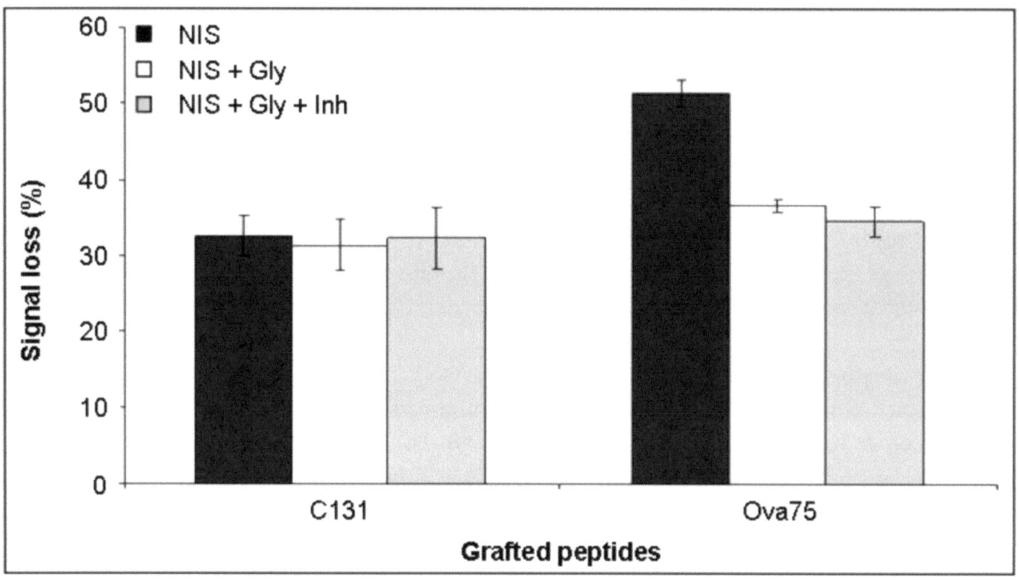

Fig. 9. Effect of glycerol and protease inhibitors on SPR signal loss. C131 and Ova75 peptides were immobilized in triplicate on the gold chip surface via pyrrole electropolymerization. Saturation of the chip surface was ensured using NIS. Successive injections of anti-C131 and anti-Ova75 serums (1/50) were performed, followed by HCl-Glycine regeneration and signal loss after 76 injections was quantified. + Gly: presence of glycerol 0.1% in the running buffer and regeneration solution. + Inh: addition of protease inhibitors in the serum samples.

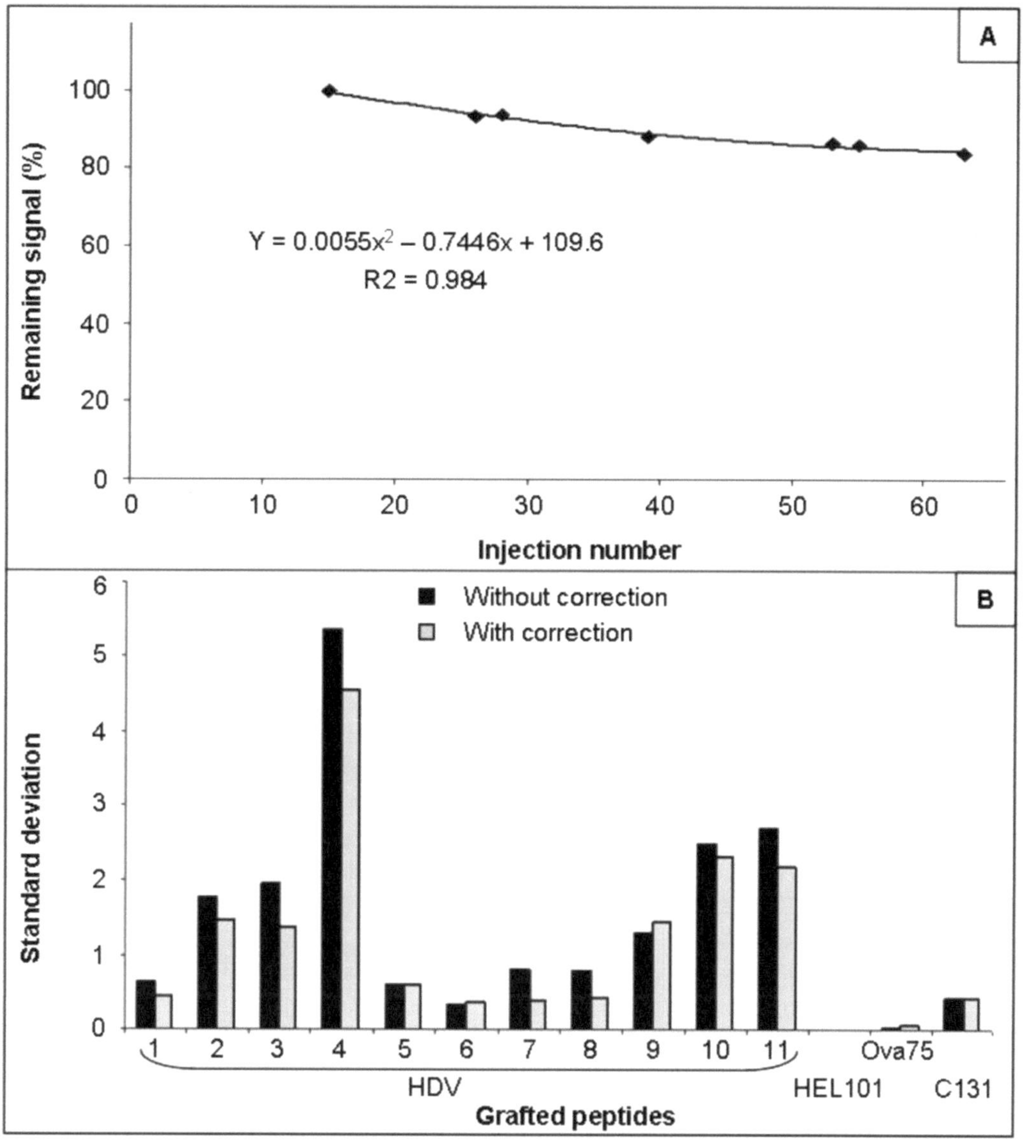

Fig. 10. SPR signal correction: 14 peptides were immobilized in triplicate on gold chips surface via pyrrole electropolymerization. Saturation of the chip surface was ensured using NIS and glycerol was added to running solutions. Successive injections of serums from HDV infected or healthy donors (1/50) were performed, followed by HCl-Glycine regeneration. Rabbit anti-C131 serum was periodically injected. A) Remaining signal obtained on C131 spot upon anti-C131 serum injections. B) Standard deviation of the SPR signal obtained before and after signal correction, for three independent sets of injections.

To validate this correction procedure, we realised the following experiments: 14 peptides (11 from HDV, 1 from HEL, 1 from Ova and the control peptide C131) were grafted in triplicates on 2 chips. A set of 17 serums (14 from HDV infected patients and 3 from healthy donors) was injected twice on the first chip and once on the second chip. Serums were in a different random order within each set. Rabbit anti-C131 serum was periodically injected (at least every 12 injections) to establish a control curve for signal decrease. The sum of the SPR signals obtained for each peptide was calculated for each injection set, before and after application of the correction, as well as the standard deviation (Fig. 10B). Indeed, as signal loss depends on the peptide/ligand pair, this correction is not optimal, but nevertheless improves the results.

4. Conclusion

Among factors susceptible to be involved in signal loss during sample screening on peptide chip, we can exclude probe proteolysis as protease inhibitors had no significant effect on peptide chip stability. Nevertheless, it is recommended to check for protease activities in samples before running large screening. The accessibility to peptides does not seem to limit the signal: steric hindrance and/or chip fouling due to partial regeneration or poor chip surface saturation would lead to an increase in base line, which is not the case.

The influence of peptide grafting process suggests a release of the probes during the experiments. However, in this case, the reduction of signal should be independent of the peptide, which is actually not the case. It is worth to notice that SPR signal is much lower when diazonium protocol was used. We suggest that this is due to a weaker grafting efficiency, leading to fewer immobilized peptide on the chip. Thus, SPR signal is lower and the competition for Ab binding favours high-affinity Ab. As discussed in § 3.7, high-affinity Ab keep their capacity to bind their epitopes even after epitope conformational changes, thus leading to a more stable signal. This is supported by the slope values obtained for the dissociation curves ($12 \pm 2.6 \times 10^{-3}$ %reflectivity/min and $4 \pm 0.9 \times 10^{-3}$ %reflectivity/min for ppy and diazonium respectively, first C131 injection).

Altogether, our results suggest that the main reason for the loss of SPR signal during the experiments is a change in the conformation of the grafted peptides, probably induced by the successive injection/regeneration cycles. This conformational change impairs peptide recognition by some Ab, selecting the Ab possessing a high affinity for the probe. The signal loss, due to a decrease in Ab amounts that bind to peptides, depends on both the peptide (according to its conformational stability) and the Ab (according to the targeted epitope via its accessibility upon conformational change of the peptide). Saturation process can also impact peptide stability, depending on the nature of peptide-chip surface interactions, some of them being able to induce epitope masking. Involvement of peptide conformational change in signal loss is strengthened by the improvement of the signal stability observed for some peptides when glycerol was added in the running solutions.

This study points out several ways to both reduce and correct the signal loss occurring on peptide chip upon multiple injection/regeneration cycles.

5. Acknowledgment

This work was supported by the "Agence Nationale de Recherches sur le Sida" (ANRS), by the "Fonds Unique Interministériel (FUI) Biotherapic and Alphavac" and the "Région Rhones-Alpes (Cluster Infectiologie)".

6. References

Barton, A.C., Davis, F. & Higson, S.P. (2008). Labeless immunosensor assay for prostate specific antigen with picogram per milliliter limits of detection based upon an ac impedance protocol. Anal Chem, Vol. 80, No. 16, pp. 6198-6205, 0003-2700 0003-2700

Benters, R., Niemeyer, C.M., Drutschmann, D., Blohm, D. & Wohrle, D. (2002). DNA microarrays with PAMAM dendritic linker systems. Nucleic Acids Res, Vol. 30, No. 2, pp. E10, 0305-1048 0305-1048

Berek, C. & Ziegner, M. (1993). The maturation of the immune response. Immunol Today, Vol. 14, No. 8, pp. 400-404, 0167-5699 0167-5699

Cherif, B., Roget, A., Villiers, C.L., Calemczuk, R., Leroy, V., Marche, P.N., Livache, T. & Villiers, M.B. (2006). Clinically related protein-peptide interactions monitored in real time on novel peptide chips by surface plasmon resonance imaging. Clin Chem, Vol. 52, No. 2, pp. 255-262, 0009-9147 0009-9147

Corgier, B.P., Bellon, S., Anger-Leroy, M., Blum, L.J. & Marquette, C.A. (2009). Protein-diazonium adduct direct electrografting onto SPRi-biochip. Langmuir, Vol. 25, No. 16, pp. 9619-9623, 0743-7463 0743-7463

Donhauser, S.C., Niessner, R. & Seidel, M. (2009). Quantification of E. coli DNA on a flow-through chemiluminescence microarray readout system after PCR amplification. Anal Sci, Vol. 25, No. 5, pp. 669-674, 0910-6340 0910-6340

Fieser, T.M., Tainer, J.A., Geysen, H.M., Houghten, R.A. & Lerner, R.A. (1987). Influence of protein flexibility and peptide conformation on reactivity of monoclonal anti-peptide antibodies with a protein alpha-helix. Proc Natl Acad Sci U S A, Vol. 84, No. 23, pp. 8568-8572, 0027-8424 0027-8424

Halperin, R.F., Stafford, P. & Johnston, S.A. (2011). Exploring antibody recognition of sequence space through random-sequence peptide microarrays. Mol Cell Proteomics, Vol. 10, No. 3, pp. M110 000786, 1535-9476 1535-9476

Lee, C. & Kim, B. (2002). Improvement of protein stability in protein microarrays. Biotechnol Lett, Vol. 24, No. pp. 839-844, 0167-5699 0167-5699

Neuman de Vegvar, H.E., Amara, R.R., Steinman, L., Utz, P.J., Robinson, H.L. & Robinson, W.H. (2003). Microarray profiling of antibody responses against simian-human immunodeficiency virus: postchallenge convergence of reactivities independent of host histocompatibility type and vaccine regimen. J Virol, Vol. 77, No. 20, pp. 11125-11138, 0022-538X 0022-538X

Vagenende, V., Yap, M.G. & Trout, B.L. (2009). Mechanisms of protein stabilization and prevention of protein aggregation by glycerol. Biochemistry, Vol. 48, No. 46, pp. 11084-11096, 0006-2960 0006-2960

Villiers, M.-B., Cortes, S., Brakha, C., Marche, P., Roget, A. & Livache, T. (2009). Polypyrrole-peptide microarray for biomolecular interaction analysis by SPR imaging, In: Peptide microarrays: methods and protocols, M. Cretich and M. Chiari, pp. 317-328, Humana Press, 978-1603273930 978-1603273930, London

Yakovleva, J., Davidsson, R., Bengtsson, M., Laurell, T. & Emneus, J. (2003). Microfluidic enzyme immunosensors with immobilised protein A and G using chemiluminescence detection. Biosens Bioelectron, Vol. 19, No. 1, pp. 21-34, 0956-5663 0956-5663